Radio Engineering and Telecommunications Waveguide Systems in the Microwave Range

Islam Islamov

Radio Engineering and Telecommunications Waveguide Systems in the Microwave Range

Modeling and Synthesis

Springer

Islam Islamov
Dept of Radio Engineering & Telecom
Azerbaijan Technical University
Baku, Azerbaijan

ISBN 978-3-031-37915-4 ISBN 978-3-031-37916-1 (eBook)
https://doi.org/10.1007/978-3-031-37916-1

This Springer imprint is published by the registered company Springer Nature Switzerland AG
The registered company address is: Gewerbestrasse 11, 6330 Cham, Switzerland

Preface

The appearance of such a book is relevant and timely, since microwave technology is now reaching the mass consumer. This is connected not only with the development of television broadcasting in the decimeter range, but also with the appearance on the market of systems for the individual reception of television signals in the range of 4–12 *GHz* from satellites located in geostationary orbit. Considering that dozens of television programs can already be transmitted from one such satellite, it is easy to imagine the possibilities of satellite television broadcasting and its effectiveness. Based on a fairly detailed presentation of the material, the reader will be able to perform not only analysis, but also the synthesis of both simple and complex microwave devices in radio engineering and telecommunications. Preference was given to devices that have found application in real practice. The book includes many practical problems with close to real values of the initial values. The book is devoted to the numerical calculation of electromagnetic fields of microwave devices in radio engineering and telecommunications and the implementation of these methods.

The effectiveness of numerical calculation methods is evaluated by final parameters: accuracy and performance of programs, ease of implementation. The author made his contribution to the problem under consideration by analyzing numerical methods, developing these methods in the analysis of the widest possible range of devices, solving related issues, and universalizing algorithms in relation to the geometry and nonlinear properties of microwave devices in radio engineering and telecommunications.

Modern applied electrodynamics, that is, mainly the electrodynamics of hollow systems (in particular, systems of complex shape containing nonlinear properties of the medium), leads to significant difficulties of a mathematical, or rather, computational nature.

For the design and improvement of modern radio engineering and telecommunication devices in the microwave range, new approaches are needed to develop methods for calculating electromagnetic fields. These methods should be based on mathematical models that should take into account the real electrophysical and geometric properties of media, such as inhomogeneity, nonlinearity, anisotropy, specific features, complexity of boundary configuration, wave types, and boundary conditions.

The desire to take into account the largest possible number of factors that significantly affect the formation of an electromagnetic field in microwave devices necessitates the choice of the most rational approaches to the implementation of models based on the use of effective numerical methods for solving boundary value problems.

Currently, engineers and researchers who solve specific problems of modeling and calculating an electromagnetic field during design and research often need recommendations on the formulation of boundary value problems, the choice of a solution method, the construction of an algorithm for calculating an electromagnetic field, and its effective software implementation.

The mathematical description of the processes occurring in the nonlinear medium of microwave devices leads to the equations of mathematical physics, which are quite difficult to solve even with the simplification of the original problem. At the same time, the rapid development of mathematical modeling methods in recent years provides an effective means for solving boundary value problems.

When preparing the monograph, two main goals were set: to give a systematic presentation of all stages of mathematical modeling in engineering calculations – from formulation to obtaining numerical results; to illustrate the application of the considered methods and algorithms using examples of solving specific problems of calculating electromagnetic fields, as well as to help scientific and engineering workers, who are faced with the problems of calculating electromagnetic fields in nonlinear media, master the methodology for solving these problems through mathematical modeling.

The monograph is written for specialists in radio engineering and telecommunications, as well as applied electrodynamics. Some questions may be of interest to mathematicians, which will contribute to the further development of mathematical methods.

It is known that any technique is best learned through practical examples. Therefore, most of the monograph is devoted to the analysis of specific examples of mathematical modeling of electromagnetic fields of microwave devices in radio engineering and telecommunications.

When working on the book, the author discussed emerging issues with specialists from the Department of Radio Engineering and Telecommunications of the Azerbaijan Technical University, as well as with specialists from the National Aerospace Agency (NASA) of Azerbaijan Republic and the National Defense University. The author is grateful to all the specialists of the above-mentioned institutions for valuable, principled, and benevolent remarks, which contributed to the improvement of the content of the monograph.

Baku, Azerbaijan Islam Islamov

Abstract

In the monograph, the numerical methods for the modeling and synthesis of microwave range radio engineering and telecommunication systems of the rectangular and circular waveguide type are systematized, and the theoretical bases of the analysis and optimal synthesis of these systems are given, and new research results are shown. For the first time, the HFSS complex was used for the analysis, optimization, and synthesis of microwave range systems operating in *E*-type and *H*-type waves. As a result of analysis and synthesis, various microwave range devices with improved technical and operational parameters were developed, and based on them, a new waveguide tract system connecting the transmitter and antenna in television towers was proposed. The monograph is designed for engineering and scientific workers engaged in the design of radio engineering and telecommunication systems. At the same time, it can be used by students, graduate students, doctoral students, and a wide readership.

Introduction

Since the fundamental limit of telecommunication and radio engineering systems is the mastering of increasingly higher frequency bands, information exchange is increasing day by day in the twenty-first century, which is the age of information and communication. It is known that wave-transmitting systems are widely used in the lossless transmission and reception of information in the microwave range, which has a larger information capacity and speed. Microwave wave-transmitting systems used to transmit information are of various constructions and specifications, and are used in telecommunications, television technology, radiolocation, radionavigation, medical technology, radio control, communication technology, antenna technology, and other fields of science and technology. Of such waveguides, the most widely practical are the microwave rectangular and circular waveguides. Microwave rectangular and circular waveguides have several advantages. Thus, they have high reliability, stability of parameters, and longevity in the process of operation. At the same time, since microwave rectangular and circular waveguides are completely isolated from the environment, when transmitting information with such systems, electromagnetic waves (radio waves) are hardly radiated to the external environment.

Despite the wide application of extremely microwave rectangular and circular waveguides, there are not enough scientific works that fully reflect the electromagnetic processes propagated in such transmission systems and are of experimental importance. V.A. Donchenko, G.F. Zargano, G.P. Sinyavskiy, K.V. Vdovenko, V.A. Katrich, M.V. Nesterenko, S.C. Chen, V.C. Qiu, L. Gong, K. Zhang, T. Grik, L. Nickelson, S. Asmontas, A. Zaghdani, O.S. Zakharchenko, P.Y. Stepanenko in the current works of scientists such as in this field, the medium in which electromagnetic waves (radio waves) propagate is considered as a linear medium. However, our research has shown that in real practice, during the operation of microwave rectangular and circular waveguides, the medium behaves as a non-linear medium. At the same time, in the works of the above-mentioned scientists, complex classical mathematical apparatuses were used during the modeling of such waveguides, which increased the error of calculations and made their implementation difficult. Therefore, in order to improve the electrical, magnetic, technical, structural, and operating parameters and characteristics of microwave rectangular and circular waveguides, taking into account the nonlinearity of the environment, there is a serious need to develop new mathematical models that reflect the specific characteristics of these devices and to algorithmize them based on numerical methods as it is heard; this issue has great scientific and practical importance.

On the other hand, in the works of scientists such as L.I. Babak, S.Y. Bankov, A.A. Kurshin, V.D. Razevig, M.P. Batura, A.A. Kurayev, C.J. Railton, D.L. Paul, A. Munir, optimization of rectangular and circular waveguides with microwave a one-step synthesis process was used, which was accompanied by the following negative situations:

- There are significant errors in the geometric dimensions of the devices.
- Since it is not possible to use the optimal information about the results of experimental research and development of devices, the scientific and technical efficiency has decreased.

Therefore, there is a need to solve the problem of multi-criteria optimal synthesis of rectangular and circular waveguides with microwave with high reliability, stability of parameters, and longevity in the operation process. Thus, solving the problem of multi-criteria optimal synthesis of rectangular and circular waveguides with microwave, improving the electrical, magnetic, technical, structural, and operational parameters and characteristics of these devices, synthesis and, as a result, developing new models with optimal parameters, remains an urgent scientific and technical issue.

Summarizing the above, it can be said that since microwave rectangular and circular waveguides are widely used in production, the improvement of the electrical, magnetic, technical, structural, and operational parameters and characteristics of these devices and taking into account the non-linearity of the environment, a new waveguide with more optimal structural dimensions designing such systems is important and relevant both from a scientific point of view and from a production point of view.

Contents

About the Author

Islam Islamov received his M.S. degree in Radio Engineering from Azerbaijan Technical University, Azerbaijan. He is currently a professor at the Department of Radio Engineering and Telecommunication, Azerbaijan Technical University, as well as the National Aerospace Agency (NASA) and National Defence University, Azerbaijan. His research interests include Information Technology, Radio Engineering, Antenna Engineering, Microwave Engineering, Digital Signal Processing, and Telecommunication networks and systems.

Chapter 1
Current State of Analysis and Optimal Synthesis of Microwave Waveguide Systems of Complex Structure

1.1 Classification of Microwave Devices

Microwave devices include power converters, branchers, phase shifters, circulators, filters, valves, etc. belongs to. Generally, microwave devices are divided into regular and irregular devices.

The distribution of microwave devices according to frequency ranges is shown in Table 1.1. According to the specified terminology, this frequency range covers the region from 3 kHz to 3000 GHz. The microwave range shown in Table 1.1 covers the range from meters to millimeters [1, 2].

Microwave devices are classified according to the types of electromagnetic waves emitted by them, and the following can be attributed to them: Transverse electromagnetic wave microwave devices (*TEM*-type wave), magnetic wave microwave devices (*H* or *TE* wave), electric wave microwave devices (*E* or *TH* wave), and hybrid wave microwave devices.

If we direct the *Z* axis along the microwave device in the rectangular coordinate system, the type of each wave can be determined from the conditions shown in Table 1.2. As can be seen from Table 1.2, the intensity vectors of the electric and magnetic fields in the *TEM*-type wave are located in the plane perpendicular to the wave propagation direction; in the *H*-type wave, the intensity vector of the magnetic field has longitudinal and transverse summaries, while the intensity vector of the electric field has only the transverse summation; in the *E*-type wave, the intensity vector of the electric field has longitudinal and transverse summaries, and the intensity vector of the magnetic field is located in the plane of the cross section of the microwave device; in a hybrid wave, the intensity vectors of the electric and magnetic fields have both longitudinal and transverse summations. A classification of microwave devices according to appearance is shown in Fig. 1.1 [3].

Microwave devices without elastic or plastic curves are called rigid, otherwise flexible microwave devices. microwave devices with one or more conducting

I. Islamov, *Radio Engineering and Telecommunications Waveguide Systems in the Microwave Range*, https://doi.org/10.1007/978-3-031-37916-1_1

Table 1.1 Different frequency bands of radio waves

Wave length	Term	Frequency	Term
100–1 km	Miriametric wave	3–30 kHs	Lowest frequency
10–1 km	Kilometer wave	30–300 kHz	Low frequency
1000–100 m	Hectometre wave	300–3000 kHz	Average frequency
100–10 m	Decameter wave	3–30 MHz	High frequency
10–1 m	Meter wave	30–300 MHz	Very high frequency
100–10 cm	Decimeter wave	300–3000 MHz	Ultra high frequency
10–1 cm	Centimeter wave	3–30 QHs	Microwave
10–1 mm	Millimeter wave	30–300 GHz	Peak frequency
1–0.1 mm	A decimillimeter wave	300–3000 GHz	Hyper high frequency

Table 1.2 Types of electromagnetic waves emitted in microwave devices and the conditions of their longitudinal accumulation in the field of propagation

Type of wave	Along the field the terms of those collected
T-type wave	$E_Z = 0, H_Z = 0$
H-type wave	$E_Z = 0, H_Z \neq 0$
E-type wave	$E_Z \neq 0, H_Z = 0$
Hybrid wave	$E_Z \neq 0, H_Z \neq 0$

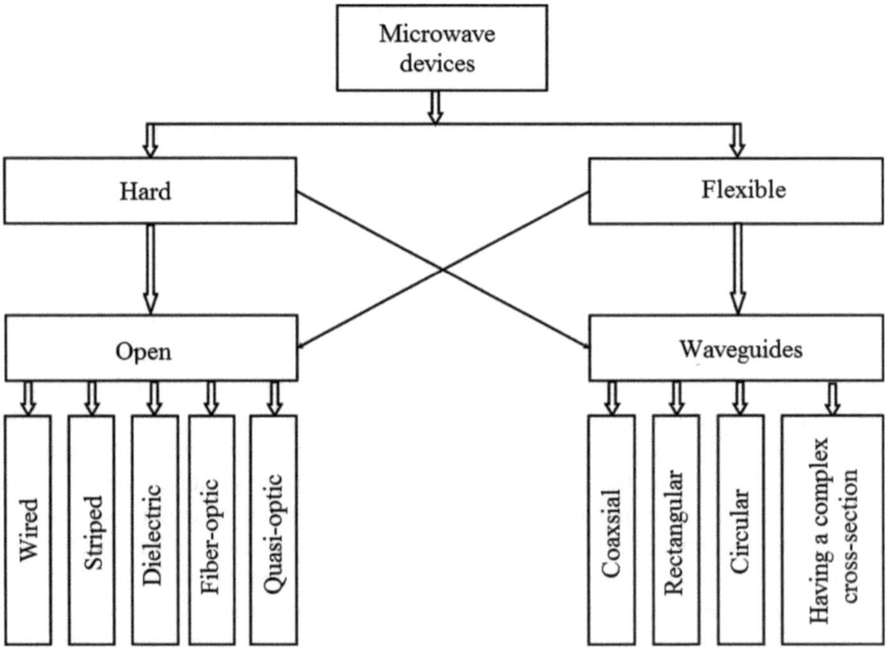

Fig. 1.1 Classification of microwave devices

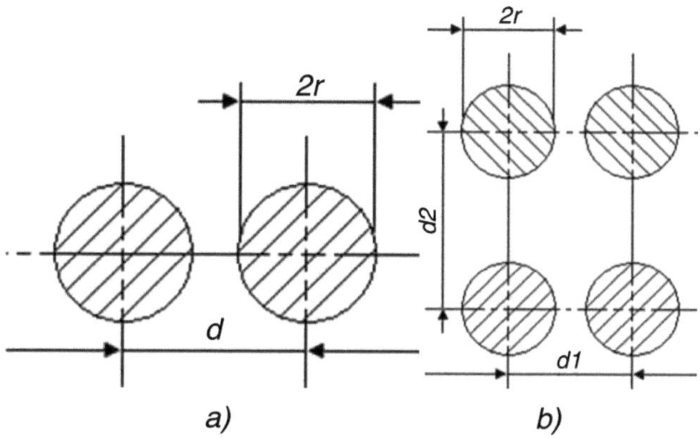

Fig. 1.2 Cross-section of wired microwave devices: (*a*) double – wired; (*b*) four – wire

surfaces and whose cross-sectional area is in the form of a closed conductor are called waveguides. If there is no conductive loop, such microwave devices are called open-type devices.

Wired microwave devices include two-wire and four-wire devices. Figure 1.2 shows the cross-sections of such devices [4].

The wires of these lines are covered with dielectric. In such microwave devices, the main type of wave is the *TEM*-type wave. In four-wire microwave units, paired wires are excited, e.g., horizontally, vertically, or diagonally. Such microwave devices are used in the hectometre, decametre, and meter wave ranges. Banded microwave devices can include symmetric and nonsymmetrical devices, as well as slotted and coplanar devices.

Thus, the cross-sectional area of microwave devices and the structures of electromagnetic fields are shown in Fig. 1.3 [5].

Such microwave devices are mainly used in decimeter, centimeter, and millimeter wave ranges. The main operating wave of an asymmetric strip line is a *TEM*-type wave. In slotted and coplanar lines, the main type of wave is the *H*-type wave.

Apart from these, there are microstrip microwave devices, in which the dielectric permeability of the substrate takes a very large value (over 10), and the loss is small. Therefore, the geometric dimensions of this device are reduced by several times. The material of the six microstrip microwave devices is mainly polycor, cital, silicon, sapphire, etc.. In banded microwave devices, it is convenient to use air as a dielectric to reduce loss. Such microwave devices are called air-filled or high-quality strip devices.

Dielectric microwave devices can be classified according to the shape of the cross-sectional area. Some of them are shown in Fig. 1.4 [6].

Such microwave devices are used in the millimeter wave range. The main type of working wave in them is hybrid *HE*-wave. In microwave devices, the presence of a metallic screen (Fig. 1.4*d*, *e*, *c*) allows maintaining the polarization structure of the propagated wave.

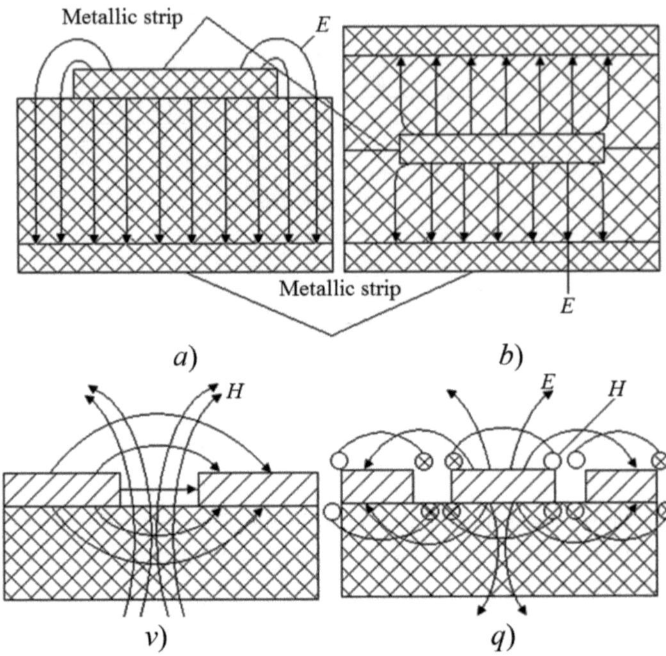

Fig. 1.3 Cross-section of strip microwave devices: (*a*) nonsymmetrical, (*b*) simmer, (*c*) split, (*q*) coplanar

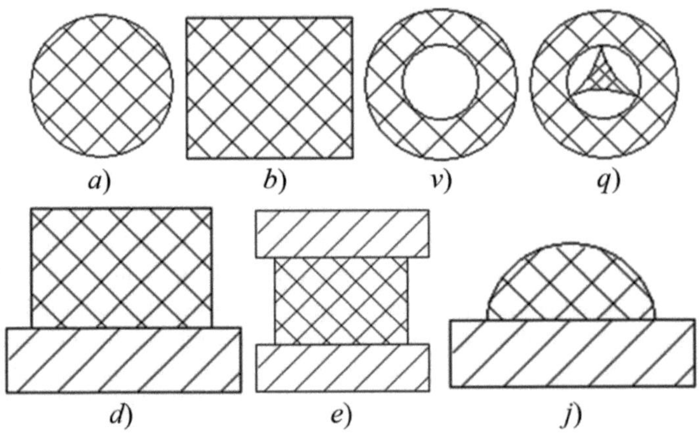

Fig. 1.4 Cross-sections of dielectric microwavedevices: (*a*) circular; (*b*) rectangle; (*v*) tubular; (*q*) star-shaped; (*d*), (*e*), (*c*) mirrored

Fiber-optic transmission lines are used in decimillimeter (submillimeter) and optical ranges. They are in the form of a dielectric line with a circular cross-section made of quartz, along which several types of waves can be propagated simultaneously. Therefore, such transmission lines are called multi-mode transmission

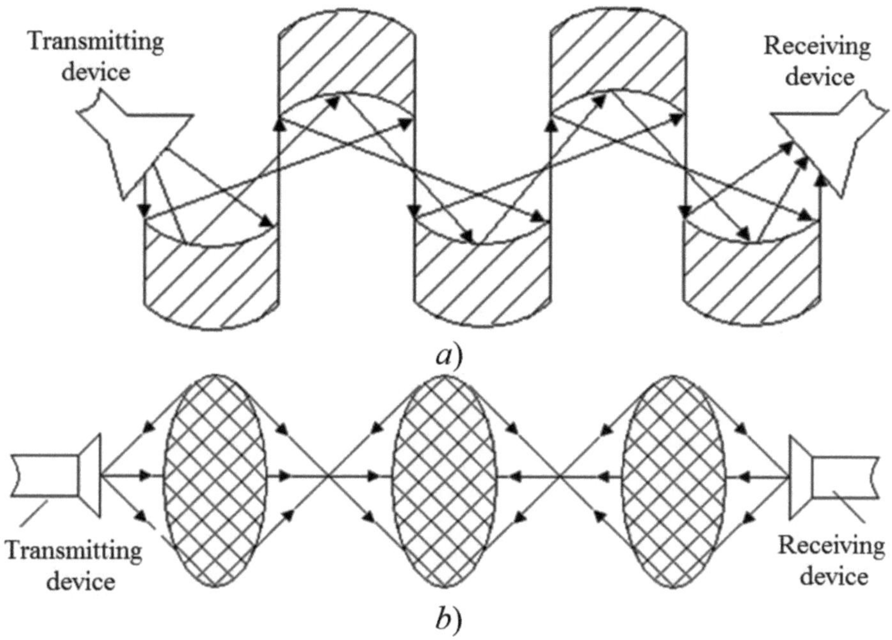

Fig. 1.5 Radiant transmission lines: (*a*) reflective type; (*b*) lens type

lines. The diameter of the circular fiber is made in the size of several wavelengths of electromagnetic oscillation. Wave propagation in optical-fiber transmission lines is based on the effect of total reflection from the dielectric-air interface. In order to reduce heat losses in such transmission lines, fibers with a variable refractive index in the cross-sectional area are used. This, in turn, leads to the reduction of the geometric path of the beam passing through the unit length of the transmission lines.

Quasi-optical (radial) transmission lines are called irregular transmission lines, and its working principle is based on the use of the optical content of radio waves. Figure 1.5 shows the construction options of these transmission lines schematically [7].

Such transmission lines are used in the millimeter and submillimeter wave ranges. Coaxial transducers come in the form of rigid or flexible coaxial cables and operate in a *TEM*-type wave. Such microwave devices can be applied from the hectometre wavelength range to the centimeter wavelength range. Cross-sections of coaxial waveguides, which are more widespread in practice, are shown in Fig. 1.6 [8].

Rectangular, circular and hollow metallic waveguides with a more complex cross-sectional area are shown in Fig. 1.7 [9]. *E*-type and *H*-type waves can be propagated in such waveguides. These transducers can operate from the decimeter wavelength range to the millimeter wavelength range. Microwave devices can also be classified according to the compatibility of their cross-sectional areas.

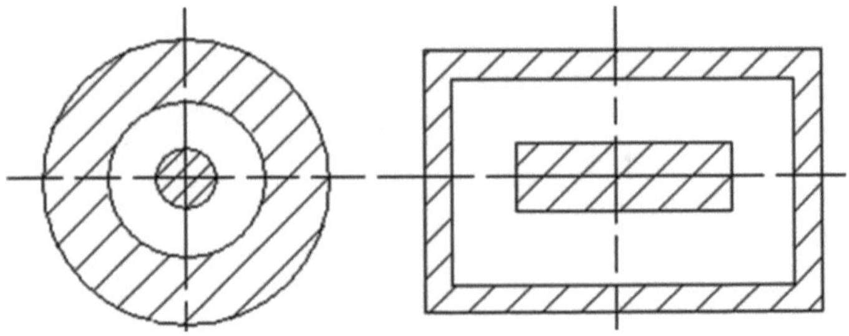

Fig. 1.6 Cross-sections of coaxial waveguides: (*a*) circular; (*b*) rectangle

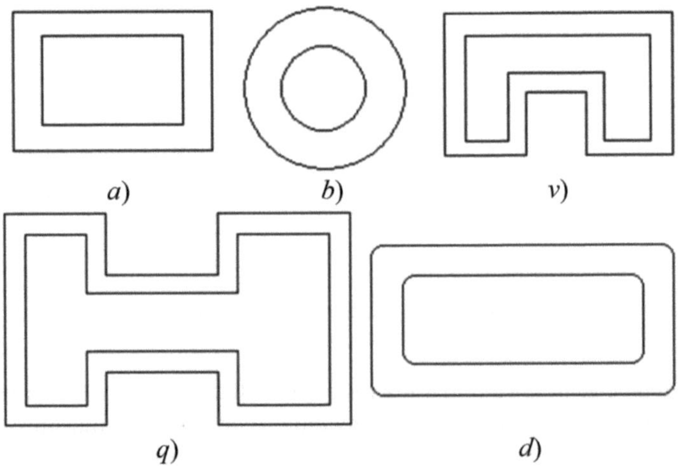

Fig. 1.7 Cross sections of metallic microwave waveguides: (*a*) rectangular; (*b*) circular; (*v*) Π – shaped; (*q*) H – shaped; (*d*) elliptical

The degree of communication is a geometric characteristic of the cross-section of microwave devices and is determined by the number of conductive surfaces. Depending on the number of conductive surfaces, microwave devices can be divided into single-connection, two-connection, three-connection, multi-connection and zero-connection devices.

Zero-coupling microwave devices do not have conductive surfaces. For example, metallic waveguides are single-coupled, coaxial waveguides are double-coupled, and dielectric waveguides are zero-coupled microwave devices.

It should be noted that air-filled rectangular and circular waveguides are more widely used in practice in microwave technology to transmit high power energy. Microwave rectangular and circular waveguides are used in telecommunications, radio communication, television, radar, etc. Its application not only reduces the costs spent on the development of these systems, but also increases their effectiveness in turn.

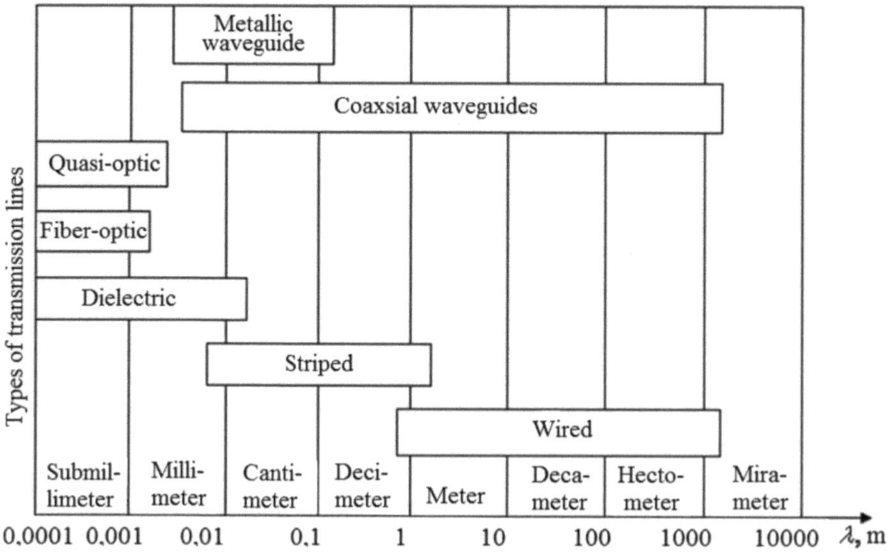

Fig. 1.8 Application scheme of transmission lines in different wavelength ranges

Figure 1.8 shows the frequency ranges where microwave can be applied [9].

Microwave rectangular and circular waveguides have several advantages. They have high reliability, stability of their parameters, and long service life in the process of operation. At the same time, since rectangular and circular waveguides are completely isolated from the environment, when transmitting information with such waveguides, radiation of electromagnetic waves (radiowaves) to the external environment almost does not occur. When transmitting information through microwave rectangular and circular waveguides, electromagnetic processes propagate under extremely complex conditions.

Until now, when designing air-filled microwave rectangular and circular waveguides in telecommunications, including calculating their parameters, the medium through which electromagnetic waves propagate has been assumed to be a linear medium. But in real practice, when transmitting information through air-filled microwave rectangular and circular waveguides, the medium behaves as a nonlinear medium. Thus, the electrophysical parameters of the environment change depending on the intensity of the electric or magnetic field of microwave rectangular and circular waveguides. Therefore, the study of air-filled microwave rectangular and circular waveguides, taking into account the nonlinearity of the medium, is an important issue for both science and practice.

1.2 State-of-the-Art Analysis of Microwave Devices

In order to clarify the current state of analysis of microwave devices, let's look at the overview of scientific research works performed in this field worldwide.

Alsuyuti et al. [10] developed an algorithm for determining the parameters of the damage areas of a stepped microwave rectangular waveguide using the Lemba wave. The developed algorithm is based on the interaction of the Lemba wave with damage areas and the phenomena of transformation of wave modes. The analysis is divided into two stages: determination of damage areas and determination of geometric dimensions of damage areas. The main advantage of this method is that the parameters of the damage areas can be determined with any accuracy in any part of the stepped rectangular waveguide. The presented method is distinguished by its universality. Thus, this method can be successfully applied to determine the parameters of damage areas in any multilayer layers.

Altufaili et al. [11] proposed an expression for a new integral equation in the spatial domain of a stepped rectangular waveguide. The importance of the proposed equation is that it takes into account the electrical conductivity in an arbitrary spatial region. The proposed integral equation is solved in the spatial domain of the stepped rectangular waveguide. In this case, it was considered that the core of potentials is located at the center of the waveguide. Since the proposed integral equation could not be solved by analytical methods, it was solved by numerical methods. At the same time, the proposed equation has also been solved numerically for bandpass filters and related dielectric resonators from other microwave band devices. The obtained results were compared with other available results and the error was determined to be satisfactory.

In a study by Amin [12], the electromagnetic parameters of the waveguide structure were determined by applying the finite difference method. It is recommended to apply the obtained results in determining the frequency characteristics of multi-level asymmetric waveguide structures. Anand and Kushwah [13] proposed a new structure for accurate measurement of the characteristics of an extremely high frequency TE_{10} wave mode rectangular waveguide to TM_{01} wave mode circular waveguide converter used in powerful microwave systems. A TM_{01} mode wave is excited in the hybrid rectangular waveguide with *E-T* structure. In order to quench the TE_{11} mode wave, the circular waveguide slits are opened asymmetrically on the rectangular waveguide. Attenuator slots on the small wall of the rectangular waveguide are designed to increase the frequency range. In this study, the experimentally measured parameters of the electromagnetic field were confirmed by the results obtained from numerical modeling. During the conducted studies, it was shown that if the size of the quenching slit increases from 5.5 mm to 11.7 mm, then the frequency range of the device will increase by 500 MHz. This allows the unit's efficiency to increase to 99%.

Annadhasan et al. [14] investigated the characteristics of an air-filled waveguide used as a polarization element transducer. At the same time, the theoretical results obtained here are compared with the experimental results, and the error is determined to be satisfactory. Numerical modeling was performed using the line method, and the results were compared with experiments. In a study by Argyropoulos et al. [15], the problem of propagation of radio waves in a rectangular waveguide with an ideal conducting wall was considered using the solution of Maxwell's equations by numerical methods. A method was developed to recover the dielectric constant in the emission band taking into account the experimental error. The experimental results for a homogeneous diaphragm were compared with the results obtained from solving Maxwell's equations by numerical methods.

It is known that thick multilayer dielectric composite materials are used in aerospace, marine, petrochemical, etc. and are widely used in industries. For these materials to be used in critical applications, they must be of high quality and free from any defects. Various test methods have been applied to check and evaluate any defects in such materials. In a study by Arnberg et al. [16], the waveguide-based NDT microwave method was used for the first time to analyze composite materials. A specially designed circular waveguide was used in the test sample for inspection of defects. The multilayer composite structure and waveguide are modeled in electromagnetic numerical modeling software. The sensitivity of the composite in the circular waveguide is checked by analyzing the change of the electric field at a frequency of 24 GHz.

Azeez et al. [17] obtained injection-molding techniques and designed and manufactured a power adapter operating in the microwave range. Both numerical modeling and experimental results have shown that using fabricated plastic adapters, low-loss broadband power transmissions can be realized with metallic and dielectric waveguides. The results show that the prepared adapter can be applied in practical microwave functional waveguide devices.

In a study by Babak [18], a rectangular waveguide with a homogeneous medium was developed. The characteristics of the developed waveguide were studied, including the propagation of electromagnetic waves in different directions along axes, and the transmission characteristics of controlled waves were also studied. Here, the results obtained when the optical axis is parallel to one of the *XYZ* axes are compared with transversal *Z*-solutions. Calculations have shown that anisotropy not only changes the field distributions but also leads to a change in the mode of the guiding wave in the waveguide. If the optical axis propagates towards TE_{0n} or TE_{m0} (Z) wave modes, then the proposed boundary condition in the resulting hybrid wave mode is found from matrix calculations. Here, hybrid waves are expressed as a combination of ordinary and extraordinary waves. An algorithm is developed to solve this problem and numerical examples are given. The accuracy of the results was confirmed by comparison with the results obtained from the experiment.

Bachiller et al. [19] studied a waveguide with planar hybrid and non-planar structure. Such a waveguide is called an integral waveguide. These transducers have several advantages. So, they work with great powers and have a low cost. Here, the integral rectangular waveguide was analyzed using the finite difference method in the time domain using the Matlab software complex.

In a study by Bakkali et al. [20], a numerical-analytical method was developed for studying the tensor conductivity and conductivity of the dielectric diaphragm in a rectangular waveguide. The developed method and obtained results are recommended to be used in optical electronic devices, nanotechnology and in the construction of microwave devices. In a study by Balbastre and Nuno [21], a graphene-based nanometer waveguide was investigated. This waveguide is coupled to a resonator with a rectangular structure placed on a sapphire substrate. Two different methods are used to achieve the PIT effect: one is based on direct exposure to radiation and dark levels and the other is a graphene nanooptical waveguide. Numerical results obtained in this study show that the acquisition of a high-transparency PIT window is achieved due to the change of the Fermi energy level of the graphene rectangular resonator. Compared with the previously proposed graphene-based PIT, the proposed scheme is structurally simpler and has better economic efficiency.

Bandler and Seviora [22] present a very flexible and efficient tool for automated design (CAD) of rectangular waveguide filters. The software tool developed based on the use of the half-wave method made it possible to accurately study the electromagnetic field of resonators. The developed CAD tool has significantly increased the efficiency of the CAD process. The developed software tool made it possible to determine the electromagnetic characteristics of the microwave waveguide. In addition, a new simple method is proposed for the effective combination of all obtained broadband matrices. In order to test the accuracy and efficiency of the new instrument, new designs of strip waveguide devices have been proposed. Checking the accuracy of the CAD tool was performed using the finite element method.

In a study by Basir and Yoo [23], the Taguchi method was used for structural optimization of a rectangular waveguide, which made it possible to reduce the power transmitted through that device to the minimum level. The waveguide under study is used to transfer the energy of the electromagnetic field from one point to another and is filled with internal air. Here, a microwave rectangular waveguide is optimized using the orthogonal Taguchi method and then modeled using the finite element method. The medium of air used to transmit the energy of the electromagnetic field from one point to another is considered as a linear medium. At the same time, the effective parameters of the rectangular waveguide were determined using the dispersion method. After optimization, the rectangular waveguide had dimensions of $L = 20$ cm and $D = 3$ cm. This allowed the extinction coefficient to be equal to 0.1.

In Belenguer et al. [24], a plasma-filled metal rectangular waveguide was studied using the two-dimensional finite element method. The result of the numerical modeling showed that the resonance in the presented structure can be created by a sharp and asymmetric Fano-line structure. An analytical model based on scattering matrix theory is used to describe and explain this phenomenon. The results obtained here contribute to the successful use in the production of nanosensors with effective physical properties, that is, a sensitivity of 1300 nm/RIU and a numerical index equal to 6838. At the same time, these plasma structures can find their application in nano-dispersed crystals.

Benassi et al. [25] analyzed a *2D* rectangular waveguide through the finite element method using first-order triangular elements. Given that the equations obtained here are large matrix equations, taking into account that a large amount of time is spent on solving them, various methods have been sought to reduce the time of calculations. Therefore, the Rayleigh method was used to solve this problem. The obtained results were compared with the results obtained by *QR*-method, *QZ*-methods and the comparison gave good results. In a study by Bird et al. [26], the method of calculating transverse electric $E(x,y)$ and transverse magnetic fields $H(x,y)$ of a rectangular waveguide was considered. The calculations were performed using Fourier transform in TEM_{pq} modes. A wave mode was used to perform the calculations. In addition, the parameters of the electric $E(x,y,z)$ and magnetic $H(x,y,z)$ fields have also been determined.

It is known that at least n independent measurements are needed to get the true value of any parameter when performing multi-parameter measurements at the same time. Therefore, in this study, the multilayer medium method was used to measure the reflection coefficient of electromagnetic waves in the microwave range. The measurement methodology of this method is based on the principle of measuring two reflection coefficients reflected from reflective sensors. The first reflection coefficient is measured using the material under study, and the second reflection coefficient is measured for a multilayered medium. At the same time, along with the measurements, the finite difference method in the time domain was used to determine the reflection coefficient and other main parameters of the electromagnetic field. The complex dielectric permittivity of the investigated device is determined by the value of the reflection coefficient determined using the finite difference method in the time domain. In addition to the time domain finite difference method, the parameters of the studied device were also determined by the Newton-Raphson method. The obtained results were compared and it was determined that the error was satisfactory [27].

In a study by Bogolyubov et al. [28], two-dimensional numerical analysis of electromagnetic waves propagating inside a rectangular waveguide was performed. The vacuum system inside the rectangular waveguide creates a monochromatic wave in the TE_{10} mode. Using Maxwell's equations and the finite difference method in the time domain, the main parameters of the electromagnetic field propagated inside the rectangular waveguide were determined. At the same time, a half-wave numerical approach was proposed for the analysis of a rectangular waveguide with a periodic interval. This approach is based on the solution of the integral equations for the magnetic field in a single slot. The presented method allows to analyze both coherent and non-hermetic wave modes in a rectangular waveguide. At the same time, the presented method can be used in the analysis of a slotted rectangular waveguide. A numerical study of the environment inside the rectangular waveguide resonator was carried out. Here, the ultra-high-frequency system excites a monochromatic wave with an operating frequency of 2.45 GHz in the fundamental wave mode (TE_{10} mode).

In a study by Bozinovic et al. [29], a thin conducting layer of a rectangular waveguide in the TE_{10} electromagnetic field regime was numerically investigated. The time domain finite difference method was used to perform the modeling. The distribution of the electric field in the thin film and the determination of the reflection coefficient from that film were performed for different lengths, shapes and conductivities. It was recommended to apply the results obtained in the study in the preparation of microwave transmitters. Bozzi et al. [30] studied the two different types of waveguides with circular and elliptical cross-sections: dielectric-coated and plasma-core waveguides. New mathematical expressions have been obtained for the transverse and longitudinal summations of the electromagnetic field in different areas of these waveguides. Expressions of dispersion of electromagnetic waves for structures of waveguides were obtained by using boundary conditions, and dispersion dependencies were established based on this. The used differential equations were solved by the fourth-order Runge-Kutta method, and the obtained numerical results were presented in graphical form.

Bozzi et al. [31] proposed a new approach for determining the propagation constant of electromagnetic waves in a circular waveguide with a periodic structure. Other main parameters of the circular waveguide were related to the propagation constant of electromagnetic waves and new expressions were obtained among these parameters.

Buesa-Zubiria and Esteban [32] analyzed the transition of the TM_{02} mode to the TE_{11} mode in a circular waveguide operating in the TE_{11} output mode. The obtained results showed that the conversion efficiency in circular waveguide operating in TE_{11} output mode and 9.4 GHz frequency is 99.6%, and the maximum power is 1.66 GWt. At the same time, it was determined that the frequency range at 90% conversion efficiency varies in the frequency range of 9.23–9.83 GHz and the relative width of the convergence band is 6.38%. In a study by Bulashenko et al. [33], the circular waveguide with a dielectric layer is modeled by the finite difference method in the time domain. For this method, the time step size depends only on the time dimension. Here, the stability of the modeling method is evaluated. The obtained numerical results showed that the time domain finite difference method applied to the modeling of the circular waveguide is more efficient than the traditional BOR-FDTD method in terms of computational time.

It is known that various problems of the electromagnetic field of microwave devices can be solved with the help of different numerical methods. One of these methods is the time domain finite difference method. Therefore, in Butt [34], an extremely high-frequency circular waveguide filled with air was modeled using the time-domain finite-difference method. At this time, the medium in which the electromagnetic waves propagate in the waveguide was considered as a linear medium. In other waveguides, the equations characterizing the electromagnetic field are written in the Cartesian coordinate system, and in the circular waveguide, they are written in the cylindrical coordinate system. In this work, a circular waveguide with a length of 100 m and a radius of 50 m is modeled in $2D$ and $3D$ systems. The wave source for the calculation is a sinusoidal Gaussian signal with a frequency of 3 MHz. From the obtained numerical results, it can be seen that the amplitude of the wave propagating inside the waveguide is reduced due to attenuation in both $2D$ and $3D$ systems.

In a study by Calignanoa et al. [35], sound propagation through a waveguide is modeled by the Webster equation. Here, it forms a plane wave front. In this study, a one-dimensional model of the sound propagated in a $2D$ circular waveguide was obtained. The model is derived using the $2D$ Helmholtz equation. In a study by Camacho et al. [36], the structural optimization of the TE_{0n} wave mode circular waveguide was considered. Here, the numerical optimization method was used and the obtained results were compared with the results obtained with the HFSS software complex and it was determined that the error was satisfactory.

Cao et al. [37] proposed orthogonal integral equations for different modal modes for a circular waveguide. Using these equations, the components of different wave modes were determined. Finally, the results of some numerical experiments are given, which show that the proposed method is effective. In a study by Carceller et al. [38], the characteristic modes of a cylindrical dielectric waveguide with a circular cross-section were analyzed. The investigated waveguide is a nonlinear medium and the nonlinearity is described by Kerr's law. Here, the nonlinear problem for the nonlinear Helmholtz equation is considered. Numerical iteration methodology was developed for solving the problem and numerical results were obtained.

In a study by Cassivi et al. [39], two-dimensional mathematical models of a photonic crystal waveguide device are given. These mathematical models were solved by the Fourier method under periodic boundary conditions. The obtained numerical results were compared with the results obtained in existing works. In Castillo et al. [40], in the problem of electromagnetic waves propagating in a circular waveguide, Maxwell's equations were solved by the method of singular integral equations and the method of partial section area. It was determined that the method of singular integral equations used in the work is more universal. Here, taking into account the boundary conditions, a circular waveguide with any configuration can be modeled using the method of singular integral equations. The Matlab environment was used to solve the problem, and the obtained results were depicted in the form of a $3D$ graph. The diameter of the circular waveguide studied in this work is equal to 5×10^{-3}m. This transceiver operates in the frequency range from 75 GHz to 115 GHz. The dependence of the phase coefficient on the extinction coefficient for this waveguide is given. The three-dimensional $3D$ electric field distribution in the fundamental wave mode was calculated at about 10,000 points.

Ceccuzi et al. [41] evaluated the scattering of microwave power inside a nonuniform waveguide using the excitation method by analytical methods. It was emphasized that the applied analytical method is more efficient than numerical methods. This method was applied during the design of the transition device from TE_{01} mode circular waveguide to TE_{20} mode rectangular waveguide. In a study by Chaiyo and Rattanadecho [42], a mathematical model of the heat loss of an electromagnetic wave propagating in a circular waveguide using the Langer transformation was given, and this model was solved using the finite element method.

Chakravarthy et al. [43] developed a method for the analysis of microwave rectangular and circular waveguides with linear media. This method is distinguished by its universality and is not limited by symmetry at the node or other restrictions. The obtained numerical results agree well with the results obtained by other methods, including experimental results. A method has been developed for the analysis of microwave rectangular and circular waveguides with linear media using the mode matching method. In a study by Chandel [44], the circular waveguide was analyzed by analytical and numerical methods. Laplace transform was used during the analysis. On the other hand, here the field parameters are calculated by the Kirchhoff method and modeled by the finite difference method in the time domain. At the same time, the frequency characteristic of the circular waveguide is given. In Chang et al. [45], the propagation characteristics of the electromagnetic field in a waveguide with a circular conductive medium were studied. The obtained propagation characteristics were solved with the help of the excitation method and the results obtained from the calculation were compared with the exact values. The comparison has shown that the approximate solution gives good results for a wide range of conductivities.

In a study by Chang and Huang [46], the characteristics of electromagnetic waves propagating in a circular waveguide were analyzed with the help of the finite difference method in the two-dimensional frequency domain ($2D$ FDTD). Using $2D$ FDTD, finite difference schemes for the electromagnetic field of a circular waveguide in a cylindrical coordinate system were drawn up, and the characteristics of the electromagnetic field were determined based on this. In order to verify the proposed method, the obtained results were compared with the existing results obtained by other research methods.

In Charles and Guo [47], the problems arising during the transition of electromagnetic waves from the TM_{01} mode to the TE_{11} mode in a circular waveguide were studied. Using the FDTD numerical modeling method, the structures of the input and output signals were investigated. At the same time, the frequency characteristic of the microwave circular waveguide was studied in the circular waveguide using the Fourier transform. In a study by Che et al. [48], the propagation characteristics of a rectangular waveguide type switch were analyzed using the mode matching method. At the same time, the propagation characteristics of the rectangular waveguide-type switch were experimentally studied and compared with theoretical results.

It is known that the analysis of diaphragm circular waveguide systems using the time domain FDTD is a very complex problem. So, sometimes it is necessary to open small slits on the metal surface of these waveguides. Therefore, in a study by Chen et al. [49], the following improvements were made to the time domain finite difference method to solve the field problem:

1. Local step conformal algorithms were used during modeling.
2. New empirical correction factors have been considered to calculate the area.
3. Compensation for numerical dispersion was implemented in the FDTD algorithm. Within the above mentioned three conditions, the analysis of diaphragm circular waveguide systems with the help of time domain FDTD has been proven to give very good results.

A study by Chen and Liu [50] analyzed the microwave rectangular and circular waveguides using the mode matching method. The impedance, attenuation and power characteristics of these devices are derived. From the results obtained in the study, it was recommended to apply it during the design of various types and purposes of microwave devices and systems. Chen et al. [51] presented the decomposition mathematical models of the resonance valve and the circulator based on magnetic nanocomposites. Mathematical models are built with the help of autonomous blocks. In Chen et al. [52], the mathematical modeling of the microwave magnetic nanostructure and nanodevice based on the solution of Maxwell's equations was performed using an electrodynamic approach. In a study by Chen and Chew [53], the modeling of H_{10} wave diffraction of a 3*D* magnetic nanostructure in a rectangular waveguide in the frequency ranges $f = 26$ GHz and 30 GHz was performed. It is shown that the results obtained by the numerical method are in good agreement with the results obtained in the experiment. Later on [54], the electrodynamic modeling of systems in applied electrodynamics was carried out. The conducted studies were performed in the example of designing complex nodes of microwave systems.

Chen, W. et al. [55] analyzed a microwave pulsed active compressor based on a circular waveguide. Chen, Z. et al. [56] reviewed the electrodynamic modeling and computer visualization of the electromagnetic field structure of an *H*-type wave rectangular waveguide with two *L*-outputs. The considered issues were partially resolved by the method of regions. Later, they [57] proposed an algorithm to solve the waveguide boundary problem. The proposed algorithm is based on coordinate transformation method, line method and block matrix extraction methods. Researches were carried out on a circular waveguide with H_{01} wave. Cheng et al. [58] performed the mathematical modeling of waveguide transitions. In Choo et al. [59], the numerical modeling of diffraction in the waveguide was carried out using the finite element method. In Chowdhury and Chaudhary [60], the powerful microwave devices were analyzed and optimization of these devices was considered.

In Ciarlet et al. [61], the powerful microwave electronic devices were analyzed, and the optimization of these devices was considered. Cole et al. [62] give a brief summary of works on the issue of calculating resonators. The methods for determining the specific frequency of the Helmholz resonator were considered, and the finite element analysis of this device was considered using the application of the COMSOL Multiphysics software complex. In a study by Collino [63], a method for calculating the complex propagation constant in slotted waveguides was developed. The obtained results were compared with the results obtained by Oliner's analytical method. Cui and Yang [64] developed a method of making a strip filter based on a rectangular waveguide made of 32 mixed materials. In a study by Cui et al. [65], the issue of new construction principles of SIW (Substrate Integrated Waveguide) integrated waveguide devices and their realization in passive and active components was considered. The calculation and design issues of SIW facilities were considered.

Dadgarpour et al. [66] considered the issue of electrodynamic modeling of the two-channel antenna splitter of the radiometric system. The conducted studies were performed in the frequency range of 8–12 GHz. Dai et al. [67] proposed a difference scheme for solving the Maxwell-Vlasov equations in an electrodynamic approach. The proposed difference scheme is applied to sonlu elementlər metodui-open waveguide systems. Datta et al. [68] considered the application of the finite difference method for the synthesis of optical fiber transmission lines. A concrete example for the synthesis of systems is shown. In a study by Dault [69], the issue of computer modeling of waveguide-slit radiator was considered. In David et al. [70], the problem of designing a bandpass filter based on a circular waveguide was solved.

Denisov and Kulygin [71] provide a methodology for calculating special microwave blocks using engineering analysis. In a study by Deslandes and Wu [72], the problem of determining the critical frequency and propagation constant of electromagnetic waves in circular and coaxial waveguides was solved. At this time, the method of partial regions was used. Dobler et al. [73] considered the modeling of a waveguide-slit radiator given a given discretization step. In this case, the HFSS software package was used. Du et al. [74] reviewed the electrodynamic modeling and optimization of a transistor microwave amplifier using the FEKO software package. In a study by Dumin et al. [75], using numerical modeling, it was shown that taking into account the ohmic attenuation in the walls of the waveguide leads to the fact that the field distribution in longitudinal periodic waveguides does not satisfy the condition of second-order periodicity.

Duvigneau [76] investigated the two types of waveguides with *H*-shaped cross sections. Numerical calculation of these waveguides was carried out in the frequency range of 7.8–9.2 GHz. The frequency dependencies of the emission and reflection coefficients were measured. In a study by Elmaozzen and Shafai [77], the problem of determining the intensity of the high-frequency electric field of the integral-optical sensor and the method of determining the sensitivity of this device were considered. Epstein et al. [78] experimentally studied the transition characteristics of a dielectric waveguide with a frequency range of 88–102 GHz. At the same time, they also carried out the numerical modeling of the rectangular waveguide-to-dielectric waveguide transition device. In a study by Famariji and Shongwe [79], the electrodynamic modeling of the diffraction structure included in the antenna-feeder devices based on the method of integral equations was considered.

In a study by Feng et al. [80], the issue of designing an antenna based on a planar dielectric waveguide was considered.

In Ghadami et al. [81], the issue of evaluating the scattering characteristics of electromagnetic waves in diffraction structures during design was considered. In Gong et al. [82], the problem of diffraction of the basic wave of a plane waveguide was solved by the projection method. In Gong and Zhang [83], the mathematical modeling of the electromagnetic field of a fractal regular waveguide with the *R*-function method was considered. Gric et al. [84] designed a microwave antenna in the Agilent EMPro three-dimensional electromagnetic modeling environment. In a study by Guerra et al. [85], the analysis of the existing constructions of the spiral antenna was carried out, and their advantages and disadvantages were clarified. In Gupta et al. [86], the analysis of the tract elements of the planar waveguide by the finite element method was considered. In a study by Han et al. [87], various options for the construction of a microwave waveguide were considered. Han et al. [88] studied a double-layer dielectric waveguide. Hassan [89] proposed a method of measuring the dielectric parameters of materials in the microwave range. In He et al. [90], a rectangular dielectric waveguide was studied and the main parameters of this device were determined. He et al. [91] analyzed losses in a rectangular waveguide filled with dielectric. He et al. [92] modeled a micro-strip line in the microwave range. In a study by Heidari and Ahmadi [93], the issue of realizing a cylindrical dielectric waveguide with a microwave range was solved. Helfert et al. [94] reviewed the electrodynamic analysis of the field distribution to develop a new microwave device and antenna. Hess et al. [95] performed an electrodynamic modeling of a waveguide diplexer on *E*-plane section and diaphragms. In a study by Hong [96], it was clarified that the use of an automated design system at the design stage has a positive effect on the quality of radio electronic devices.

In a study by Hozen [97], the solution of wave dispersion equations in guiding electrodynamic structures was considered [98]. analyzed waveguide transducers for controlling dielectric permittivity of materials and media. In Hu et al. [99], mathematical models of complex waves are given, and their solution algorithms are shown. In a study by Hui [100], the methods of measurements in the microwave technique are given and the characteristics of the measuring devices are indicated. In Hussain et al. [101], the generalized form of the Helmholtz equation for hybrid waves in regular gyrotropic waveguides was obtained.

Hwi-Min et al. [102] discussed the development and experimental study of axial waveguide transitions at a wavelength of 8 mm. In a study by Ikram et al. [103], the problem of diffraction was solved using the projection method for the H_{10} fundamental type wave of a rectangular waveguide. Iqbal et al. [104] analyzed a circular waveguide with ferrite layer and metamaterial. In a study by Islamov and Humbataliyev [105], the parameters of the emitter in a circular dielectric waveguide are modeled.

Islamov and Ismibayli [106] proposed a new type of lines for radio communication in microwave bands. In Islamov et al. [107], the dispersion characteristics of strip transmission lines in the centimeter and millimeter wave ranges were studied. They [108] proposed a no-contact method for measuring the parameters of a microwave signal in a rectangular waveguide. Earlier, they [109] discussed the issue of designing a device for measuring microwave power.

1.3 State of the Art of Optimal Synthesis of Microwave Devices

Currently, despite the wide application of microwave transmission lines, very little research work has been devoted to the issue of their synthesis. Existing [110] in scientific works, only separate aspects of the construction process of real microwave transmission lines were considered. Taking these into account, the relevance and expediency of considering the comprehensive synthesis process of microwave transmission lines operating in *E* and *H*-type waves has arisen. Therefore, the optimal synthesis of microwave transmission lines operating in *E* and *H*-type waves was considered in the monograph, taking into account the above-mentioned factors.

Since 1990, the issue of synthesis of microwave transmission lines has been continuously improved. If at the beginning the object of research was elementary transmission lines, then researches related to the synthesis of multi-element transmission lines, which are considered microwave transmission lines with a more complex structure, and even combined transmission lines with several structures began to be conducted. Thus, the problem of unconstrained single-criteria synthesis was replaced by the problem of multi-constrained multi-criteria synthesis.

The process of obtaining the final product of the synthesis issue is constantly being updated. Previously, the process of synthesis of real devices in a number of research works was considered satisfactory only according to the concrete results of one stage, but now this indicator is not justified. Because users demand more reliable and perfect microwave transmission lines in real practice. From this point of view, users are more interested in their parameters when designing real microwave transmission lines. When creating microwave transmission lines, in addition to optimal parameters, the following information about its structure must be considered: type of microwave transmission lines; the geometric size of its elements; optimal values of allowable dimensions; results of development and experimental study of microwave transmission lines.

Thus, the process of synthesis of microwave transmission lines can be considered satisfactory when the above mentioned factors are considered together. The physical model prepared by this method can be considered suitable for production. Therefore, considering the synthesis as a multi-stage process, developing real devices is not only an urgent issue, but also of great practical importance. From this point of view, the importance of the presented monograph is beyond doubt.

As mentioned above, in the scientific works in this field by Islamov et al. [111] as a one-step process for the optimal synthesis of some microwave transmission lines, in some scientific works Ismibayli and Islamov [112] considered the optimal synthesis of elementary microwave transmission lines.

In the presented work, the optimal synthesis of microwave transmission lines with a more complex structure operating in E and H-type waves is considered, which further increases the value of the dissertation.

During the development of microwave transmission lines, the issue of single-stage synthesis leads to the following negative situations:

- Noticeable errors occur in the geometric dimensions of the device.
- The economic and scientific-technical effectiveness decreases due to the fact that it is not possible to use the optimal information about the development of the device and the results of experimental studies.

1.4 The Role of Mathematical Models in the Process of Synthesis

The role of mathematical models in the process of optimal synthesis is very large. In practice, the time and economic cost to develop real devices is usually severely limited. This prompts the researcher to develop a more optimal version of the mathematical model. The main criteria of an optimal mathematical model are its accuracy and adequacy to the device or physical model. In principle, the possibilities of specifying the mathematical model of the device are not limited. However, as soon as the time and economic value are limited, the solution of the synthesis problem, the accuracy of the model takes the second place in the criterion of optimality, and the time spent on the numerical experiment comes to the fore. Indeed, the main stage of the synthesis process is the solution (in many cases approximation) of the optimization problem, which ultimately leads to multiple numerical experiments. Thus, researchers face an important issue – it is necessary to build a mathematical model of a microwave device that includes the factors of accuracy, preparation, numerical experiment, and time spent on solving the optimization problem. Such mathematical models, first introduced by Ismo, Janjan et al., Javanbakht et al., Jeong et al., and Ji et al. [113–117], were developed.

It should be noted that in order for the research to be successful, the optimization operation should be operational as well as economically efficient.

1.5 State of the Art of Microwave Range Device Research

To date, hundreds of journal articles and dozens of monographs have been published on the calculation of microwave transmission lines. Therefore, it is appropriate to highlight the most important ones by separating the scientific and research works done in this field.

The following can be attributed to these stages:

1. Study of simple devices (one-element). These devices are homogeneous transmission lines.
2. Study of class I multi-element (multi-stage) devices. These devices are connected to each other by cascade connections.
3. Study of nonhomogeneous transmission lines and devices based on them.
4. Study of class II multi-stage devices.
5. Optimum synthesis of devices with complex compounds taking into account inhomogeneity.

As you can see, the above stages are arranged in a certain time sequence with each other. Theoretical and experimental studies conducted at various stages are still ongoing. These stages can be briefly divided into the following.

I Stage The publication of the first scientific works related to the study of the composition of long lines dates back to the end of the nineteenth century. Soon, the process of researching facilities based on transmission lines was started. The simplest of such devices can be considered a short section of homogeneous transmission lines. This device made of transmission lines is connected to each other by two cascade connections. Depending on the ripple resistance, the combination of three such transmission lines can act as a ripple resistance transformer or filter. The study of such devices is hardly a priority in the current scientific researches.

In parallel with the study of homogeneous transmission lines, the study of multi-wire transmission lines has also been started. The first stage of such studies was the minimization of electromagnetic coupling between transmission lines. This connection between transmission lines was considered undesirable and was not even considered suitable for technical use. Scientists A.A.Pistolkors, M.S.Neyman, A. R.Wolpert were the first to use the electromagnetic effect of transmission lines for technical purposes. This prompted the creation of an extremely important microwave range device-directional splitter. Directional branchers are still widely used [118–120].

II Stage The main drawback of single-stage (single-element) devices was that their electrical characteristics were incomplete. The continuous expansion of the operating frequency range and the improvement of other parameters of devices with the microwave range have led to simpler, but more effective solutions. Thus, it was determined that the cascade connection of several similar transmission lines allows to significantly increase the amplitude-frequency characteristic or the phase-frequency characteristic of the device. However, at the same time, the complexity of the structure of this device and the increase in the number of parameters made its analysis, including its optimization, even more complicated. Therefore, the problem of evaluating the optimization of the obtained results appeared. The use of polynomials with optimal composition as a way out of this problem has paid off. It should be noted that at these stages the issue of parametric synthesis (parametric optimization) was solved only through analytical methods. Thus, the degree of idealization of the studied objects has increased. Therefore, this stage of the research of microwave devices can be considered a revolutionary stage.

First, a study of multi-stage devices built on homogeneous transmission lines was carried out. Many magazine articles were devoted to this issue. However, more serious and effective work in this field was devoted to the parametric synthesis of multistage devices, and this work belonged to Kohn and Collin. Unknown to each other, they solved the problem of the synthesis of a wave resistance transformer. Later, it was Rible who continued and expanded the work of these authors. He gave a solution to the problem of the synthesis of a multi-stage transformer with a more complex structure. Later, the device with this multi-element structure found its application as different types of filters [121]. Similar to the problem of analysis of a multilevel system based on homogeneous transmission lines, the problem of analysis of a multilevel system based on connected transmission lines used as a directional brancher was solved by Shimizu and Johnson. These researchers analyzed the three-stage symmetrical system and came to the conclusion that it is possible to increase the operating frequency range of the device. A.L.Feldstein first gave the solution to expand the working frequency band of the step brancher. A little later, a solution to the problem of synthesis of asymmetric and symmetric branchers built on connected transmission lines was given [122]. Stepped systems laid the foundation for the creation of other types of micowave devices. Examples of these are micowave band filters, wide band phase shifters, power dividers, amplitude regulators.

III Stage This stage can be characterized as the stage of increasing the working frequency band of stepped systems. So, as a result of the conducted research, the working frequency range of the specified devices was increased, and as a result, nonhomogeneous transmission lines were created. It is for this reason that in the last 10 years it has been possible to synthesize ultra-broadband microwave range devices. Examples of these devices include splitters, filters, transformers, dividers, and regulators. In parallel, the theory of inhomogeneous transmission lines and the theory of synthesis of microwave devices based on transmission lines were developed. Problems related to the theory of inhomogeneous transmission lines can be found in the works of Khevisayd. However, the theory of nonlinear transmission lines was more deeply dealt with by M.S.Neiman, I.I.Volman, A.R.Volpert, P.I. Kuznetsov, P.L.Stratonovich, O.N.Litvinenko, V.I.Soshnikova, A.L.Feldstein and others. The following equation was derived by them for the input reflection coefficient of nonhomogeneous transmission lines:

$$S_{11}^{1} + 2\gamma S_{11} - N\left(1 - S_{11}^{2}\right) = 0. \tag{1.1}$$

For smaller reflections

$$S_{11} = \int_{o}^{l} N \exp\left(-2\int_{z}^{l} \gamma d\varepsilon\right) dz. \tag{1.2}$$

Equations (1.1 and 1.2) were derived independently by Feldstein and Bolinder. Later, the differential equation for element S_{12} was derived and an approximate analytical solution for the weak coupling was shown. The indicated equation laid the basis for solving the problems of analysis and synthesis of microwave devices.

The device built on inhomogeneous transmission lines, first introduced by Kerim and Suad [123], was synthesized. These results were later confirmed by Kim, J. et al. [124]. Later, inhomogeneous transmission lınes, whose wave resistance varies according to a special law, were successfully used as amplitude regulators. In a study by Kim, S. et al. [125], the effective use of multi-element inhomogeneous transmission lines for those purposes was determined.

As for related inhomogeneous transmission lines, directional branchers were first synthesized based on them [126]. This topic is still developing intensively. In recent years, coupled nonlinear transmission lines have been used in filter synthesis problems. In the first works in this field, single-element inhomogeneous transmission lines were studied. Recently, multi-element connected inhomogeneous transmission lines have been successfully studied [127]. In addition to branching and filters, it was effectively applied in the matter of the synthesis of power dividers from connected nonhomogeneous transmission lines [128].

IV Stage The three stages described above differ from each other in one common aspect: the synthesized device consisted of one or more transmission lines connected in cascade with each other. Thus, only the amplitude principle of shaping the given frequency characteristics was used. The research conducted at the indicated stages later became the basis for the creation of a new type of microwave devices with a more complex structure.

The first studies of the fourth stage [129] was performed. This work is devoted to the analysis of the system of connected transmission lines of different lengths, which perform the function of directional branching. Later, this issue [130] was more successfully solved as a problem of parameric synthesis. As a result of these studies, filters and class II broadband rotators began to be created. At the same time, harmonic filters and surge resistance transformers were synthesized.

Microwave devices based on non-equidistant section transmission lines have surpassed classical devices in a number of aspects. The research of such devices is still ongoing. Their characteristic feature is that, unlike class 1 step devices, the problem of approximation of the given characteristics of these devices can be solved only by applying numerical methods. The global optimal evaluation of found parts remained an unsolved problem. Thus, in the fourth stage of research, structural devices were synthesized with the application of new mathematical methods and modern computing techniques.

V Stage The basis of this stage is the issue of multi-criteria synthesis of uni-functional and multi-functional microwave devices. This stage is characterized by the study of microwave devices with more complex structures. These devices differed not only in cascade connections of their elements, but also in type connections. Examples of these are 4-pole and 8-pole. In the issue of synthesis of this stage, constructive, economic, technological and other restrictions were imposed on devices. At this stage, the basis of solving the optimization problem is mathematical programming methods [131]. Later, new approaches were developed to solve the optimization problem. The most important of them, the optimal management method [132] is an example. Therefore, with the help of the optimal control method, microwave devices [133] began to be synthesized and its effectiveness was justified.

The research of the fifth stage can be divided into two main directions. More studies are devoted to the first direction and they aim to improve the obtained solutions. As a result of scientific research conducted in this direction, a new local optimal solution for the synthesis of microwave devices was developed. In the second stage, the issue of searching for global optimal solutions was put forward. As a result of this direction, the theory of circuits began to develop as a science, and the physical processes occurring in electric circuits began to be studied in depth.

VI Stage This stage has been continued since 1990. At this stage, more complex microwave range transmission lines began to be synthesized. These include directional splitters with a complex structure, channel converters, variable attenuators, matching switches, two-channel power dividers, etc. an example can be given. In their work, Larer and Tsvetrovskaya [134] dealt with the issues of optimal synthesis of various transmission lines operating in the centimeter, millimeter, and submillimeter wave ranges, and as a result developed new passive functional microwave devices. The study by Li et al. [135], looking at the synthesis of gradient-type optical-fiber transmission lines, used the finite element method to solve the synthesis problem.

Thus, by analyzing the development path of the research of microwave range of transmission lines, the following conclusions can be reached. Researches have been carried out by complexing the structures of transmission lines in the microwave range. At the initial stage of research, transmission lines with a multi-level structure were composed of homogeneous transmission lines. Later, devices with more complex structures began to be synthesized. Currently, the issue of the synthesis of transmission lines with a homogeneous and nonhomogeneous structure is almost solved. At the same time, it should be noted that the issue of optimal synthesis of some microwave range transmission lines has not been resolved considering multi-criteria factors. Examples of such devices are the metallic microwave waveguides shown in Fig. 1.8, which operate at frequencies of 300 MHz–300 GHz in the millimeter wave range. The most widely used metallic waveguides are microwave rectangular and circular waveguides.

Microwave rectangular and circular waveguides, which have high reliability, parameter stability and long service life, are widely used in communication technology, antenna technology and other fields of science and technology. Despite the wide field of application of these transmission lines, currently, quite few scientific studies have been devoted to the issue of their multi-criteria optimal synthesis.

Further improvement of electrical, magnetic, technical, constructive and operating parameters and characteristics of microwave rectangular and circular waveguides is one of the actual problems of microwave technique. Therefore, as a result of solving the problem of multi-criteria optimal synthesis of microwave rectangular and circular waveguides with modern numerical methods and the application of computer techniques, it will allow to get new physical models with optimal parameters, as well as improving the electrical, magnetic, technical, constructive and operational parameters and characteristics of these devices. It is from this point of view that the issue of multi-criteria optimal synthesis of devices shown in the monograph was considered.

1.6 Conclusions

1. Classification of microwave devices according to frequency ranges and constructions is given.
2. Microwave devices are classified according to the types of electromagnetic waves emitted, and the following are clarified: transverse electromagnetic wave microwave devices (*TEM*-type wave); magnetic wave microwave devices (*H* or *TE* wave); electric wave microwave devices (*E* or *TH* wave); hybrid wave microwave units.

3. *E*-type and *H*-type waves have been determined to propagate in microwave rectangular, circular and hollow metallic waveguides with a more complex cross-sectional area.
4. Microwave devices are divided into classifications according to the compatibility of cross-sectional areas. It is determined that the degree of communication is a geometric characteristic of the cross section of the microwave devices and is determined by the number of conductive surfaces. Depending on the number of conductive surfaces, microwave devices are divided into single-connection, two-connection, three-connection, multi-connection and zero-connection devices. It has been determined that there are no conducting surfaces in zero-coupling microwave devices, that metallic waveguides are single-coupling, coaxial waveguides are double-coupling, and dielectric waveguides are zero-coupling microwave devices.
5. It has been clarified that microwave rectangular and circular waveguides have high reliability, stability of parameters and long life in the operation process. Thus, since it is completely isolated from the environment, electromagnetic waves (radio waves) are not radiated to the external environment when transmitting information with such waveguides.
6. The wider use of air-filled rectangular and circular waveguides in microwave technology to transmit high-power energy in practice has been clarified, they are used in telecommunications, radio communication, television, radar, etc. it has been determined that the application of these systems allows to reduce the costs and increase their efficiency.
7. In order to clarify the modern state of the analysis of microwave devices, an overview of the scientific research works performed in this field was given, and as a result, it was determined that the environment is nonlinear during the study of the electromagnetic field of air-filled microwave rectangular and circular waveguides. has not been reviewed. Elucidation of the nonlinearity of the medium in which electromagnetic waves propagate has created the need to develop a new theory of microwave rectangular and circular waveguides.
8. In order to clarify the modern state of the optimal synthesis of microwave devices, an overview of the scientific research works performed worldwide in this field was given, and as a result, it was determined that the issue of their multi-criteria optimal synthesis has not yet been resolved, considering the nonlinearity of the environment. Therefore, solving the problem of multi-criteria optimal synthesis of microwave rectangular and circular waveguides with modern numerical methods and the application of computer techniques will allow the design and preparation of new physical models with optimal parameters while improving the electrical, magnetic, technical, structural and operational parameters and characteristics of these devices.

References

1. Abderrahmane, I., et al. (2023). Efficient 4 and 8 plasmonic wavelength DEMUX with ultra high Q-factor and low FWHM based on nano-rectangular resonators. *Optical and Quantum Electronics, 55*(2), 1–20.
2. Abdoli-Arani, A., & Montazeri, M. M. (2018). Dispersion relation and electron acceleration in the combined circular and elliptical metallic-dielectric waveguide filled by plasma. *Physics of Plasmas, 25*(4), 789–795.
3. Abdulhameed, M. K., et al. (2018). Controlling the radiation pattern of patch antenna using switchable EBG. *Telecommunication, Computing, Electronics and Control, 16*(5), 2014–2022.
4. Abir, T., et al. (2022). Second-harmonic enhancement from a nonlinear plasmonic metasurface coupled to an optical waveguide. *Nano Letters, 22*(7), 2712–2717.
5. Adnan, F., et al. (2014). *Numerical analysis for 2D waveguide by applying raleigh quotient iteration method.* In International conference on mathematics, engineering and industrial applications (Vol. 1660, pp. 437–445)
6. Alhassoon, K. A., et al. (2021). Complex permittivity and permeability extraction of ferromagnetic materials for magnetically tuned microwave circuits. *IEEE Journal of Microwaves, 1*(2), 639–645.
7. Ali, A. K. S., et al. (2022). Stabilization of light bullets in nonlinear metamaterial waveguides. *Physical Review A, 105*, 033516.
8. Ali, M. M. M., et al. (2018). Rectangular waveguide cross-guide couplers: Accurate model for full-band operation. *Institute of Electrical and Electronics Engineers Microwave and Wireless Components Letters, 28*(7), 561–563.
9. Al-Sabea, Z. S., et al. (2022). Plasmonic logic gates at optimum optical communications wavelength. *Advanced Electromagnetics, 11*(4), 10–21.
10. Alsuyuti, M. M., et al. (2019). Modified Galerkin algorithm for solving multitype fractional differential equations. *Mathematical Methods in the Applied Sciences, 42*(5), 1389–1412.
11. Altufaili, M. M. S., et al. (2022). Design of circular-shaped microstrip patch antenna for 5G applications. *Telecommunication, Computing, Electronics and Control, 20*(1), 19–26.
12. Amin, A. K. (2022). A proposed method for synthesizing the radiation pattern of linear antenna arrays. *Journal of Communication, 1*(7), 1–6.
13. Anand, J. K., & Kushwah, H. (2020). Behaviour of poynting vector for dielectric-metal-dielectric optical waveguides and applications. *Optical and Quantum Electronics, 52*, 410.
14. Annadhasan, M., et al. (2020). Micromanipulation of mechanically compliant organic single crystal optical microwaveguides. *Angewandte Chemie, International Edition, 59*, 13821–31380.
15. Argyropoulos, C., et al. (2012). Enhanced nonlinear effects in metamaterials and plasmonics. *Advanced Electromagnetics, 1*(1), 46–51.
16. Arnberg, P., et al. (2020). High refractive index electromagnetic devices in printed technology based on glide-symmetric periodic structures. *Applied Sciences, 10*, 3216.
17. Azeez, Y. F., et al. (2019). Establishing a new form of primary impedance standard at millimeter-wave frequencies. *IEEE Transactions on Instrumentation and Measurement, 68*(1), 294–296.
18. Babak, L. I. (2009). *Decomposition synthesis approach to design of RF and microwave active circuits*. In IEEE MTT-S International microwave symposium digest (Vol. 2, pp. 1167–1170).
19. Bachiller, C., et al. (2016). Radio-frequency performance comparison of several H-plane rectangular waveguide filters loaded with circular dielectric posts. *IET Microwaves, Antennas and Propagation, 10*(5), 536–545.
20. Bakkali, M. A., et al. (2022). Design and experimental validation of a multifunction single layer UHF-RFID tag antenna. *Advanced Electromagnetics, 11*(1), 22–29.

21. Balbastre, J. V., & Nuno, L. (2018). Modelling the propagation of electromagnetic waves across complex metamaterials in closed structures. *Journal of Computational and Applied Mathematics, 352*, 40–49.
22. Bandler, J. W., & Seviora, R. T. (1972). Wave sensitivities of networks. *IEEE Transactions on Microwave Theory and Techniques, MTT-20*, 138–147.
23. Basir, A., & Yoo, H. (2019). A stable impedance-matched ultrawideband antenna system mitigating detuning effects for multiple biotelemetric applications. *IEEE Transactions on Antennas and Propagation, 67*(5), 3416–3421.
24. Belenguer, A., et al. (2014). Novel empty substrate integrated waveguide for high-performance microwave integrated circuits. *IEEE Transactions on Microwave Theory and Techniques, 62*(4), 832–839.
25. Benassi, et al. (2021). A wearable flexible energy-autonomous filtenna for ethanol detection at 2.45 GHz. *IEEE Transactions on Microwave Theory and Techniques, 69*(9), 4093–4106.
26. Bird, T. S., et al. (2017). Improved finite-range gain formula for open-ended rectangular waveguides and pyramidal horns. *IET Microwaves, Antennas and Propagation, 11*(14), 2054–2058.
27. Bogle, A., et al. (2005). Electromagnetic material characterization using a partially-filled rectangular waveguide. *Journal of Electromagnetic Waves and Applications, 19*(3), 1291–1306.
28. Bogolyubov, A. N., et al. (2018). Analysis of a rectangular waveguide with allowance for losses in the walls. *Moscow University Physics Bulletin, 473*(6), 579–582.
29. Bozinovic, N., et al. (2013). Terabit-scale orbital angular momentum mode division multiplexing in fibers. *Science, 340*(6140), 1545–1548.
30. Bozzi, M., et al. (2014). *Modeling of losses in substrate integrated waveguide components*. In Proceedings of the IEEE international conference on numerical electromagnetic modeling and optimization for RF, microwave, and terahertz applications (NEMO) (pp. 1–4).
31. Bozzi, M., et al. (2019). *A review of compact substrate integrated waveguide (SIW) interconnects and components*. In Proceedings of the IEEE 23rd workshop on signal and power integrity (SPI) (pp. 1–4)
32. Buesa-Zubiria, A., & Esteban, J. (2017). Design of five-way Bagley polygon power dividers in rectangular waveguide. *Institute of Electrical and Electronics Engineers Transactions on Microwave Theory and Techniques, 66*(1), 116–127.
33. Bulashenko, A., et al. (2022). Accuracy and agreement of FDTD, FEM and wave matrix methods for the electromagnetic simulation of waveguide polarizers. *Advanced Electromagnetics, 11*(3), 1–9.
34. Butt, M. (2022). Simple and improved plasmonic sensor configuration established on MIM waveguide for enhanced sensing performance. *Plasmonics, 17*, 1305–1314.
35. Calignanoa, F., et al. (2019). High-performance microwave waveguide devices produced by laser powder bed fusion process. *Proceedings of the CIRP, 79*, 85–88.
36. Camacho, M., et al. (2020). Extraordinary transmission and radiation from finite by infinite arrays of slots. *IEEE Transactions on Antennas and Propagation, 68*, 581–586.
37. Cao, W.-K., et al. (2019). Asymmetric transmission of acoustic waves in a waveguide via gradient index metamaterials. *Science Bulletin, 64*(12), 808–813.
38. Carceller, C., et al. (2016). Design of hybrid folded rectangular waveguide filters with transmission zeros below the passband. *Institute of Electrical and Electronics Engineers Transactions on Microwave Theory and Techniques, 64*(2), 475–485.
39. Cassivi, Y., et al. (2002). Dispersion characteristics of substrate integrated rectangular waveguide. *IEEE Microwave and Wireless Components Letters, 12*, 333–335.
40. Castillo, R., et al. (2021). Platform-based electrically-small HF antenna with switchable directional radiation patterns. *IEEE Transactions on Antennas and Propagation, 69*(8), 4370–4379.

41. Ceccuzzi, S., et al. (2017). Physical mechanisms and design principles in mode filters for oversized rectangular waveguides. *IEEE Transactions on Microwave Theory and Techniques, 65*(8), 2726–2733.
42. Chaiyo, K., & Rattanadecho, P. (2013). Numerical analysis of heat-mass transport and pressure buildup of unsaturated porous medium in a rectangular waveguide subjected to a combined microwave and vacuum system. *International Journal of Heat and Mass Transfer, 65*, 826–844.
43. Chakravarthy, M. V. V. S. S. S., et al. (2022). Elliptical antenna array synthesis using evolutionary computing tools. *Arabian Journal for Science and Engineering, 47*, 2807–2824.
44. Chandel, R. (2018). Tapered fed compact UWB MIMO-diversity antenna with dual band-notched characteristics. *IEEE Transactions on Antennas and Propagation, 66*(4), 1677–1684.
45. Chang, C.-P., et al. (2018). Waveport modeling for the DGTD simulation of electromagnetic devices. *International Journal of Numerical Modelling: Electronic Networks, Devices and Fields, 66*(4), 1701–1712.
46. Chang, R.-J., & Huang, C.-C. (2022). Simulation of a high-performance polarization beam splitter assisted by two-dimensional metamaterials. *Nanomaterials, 12*, 1852.
47. Charles, A. G., & Guo, Y. (2022). A general approach for synthesizing multibeam antenna arrays employing generalized joined coupler matrix. *IEEE Transactions on Antennas and Propagation, 256*, 1–10.
48. Che, W., et al. (2010). Investigations on propagation and the band broadening effect of ridged rectangular waveguide integrated in a multilayer dielectric substrate. *IET Microwaves, Antennas and Propagation, 4*, 674–684.
49. Chen, H.-H., et al. (2022). Extraordinary optical characteristics of one-dimensional double anti-PT-symmetric ring optical waveguide networks. *Chinese Journal of Physics, 77*, 816–825.
50. Chen, J., & Liu, Q. (2013). Discontinuous Galerkin time-domain methods for multiscale electromagnetic simulations. A review. *Proceedings of the IEEE, 101*, 242–254.
51. Chen, J., et al. (2015). Using weakly conditionally stable-body of revolution-finite-difference time-domain met-hod to simulate dielectric film-coated circular waveguide. *IET Microwaves, Antennas and Propagation, 9*(9), 853–860.
52. Chen, S., et al. (2021). Optical waveguides based on one-dimensional organic crystals. *PhotoniX, 2*(2), 1–24.
53. Chen, S. C., & Chew, W. C. (2016). Discrete electromagnetic theory with exterior calculus. *PIERS Proceedings*, 896–897.
54. Chen, S. C., & Chew, W. C. (2017). Electromagnetic theory with discrete exterior calculus. *Progress in Electromagnetics Research, 159*, 59–78.
55. Chen, W., et al. (2019). Generalized coupled-mode formalism in reciprocal waveguides with gain loss anisotropy or bianisotropy. *Physical Review B, 99*(19), 345–357.
56. Chen, Z., et al. (2015). High sensitivity plasmonic sensing based on Fano interference in a rectangular ring waveguide. *Optics Communications, 340*, 1–4.
57. Chen, Z., et al. (2022). Plasmon-induced transparency for tunable atom trapping in a chiral metamaterial structure. *Nanomaterials, 12*, 516.
58. Cheng, Y.-F., et al. (2017). A novel wide-angle scanning phased array based on dual-mode pattern reconfigurable elements. *IEEE Antennas and Wireless Propagation Letters, 16*, 396–399.
59. Choo, J., et al. (2019). Evaluation of electromagnetic interference from axially ruptured coaxial cable with multiple dielectrics used in nuclear power plants. *IEEE Transactions on Electromagnetic Compatibility, 61*(3), 860–869.
60. Chowdhury, R., & Chaudhary, R. K. (2020). Investigation of new sectored hemispherical dielectric resonator antennas operating at TM101 and TE111 mode for circular polarization. *Progress in Electromagnetics Research, 167*, 95–109.

61. Ciarlet, P., et al. (2017). On the approximation of electromagnetic fields by edge finite elements. Part 2: A heterogeneous multiscale method for Maxwell's. *Computers & Mathematics with Applications, 73*(9), 1900–1919.
62. Cole, A. J., et al. (2018). The variation of the attenuation constant of low-loss transmission lines in the presence of standing waves. *IEEE Microwave and Wireless Components Letters, 28*(8), 639–641.
63. Collino, F. (2009). Conservative space-time mesh refinement methods for the FDTD solution of Maxwell's equations. *Journal of Computational Physics, 211*(1), 9–35.
64. Cui, L. C., & Yang, H. S. (2003). Rectangular-to-circular groove waveguide junction. *Science in China Series E: Technological Sciences, 46*(4), 343–348.
65. Cui, X., et al. (2018). High-efficiency, broadband converter from a rectangular waveguide TE10 mode to a circular waveguide TM01 mode for overmoded device measurement. *IEEE Access, 6*, 14996–15003.
66. Dadgarpour, A., et al. (2017). High-efficient circularly polarized magnetoelectric dipole antenna for 5G applications using dual-polarized split-ring resonator lens. *IEEE Transactions on Antennas and Propagation, 65*(8), 4263–4267.
67. Dai, L., et al. (2020). Ultrawideband low-profile and miniaturized spoof plasmonic vivaldi antenna for base station. *Applied Sciences, 10*, 2429.
68. Datta, S., et al. (2017). Fractal shaped antenna based triband energy harvester. *Advanced Electromagnetics, 6*(4), 22–26.
69. Dault, D. (2014). The generalized method of moments for electromagnetic boundary integral equations. *IEEE Transactions on Antennas and Propagation Antennas and Propagation, 62*(6), 3174–3188.
70. David, P., et al. (2022). Performance comparison of quantized control synthesis methods of antenna arrays. *Electronics, 11*(7), 994.
71. Denisov, G. G., & Kulygin, M. L. (2010). Numerical simulation of waveguide TM_{01}-TE_{11} mode converter using FDTD method. *International Journal of Infrared and Millimeter Waves, 26*(3), 341–361.
72. Deslandes, D., & Wu, K. (2002). *Design consideration and performance analysis of substrate integrated waveguide components*. In Proceedings of the 32nd European Microwave Conference (EuMC) (pp. 23–26)
73. Dobler, A., et al. (2020). An omnidirectional radial parallel-plate waveguide antenna using wideband mode converters. *IEEE Transactions on Antennas and Propagation, 68*(5), 3348–3357.
74. Du, S., et al. (2021). Optical waveguide and photoluminescent polarization in organic cocrystal polymorphs. *Journal of Physical Chemistry Letters, 12*(38), 9233–9238.
75. Dumin, O. M., et al. (2009). *Transient radiation from the open end of circular waveguide*. In 6th international conference on antenna theory and techniques (pp. 181–184).
76. Duvigneau, R. (2020). CAD consistent adaptive refinement using a NURBS based discontinuous Galerkin method. *International Journal for Numerical Methods in Fluids, 92*(9), 1096–1117.
77. Elmoazzen, Y. E., & Shafai, L. (1975). Numerical solution of coupling between two collinear parallel – Plate waveguides. *IEEE Transactions on Microwave Theory and Techniques, no. 11*, 871–876.
78. Epstein, C. L., et al. (2015). Debye sources, Beltrami fields, and a complex structure on Maxwell fields. *Communications on Pure and Applied Mathematics, 68*(12), 2237–2280.
79. Famoriji, O. J., & Shongwe, T. (2021). An effective antenna array diagnosis method via multivalued neural network inverse modeling approach. *Advanced Electromagnetics, 10*(3), 58–70.
80. Feng, L., et al. (2013). Experimental demonstration of a unidirectional reflectionless parity-time metamaterial at optical frequencies. *Nature Materials, 12*, 108–113.
81. Ghadami, A., et al. (2018). Damage identification in multi-step waveguides using lamb waves and scattering coefficients. *Archive of Applied Mechanics, 88*(6), 1009–1026.

82. Gong, L., et al. (2019). Optical orbital-angular-momentum-multiplexed data transmission under high scattering. *Light: Science and Applications, 8*(1), 1–11.
83. Gong, L., & Zhang, K. (1997). A finite element analysis on electromagnetic waves in rectangular dielectric chirowaveguide. *Journal of Infrared and Millimeter Waves, 6*, 1353–1362.
84. Gric, T., et al. (2010). Electrodynamical characteristic particularity of open metamaterial square and circular waveguides. *Progress in Electromagnetics Research, 109*, 361–379.
85. Guerra, G., et al. (2022). Unified coupled-mode theory for geometric and material perturbations in optical waveguides. *Journal of Lightwave Technology, 40*(14), 4714–4727.
86. Gupta, A., et al. (2013). Sound propagation in two-dimensional waveguide with circular wavefront. *Acoustical Physics, 59*(5), 493–497.
87. Han, X., et al. (2015). Dynamically tunable plasmon induced transparency in a graphene-based nanoribbon waveguide coupled with graphene rectangular resonators structure on sapphire substrate. *Optics Express, 23*(25), 31945–31955.
88. Han, Y., et al. (2020). Tunable metamaterial-based silicon waveguide. *Optics Letters, 45*(24), 6619–6622.
89. Hassan, A. A. (2014). *Multilayer medium technique for nondestructive EM-properties measurement of radar absorbing materials using flanged rectangular waveguide sensor and FDTD method*. In IEEE international symposium on innovations in intelligent systems and applications (pp. 68–75)
90. He, B., et al. (2017). Time decomposition method for the general transient simulation of low-frequency electromagnetics. *Progress in Electromagnetics Research, 160*, 1–8.
91. He, S., et al. (2015). Can Maxwell's fish eye lens really give perfect imaging? Part III. A careful reconsideration of the evidence for subwavelength imaging with positive refraction. *Progress in Electromagnetics Research, 152*, 1–15.
92. He, Y., et al. (2015). Hamiltonian time integrators for Vlasov-Maxwell equations. *Physics of Plasmas, 22*(12), 234–239.
93. Heidari, M., & Ahmadi, V. (2020). Graphene-based mid-infrared plasmonic isolator with multimode interferometer. *Optics Letters, 45*(20), 5764–5767.
94. Helfert, S. F., et al. (2017). Numerical simulation of hollow waveguide arrays as polarization converting elements and experimental verification. *Optical and Quantum Electronics, 4*(9), 123–132.
95. Hess, M. W., et al. (2015). Estimating the Inf-sup constant in reduced basis methods for time-harmonic Maxwell's equations. *IEEE Transactions on Microwave Theory and Techniques, 63*(11), 3549–3557.
96. Hong, W. (2017). Solving the 5G mobile antenna puzzle: Assessing future directions for the 5G mobile antenna paradigm shift. *IEEE Microwave Magazine, 18*(7), 86–102.
97. Hozen, T. (2019). Design of a planar array antenna on a narrow wall of the rectangular waveguide for linear polarization perpendicular to the axis with standing-wave excitation. *IEICE Technical Report, 119*(228), 49–54.
98. Htet, A. T. (2019). Comparative performance of the finite element method and the boundary element fast multipole method for problems mimicking transcranial magnetic stimulation TMS. *Journal of Neural Engineering, 16*(2), 52–65.
99. Hu, H.-T., et al. (2022). 60 GHz Fabry–Perot cavity filtering antenna driven by an SIW-fed filtering source. *IEEE Transactions on Antennas and Propagation, 70*, 823–834.
100. Hui, L. (2022). Synthesis, control, and excitation of characteristic modes for platform-integrated antenna designs: A design philosophy. *IEEE Antennas and Propagation Magazine, 64*(2), 41–48.
101. Hussain, N., et al. (2020). Metasurface-based single-layer wideband circularly polarized MIMO antenna for 5G millimeter-wave systems. *IEEE Access, 8*, 130293–130304.
102. Hwi-Min, K., et al. (2018). Monolithic integration of self-aligned nanoisland laser with shifted-air-hole waveguide. *Optics Express, 26*(10), 12569–12578.

103. Ikram, M., et al. (2020). Common-aperture Sub-6 GHz and millimeter-wave 5G antenna system. *IEEE Access, 8*, 199415–199423.
104. Iqbal, A., et al. (2019). Tunable substrate integrated waveguide diplexer with high isolation and wide stopband. *IEEE Microwave and Wireless Components Letters, 29*, 456–458.
105. Islamov, I., & Humbataliyev, E. (2022). General approaches to solving problems of analysis and synthesis of directional properties of antenna arrays. *Advanced Electromagnetics, 11*(4), 22–33.
106. Islamov, I. J., & Ismibayli, E. G. (2018). Experimental study of characteristics of microwave devices transition from rectangular waveguide to the megaphone. *IFAC-PapersOnLine, 51*(30), 477–479.
107. Islamov, I. J., et al. (2021). *Simulation of electrodynamic processes in a cylindrical-rectangular microwave waveguide systems transmitting information.* In 11th International conference on theory and application of soft computing, computing with words, perception and artificial intelligence (Vol. 362, pp. 246–253)
108. Islamov, I. J., et al. (2021). Numerical simulation of characteristics of propagation of symmetric waves in microwave circular shielded waveguide with a radially inhomogeneous dielectric filling. *International Journal of Microwave and Wireless Technologies, 14*(6), 761–767.
109. Islamov, I. J., et al. (2019). Modeling of the electromagnetic feld of a rectangular waveguide with side holes. *Progress in Electromagnetics Research, 81*, 127–132.
110. Islamov, I. J., et al. (2018). Electrodynamics characteristics of the no resonant system of transverse slits located in the wide wall of a rectangular waveguide. *Progress in Electromagnetics Research Letters, 8*, 23–29.
111. Islamov, I. J., et al. (2019). Calculation of the electromagnetic field of a rectangular waveguide with chiral medium. *Progress in Electromagnetics Research, 84*, 97–114.
112. Ismibayli, E. G., & Islamov, I. J. (2018). New approach to definition of potential of the electric field created by set distribution in space of electric charges. *IFAC-PapersOnLine, 51*(30), 410–414.
113. Ismo, V. L. (2015). Plane-wave propagation in electromagnetic PQ medium. *Progress in Electromagnetics Research, 154*, 23–33.
114. Janjan, B., et al. (2022). Efficient second-harmonic generation in an embedded ABC-metamaterial waveguide. *IEEE Journal of Quantum Electronics, 58*(3), 234–248.
115. Javanbakht, N., et al. (2021). Side-lobe level reduction of half-mode substrate integrated waveguide leaky-wave antenna. *IEEE Transactions on Antennas and Propagation, 69*, 3572–3577.
116. Jeong, M. J., et al. (2019). Millimeter-wave microstrip patch antenna using vertically coupled split ring metaplate for gain enhancement. *Microwave and Optical Technology Letters, 6*(10), 2360–2365.
117. Ji, L., et al. (2021). A slow wave ridged half-mode substrate integrated waveguide with spoof surface plasmon polaritons. *IEEE Transactions on Plasma Science, 49*, 1818–1825.
118. Ji, Z. Q., et al. (2018). Circularly polarized dielectric rod waveguide antenna for millimeter-wave applications. *IEEE Transactions on Antennas and Propagation, 66*(10), 5080–5087.
119. Jilani, S. F., & Alomainy, A. (2018). Millimetre-wave T-shaped MIMO antenna with defected ground structures for 5G cellular networks. *IET Microwaves, Antennas and Propagation, 12*(5), 672–677.
120. Kamada, S., & Okamoto, T. (2018). Design optimization and resonance modes of a plasmonic sensor based on a rectangular resonator. *Optics Communications, 427*, 220–225.
121. Kancleris, Z., et al. (2010). *Interaction of thin conductive sheets with TE_{10} electromagnetic wave in rectangular waveguide*. In 4th Microwave and Radar Week (MRW). 18th International conference on microwaves, radar and wireless communications (Vol. 1 & 2, pp. 35–43)
122. Kerim, G., & Suad, B. (2015). A quantized water cycle optimization algorithm for antenna array synthesis by using digital phase shifters. *International Journal of RF and Microwave Computer-Aided Engineering, 25*(1), 21–29.

123. Kesari, V., & Keshari, J. P. (2019). Hybrid-mode analysis of circular waveguide with chiral dielectric lining for dispersion characteristics for potential application in broadbanding a gyro-traveling-wave tube. *Journal of Electromagnetic Waves and Applications, 33*(2), 204–214.
124. Kim, J., et al. (2018). Radiation from a millimeter-wave rectangular waveguide slot array antenna enclosed by a von karman radome. *Journal of Electromagnetic Engineering Sciences, 18*(3), 154–159.
125. Kim, S., et al. (2020). Computational characterization of microwave planar cutoff probes for non-invasive electron density measurement in low-temperature plasma: Ring- and bar-type cutoff probes. *Applied Sciences, 10*, 7066.
126. Klinaku, S. (2019). The general Galilean transformation versus Lorentz transformation: The motion in an isosceles triangle. *Physics Essays, 32*(2), 253–254.
127. Koenen, C., & Hamberger, G. F. (2018). A low-reflectivity vacuum window for rectangular hollow waveguides. *IEEE Transactions on Microwave Theory and Techniques, 66*, 128–135.
128. Krysl, P. (2003). Natural hierarchical refinement for finite element methods. *International Journal for Numerical Methods in Engineering, 56*(8), 1109–1124.
129. Kuhler, L., et al. (2018). The propagation characteristics of 2-D metamaterial waveguides using the modal expansion theory. *Institute of Electrical and Electronics Engineers Transactions on Microwave Theory and Techniques, 66*(10), 4319–4326.
130. Kurachka, K. S. (2018). Numerical modeling of a influence of a nanoparticle pair on the electromagnetic field in the near zone by the vector finite elements method. *Computer Optics, 42*(4), 542–549.
131. Lan, F., et al. (2012). Study on TE_{0n} nonuniform ripple-wall mode converter in circular waveguide. *Acta Physica Sinica, 61, nio. 15*, 567–574.
132. Lan, J., et al. (2020). An aperture-sharing array for 3.5/28 GHz terminals with steerable beam in millimeter wave band. *IEEE Transactions on Antennas and Propagation, 68*(5), 4114–4119.
133. Larer, A., et al. (1996). The sonlu elementlər metodui inversion method for cylindrical microwave structures. *Journal of Electromagnetic Waves and Application, 10*(6), 765–790.
134. Larer, A., & Tsvetrovskaya, S. (1997). Universal method of the a of analysis multilayered planer lines and complex waveguides. *International Journal of Microwave and Millimeter Wave Computer Aided Engineering, 7*(6), 483–494.
135. Li, F., et al. (2017). Design and microwave measurement of a Ka-band HE11 mode corrugated horn for the Faraday rotator. *IET Microwaves, Antennas and Propagation, 11*(1), 75–80.

Chapter 2
Modeling of Microwave Waveguide Systems of Complex Structure in Nonlinear Media

2.1 Mathematical Models of a Microwave Rectangular Waveguide in a Nonlinear Medium

Air-filled rectangular waveguides are widely used in practice in microwave techniques to transmit large power. Microwave rectangular waveguides are used in telecommunications, radio communication, television, radar, etc. Its application not only reduces the cost spent on the development of these systems, but also increases their effectiveness in turn. When transmitting information through microwave rectangular waveguides, electromagnetic processes propagate under extremely complex conditions.

Until now, in the art, when designing air-filled microwave rectangular waveguides, including calculating their parameters, the medium through which electromagnetic waves propagate has been considered as a linear medium. But in real practice, when transmitting information through air-filled microwave rectangular waveguides, the medium behaves as a nonlinear medium. Thus, the electrophysical parameters of the environment change depending on the intensity of the electric or magnetic field of the microwave rectangular waveguide.

Therefore, the issue of developing new mathematical models of microwave rectangular waveguide filled with air, which is important for practice, taking into account the nonlinearity of the environment, is very relevant.

Now let's look at the issue of obtaining new mathematical models of an air-filled microwave rectangular waveguide, taking into account the nonlinearity of the medium. Let's use the Cartesian coordinate system for this. Let's place the origin of the Cartesian coordinate system at the center of the rectangular waveguide and take the axes in the direction of the sides of the rectangular waveguide, as shown in Fig. 2.1.

I. Islamov, *Radio Engineering and Telecommunications Waveguide Systems in the Microwave Range*, https://doi.org/10.1007/978-3-031-37916-1_2

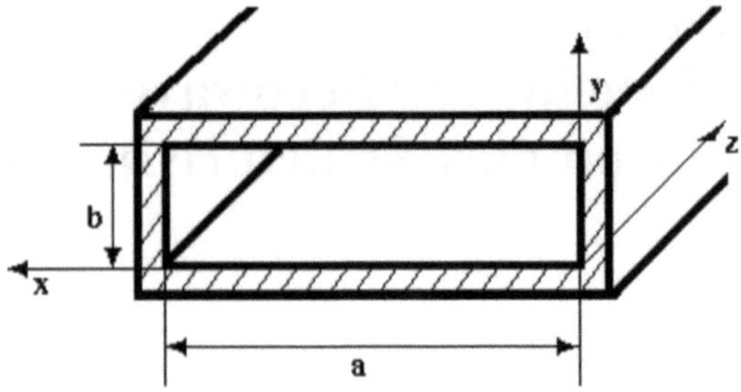

Fig. 2.1 Microwave rectangular waveguide in Cartesian coordinate system

It is known that [1]:

$$rot\vec{H} = \vec{J} + \frac{\partial \vec{D}}{\partial t}. \tag{2.1}$$

Equation 2.1 in the vectorial form is equivalent to the equations in the following three scalar forms in the *XYZ* Cartesian coordinate system:

$$\begin{cases} \frac{\partial H_z}{\partial y} - \frac{\partial H_y}{\partial z} = j_x + \frac{\partial D_x}{\partial t}, \\ \frac{\partial H_x}{\partial z} - \frac{\partial H_z}{\partial x} = j_y + \frac{\partial D_y}{\partial t}, \\ \frac{\partial H_y}{\partial x} - \frac{\partial H_x}{\partial y} = j_z + \frac{\partial D_z}{\partial t}. \end{cases} \tag{2.2}$$

On the other hand, it is known that Maxwell's differential second equation is as follows [2]:

$$rot\vec{E} = -\frac{\partial \vec{B}}{\partial t}. \tag{2.3}$$

Equation 2.3 in the vectorial form is equivalent to the equations in the following three scalar forms in the *XYZ* Cartesian coordinate system:

$$\begin{cases} \frac{\partial E_z}{\partial y} - \frac{\partial E_y}{\partial z} = -\frac{\partial B_x}{\partial t}, \\ \frac{\partial E_x}{\partial z} - \frac{\partial E_z}{\partial x} = -\frac{\partial B_y}{\partial t}, \\ \frac{\partial E_y}{\partial x} - \frac{\partial E_x}{\partial y} = -\frac{\partial B_z}{\partial t}. \end{cases} \tag{2.4}$$

Transverse electric (*TM* or *E*-type) type waves and transverse magnetic (*TE* or *H*-type) type waves can be propagated in a rectangular waveguide. For *E*-type waves, there are E_x, E_y, E_z, H_x, H_y, $H_z = 0$ components of the electromagnetic field, and for H-type waves E_x, E_y, $E_z = 0$, H_x, H_y, H_z components. So let's look at both the above cases:

(a) The system of differential Eqs. 2.2 and 2.4 for the *E*-type wave is as follows:

$$\begin{cases} -\dfrac{\partial H_y}{\partial z} = j_x + \dfrac{\partial D_x}{\partial t}, \\ \dfrac{\partial H_x}{\partial z} = j_y + \dfrac{\partial D_y}{\partial t}, \\ \dfrac{\partial H_y}{\partial x} - \dfrac{\partial H_x}{\partial y} = j_z + \dfrac{\partial D_z}{\partial t}, \\ \dfrac{\partial E_z}{\partial y} - \dfrac{\partial E_y}{\partial z} = -\dfrac{\partial B_x}{\partial t}, \\ \dfrac{\partial E_x}{\partial z} - \dfrac{\partial E_z}{\partial x} = -\dfrac{\partial B_y}{\partial t}, \\ \dfrac{\partial E_y}{\partial x} - \dfrac{\partial E_x}{\partial y} = -\dfrac{\partial B_z}{\partial t}. \end{cases} \tag{2.5}$$

Considering that, $\vec{j} = \sigma \vec{E}, \vec{D} = \varepsilon_a \vec{E}$ [3] and $\vec{B} = \mu_a \vec{H}$ [4], (2.5) becomes as follows:

$$\begin{cases} -\dfrac{\partial H_y}{\partial z} = \sigma E_x + \left(\varepsilon_a + E_x \dfrac{\partial \varepsilon_a}{\partial E_x} \right) \dfrac{\partial E_x}{\partial t}, \\ \dfrac{\partial H_x}{\partial z} = \sigma E_y + \left(\varepsilon_a + E_y \dfrac{\partial \varepsilon_a}{\partial E_y} \right) \dfrac{\partial E_y}{\partial t}, \\ \dfrac{\partial H_y}{\partial x} - \dfrac{\partial H_x}{\partial y} = \sigma E_z + \left(\varepsilon_a + E_z \dfrac{\partial \varepsilon_a}{\partial E_z} \right) \dfrac{\partial E_z}{\partial t}, \\ \dfrac{\partial E_z}{\partial y} - \dfrac{\partial E_y}{\partial z} = -\mu_a \dfrac{\partial H_x}{\partial t}, \\ \dfrac{\partial E_x}{\partial z} - \dfrac{\partial E_z}{\partial x} = -\mu_a \dfrac{\partial H_y}{\partial t}, \\ \dfrac{\partial E_y}{\partial x} - \dfrac{\partial E_x}{\partial y} = -\mu_a \dfrac{\partial H_z}{\partial t} = 0. \end{cases} \tag{2.6}$$

(b) The system of differential Eqs. 2.2 and 2.4 for the *H*-type wave is as follows:

$$\begin{cases} \dfrac{\partial H_z}{\partial y} - \dfrac{\partial H_y}{\partial z} = \sigma E_x + \left(\varepsilon_a + E_x \dfrac{\partial \varepsilon_a}{\partial E_x} \right) \dfrac{\partial E_x}{\partial t}, \\ \dfrac{\partial H_x}{\partial z} - \dfrac{\partial H_z}{\partial x} = \sigma E_y + \left(\varepsilon_a + E_y \dfrac{\partial \varepsilon_a}{\partial E_y} \right) \dfrac{\partial E_y}{\partial t}, \\ \dfrac{\partial H_y}{\partial x} - \dfrac{\partial H_x}{\partial y} = 0, \\ \dfrac{\partial E_y}{\partial z} = \mu_a \dfrac{\partial H_x}{\partial t}, \\ \dfrac{\partial E_x}{\partial z} = -\mu_a \dfrac{\partial H_y}{\partial t}, \\ \dfrac{\partial E_y}{\partial x} - \dfrac{\partial E_x}{\partial y} = -\mu_a \dfrac{\partial H_z}{\partial t}. \end{cases} \tag{2.7}$$

If we perform some mathematical operations on the system of Eqs. 2.6 and 2.7, we get the following.

Thus, the system of differential equations with the second-order special formulation for the *E*-type wave is obtained as follows:

$$\begin{cases} \sigma \dfrac{\partial E_x}{\partial t} + \left(\varepsilon_a + E_x \dfrac{\partial \varepsilon_a}{\partial E_z} \right) \dfrac{\partial^2 E_x}{\partial t^2} = \dfrac{1}{\mu_a} \left(\dfrac{\partial^2 E_x}{\partial z^2} - \dfrac{\partial^2 E_z}{\partial x \partial z} \right), \\ \sigma \dfrac{\partial E_y}{\partial t} + \left(\varepsilon_a + E_y \dfrac{\partial \varepsilon_a}{\partial E_y} \right) \dfrac{\partial^2 E_y}{\partial t^2} = \dfrac{1}{\mu_a} \left(\dfrac{\partial^2 E_y}{\partial z^2} - \dfrac{\partial^2 E_z}{\partial y \partial z} \right), \\ \sigma \dfrac{\partial E_z}{\partial t} + \left(\varepsilon_a + E_z \dfrac{\partial \varepsilon_a}{\partial E_z} \right) \dfrac{\partial^2 E_z}{\partial t^2} = \dfrac{1}{\mu_a} \left(\dfrac{\partial^2 E_y}{\partial y \partial z} + \dfrac{\partial^2 E_z}{\partial y^2} + \dfrac{\partial^2 E_z}{\partial x^2} - \dfrac{\partial^2 E_x}{\partial x \partial z} \right), \\ \left(\varepsilon_a + E_y \dfrac{\partial \varepsilon_a}{\partial E_y} \right) \dfrac{\partial^2 E_y}{\partial x \partial t} - \left(\varepsilon_a + E_x \dfrac{\partial \varepsilon_a}{\partial E_x} \right) \dfrac{\partial^2 E_x}{\partial y \partial t} = \dfrac{\partial^2 H_x}{\partial x \partial z} + \dfrac{\partial^2 H_y}{\partial y \partial z}, \\ \sigma \mu_a \dfrac{\partial H_y}{\partial t} - \left(\varepsilon_a + E_x \dfrac{\partial \varepsilon_a}{\partial E_x} \right) \dfrac{\partial^2 E_x}{\partial z \partial t} + \left(\varepsilon_a + E_z \dfrac{\partial \varepsilon_a}{\partial E_z} \right) \dfrac{\partial^2 E_z}{\partial x \partial t} = \dfrac{\partial^2 H_y}{\partial z^2} + \dfrac{\partial^2 H_y}{\partial x^2} - \dfrac{\partial^2 H_x}{\partial x \partial y}. \end{cases} \tag{2.8}$$

For the *H*-type wave, the system of differential equations with the second-order special formulation is obtained as follows:

$$
\begin{cases}
\sigma\mu_a \dfrac{\partial H_y}{\partial t} - \left(\varepsilon_a + E_x \dfrac{\partial \varepsilon_a}{\partial E_x}\right) \dfrac{\partial^2 E_x}{\partial z \partial t} = \dfrac{\partial^2 H_y}{\partial z^2} - \dfrac{\partial^2 H_z}{\partial y \partial z}, \\
\sigma\mu_a \dfrac{\partial H_x}{\partial t} + \left(\varepsilon_a + E_y \dfrac{\partial \varepsilon_a}{\partial E_y}\right) \dfrac{\partial^2 E_y}{\partial z \partial t} = \dfrac{\partial^2 H_x}{\partial z^2} - \dfrac{\partial^2 H_z}{\partial x \partial z}, \\
\sigma \dfrac{\partial E_x}{\partial t} + \left(\varepsilon_a + E_x \dfrac{\partial \varepsilon_a}{\partial E_x}\right) \dfrac{\partial^2 E_x}{\partial t^2} = \dfrac{1}{\mu_a} \left[\dfrac{\partial H_z}{\partial y} \left(\dfrac{\partial E_y}{\partial x} - \dfrac{\partial E_x}{\partial y} \right) + \dfrac{\partial H_y}{\partial z} \dfrac{\partial E_x}{\partial z} \right], \\
\sigma \dfrac{\partial E_y}{\partial t} + \left(\varepsilon_a + E_y \dfrac{\partial \varepsilon_a}{\partial E_y}\right) \dfrac{\partial^2 E_y}{\partial t^2} = \dfrac{1}{\mu_a} \left[\dfrac{\partial H_z}{\partial z} \dfrac{\partial E_y}{\partial z} + \dfrac{\partial H_z}{\partial x} \left(\dfrac{\partial E_y}{\partial x} - \dfrac{\partial E_x}{\partial y} \right) \right], \\
\sigma\mu_a \dfrac{\partial H_z}{\partial t} - \left(\varepsilon_a + E_y \dfrac{\partial \varepsilon_a}{\partial E_y}\right) \dfrac{\partial^2 E_y}{\partial x \partial t} + \left(\varepsilon_a + E_x \dfrac{\partial \varepsilon_a}{\partial E_x}\right) \dfrac{\partial^2 E_x}{\partial y \partial t} = \\
= \dfrac{\partial^2 H_y}{\partial y \partial z} + \dfrac{\partial^2 H_z}{\partial y^2} + \dfrac{\partial^2 H_z}{\partial x^2} - \dfrac{\partial^2 H_x}{\partial x \partial z}.
\end{cases}
\tag{2.9}
$$

2.2 Analytical Expression Between Relative Dielectric Permittivity and Electric Field Intensity in a Nonlinear Medium

It should be noted that in extremely high-intensity strong fields, the medium behaves as a nonlinear medium. In turn, the dependence between the relative dielectric permittivity ε and the intensity of the electric field E in a nonlinear environment will also be nonlinear. During the propagation of microwave waves inside rectangular waveguides, because the field is strong, water droplets form on its inner walls. At this time, the energy of the electromagnetic field is extinguished. This creates a serious obstacle to the transmission of information through rectangular waveguides. Extinction of the energy of the electromagnetic field in water droplets is accompanied by the generation of polarization currents in each droplet when electromagnetic waves pass through the water droplets. Therefore, the density of polarization currents increases significantly and increases as the frequency of the emitted electromagnetic waves increases.

Due to the damping of electromagnetic waves in water droplets, the field of the propagated wave is also subject to damping. The physical meaning of this process is that radio waves formed in water droplets radiate electromagnetic waves equally in all directions. This leads to power dissipation. The larger the water droplets, the greater the scattering. In addition, quenching also occurs in water molecules. This phenomenon can be explained by the fact that H_2O molecules have a constant electric moment, and O_2 molecules have a magnetic moment [5]. The electromagnetic field of the propagating wave affects the electron cloud of molecules. At this time, if the frequency of the wave coincides with the specific frequency of the molecules, a resonance phenomenon occurs and the energy of the waves is converted

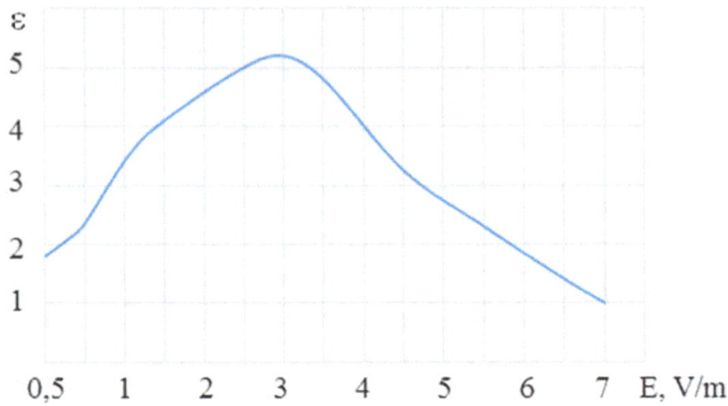

Fig. 2.2 Dependence between relative dielectric permittivity ε and electric field intensity E

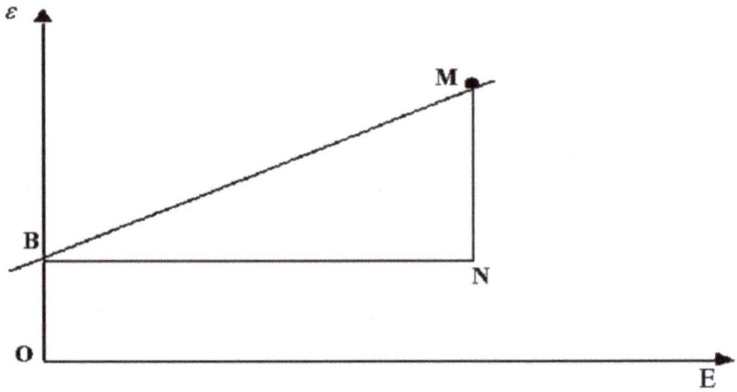

Fig. 2.3 ε and on the determination of the dependence between parameters E

into intramolecular energy. Therefore, the energy of the electromagnetic wave is subject to extinction. This is a strong proof that the medium in which electromagnetic waves propagate behaves as a nonlinear medium.

For this considered case, let's determine the dependence between the relative dielectric permeability ε and the intensity of the electric field E. For this, let's use the graph shown in Fig. 2.2 [6].

As can be seen from Fig. 2.2, the dependence between the relative dielectric permeability ε and the intensity of the electric field E behaves linearly in the region ab. Then the following relation is true:

$$\varepsilon = \varepsilon_1 + kE = 1,8 + kE. \tag{2.10}$$

Now let's choose the piece $OB = 1{,}8$ on $o\varepsilon$ (Fig. 2.3). Let's draw the straight line given by the Eq. 2.10. For this, let's choose a piece parallel to the axis OE from point B and whose length is equal to $BN = 3 \times 105$. At the same time, let's choose a piece that is parallel to the N $o\varepsilon$ axis and whose length is equal to $NM = 3{,}2$.

Then select the desired *BM*. It has an angle coefficient of $\kappa = 3,\ 2/3 \cdot 10^5$and intersects the $o\varepsilon$ axis at $\varepsilon_1 = 1,\ 833$.

Thus, considering the value of *k* in (2.10), we get

$$\varepsilon = 1,8 + 1,066 \cdot 10^{-5}E. \tag{2.11}$$

Then

$$\varepsilon_a = \varepsilon_0\varepsilon = \varepsilon_0\left(1,8 + 1,066 \cdot 10^{-5}E\right). \tag{2.12}$$

Since there are $\frac{\partial \varepsilon_a}{\partial E} = k,$ the sum of $\varepsilon_a + E\frac{\partial \varepsilon_a}{\partial E}$ will be as follows:

$$\varepsilon_a + E\frac{\partial \varepsilon_a}{\partial E} = 1,8\varepsilon_0 + 1,066 \cdot 10^{-5}(\varepsilon_0 + 1)E. \tag{2.13}$$

Thus, if we consider (2.13) in (2.8) and (2.9), the new mathematical models characterizing the electromagnetic field of the air-filled microwave rectangular waveguide will have the following form:

(a) For an *E*-type wave (*TM*-wave) of a microwave rectangular waveguide:

$$\begin{cases} \sigma\frac{\partial E_z}{\partial t} + \left[1,8\varepsilon_0 + 1,066 \cdot 10^{-5}(\varepsilon_0 + 1)E_z\right]\frac{\partial^2 E_z}{\partial t^2} = -\frac{1}{\mu_a}\left(\frac{\partial^2 E_x}{\partial z^2} - \frac{\partial^2 E_z}{\partial x \partial z}\right), \\ \sigma\frac{\partial E_y}{\partial t} + \left[1,8\varepsilon_0 + 1,066 \cdot 10^{-5}(\varepsilon_0 + 1)E_y\right]\frac{\partial^2 E_y}{\partial t^2} = \frac{1}{\mu_a}\left(\frac{\partial^2 E_y}{\partial z^2} - \frac{\partial^2 E_z}{\partial y \partial z}\right), \\ \sigma\frac{\partial E_x}{\partial t} + \left[1,8\varepsilon_0 + 1,066 \cdot 10^{-5}(\varepsilon_0 + 1)E_z\right]\frac{\partial^2 E_z}{\partial t^2} = \\ = \frac{1}{\mu_a}\left(\frac{\partial^2 E_y}{\partial y \partial z} + \frac{\partial^2 E_z}{\partial y^2} + \frac{\partial^2 E_z}{\partial x^2} - \frac{\partial^2 E_z}{\partial x \partial z}\right), \\ \left[1,8\varepsilon_0 + 1,066 \cdot 10^{-5}(\varepsilon_0 + 1)E_y\right]\frac{\partial^2 E_y}{\partial x \partial t} - \left[1,8\varepsilon_0 + 1.066 \cdot 10^{-5}(\varepsilon_0 + 1)E_x\right]\frac{\partial^2 E_x}{\partial y \partial t} = \\ = \frac{\partial^2 H_x}{\partial x \partial z} + \frac{\partial^2 H_y}{\partial y \partial z}, \\ \sigma\mu_a\frac{\partial H_y}{\partial t} - \left[1,8\varepsilon_0 + 1,066 \cdot 10^{-5}(\varepsilon_0 + 1)E_x\right]\frac{\partial^2 E_x}{\partial z \partial t} + \\ + \left[1,8\varepsilon_0 + 1,066 \cdot 10^{-5}(\varepsilon_0 + 1)E_z\right]\frac{\partial^2 E_z}{\partial x \partial t} = \frac{\partial^2 H_y}{\partial z^2} + \frac{\partial^2 H_y}{\partial x^2} + \frac{\partial^2 H_x}{\partial x \partial y}. \end{cases} \tag{2.14}$$

(b) For an *H*-type wave (*TE*-wave) of a microwave rectangular waveguide:

$$\begin{cases} \sigma\mu_a\dfrac{\partial H_y}{\partial t}-\left[1{,}8\varepsilon_0+1{,}066\cdot10^{-5}(\varepsilon_0+1)E_x\right]\dfrac{\partial^2 E_x}{\partial z\partial t}=\dfrac{\partial^2 H_y}{\partial z^2}-\dfrac{\partial^2 H_z}{\partial y\partial z}, \\ \sigma\mu_a\dfrac{\partial H_x}{\partial t}+\left[1{,}8\varepsilon_0+1{,}066\cdot10^{-5}(\varepsilon_0+1)E_y\right]\dfrac{\partial^2 E_y}{\partial z\partial t}=\dfrac{\partial^2 H_x}{\partial z^2}-\dfrac{\partial^2 H_z}{\partial{\scriptstyle >}\partial z}, \\ \sigma\dfrac{\partial E_x}{\partial t}+\left[1{,}8\varepsilon_0+1{,}066\cdot10^{-5}(\varepsilon_0+1)E_x\right]\dfrac{\partial^2 E_x}{\partial t^2}=\dfrac{1}{\mu_a}\left[\dfrac{\partial H_z}{\partial y}\left(\dfrac{\partial E_x}{\partial x}-\dfrac{\partial E_x}{\partial y}\right)+\dfrac{\partial H_y}{\partial z}\dfrac{\partial E_x}{\partial z}\right], \\ \sigma\dfrac{\partial E_y}{\partial t}+\left[1{,}8\varepsilon_0+1{,}066\cdot10^{-5}(\varepsilon_0+1)E_y\right]\dfrac{\partial^2 E_y}{\partial t^2}=\dfrac{1}{\mu_a}\left[\dfrac{\partial H_z}{\partial z}\dfrac{\partial E_y}{\partial z}+\dfrac{\partial H_z}{\partial x}\left(\dfrac{\partial E_y}{\partial x}-\dfrac{\partial E_x}{\partial y}\right)\right], \\ \sigma\mu_a\dfrac{\partial H_z}{\partial t}-\left[1{,}8\varepsilon_0+1{,}066\cdot10^{-5}(\varepsilon_0+1)E_y\right]\dfrac{\partial^2 E_y}{\partial x\partial t}+\left[1{,}8\varepsilon_0+1{,}066\cdot10^{-5}(\varepsilon_0+1)E_x\right]\dfrac{\partial^2 E_x}{\partial y\partial t}= \\ =\dfrac{\partial^2 H_y}{\partial y\partial z}+\dfrac{\partial^2 H_z}{\partial y^2}+\dfrac{\partial^2 H_z}{\partial x^2}-\dfrac{\partial^2 H_x}{\partial x\partial z}. \end{cases} \tag{2.15}$$

As you can see, these mathematical models are a system of differential equations of the second order with a special formulation. It is impossible to solve these equations by analytical methods. Therefore, numerical methods should be applied to solve these equations.

2.3 Calculation of the Electric Field of a Microwave Rectangular Waveguide by the Finite Difference Method

The finite difference method is one of the most effective methods for solving area problems. When calculating the electric field inside any volume *V* of the microwave rectangular waveguide using the finite difference method, its solution must satisfy the boundary conditions on the surface S. The equations in the system of Eqs. 2.14 and 2.15 are elliptic type equations. The ellipticity of Eqs. 2.14 and 2.15 is true under the given conditions [7]. Therefore, the simple boundary condition in region V for calculating the area is the Dirichlet condition:

$$\phi(x,y,z)=\phi_0(x,y,z). \tag{2.16}$$

At the same time, in order to solve the elliptic type equations in the system of Eqs. 2.14 and 2.15, it is necessary to add the condition of regularity of the potential at infinity to the condition (2.16). I mean,

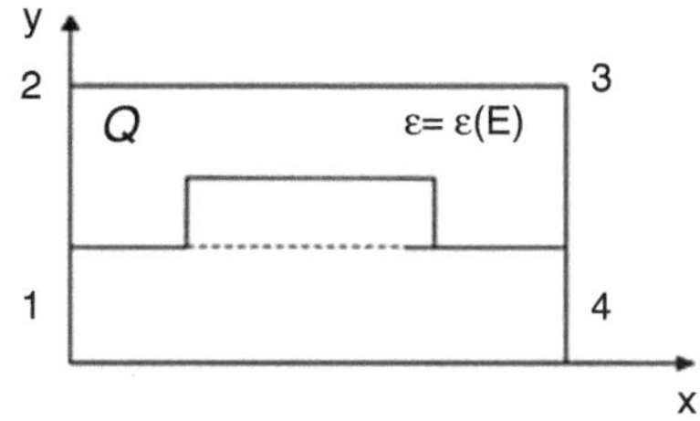

Fig. 2.4 Calculation domain: 1–4 knots

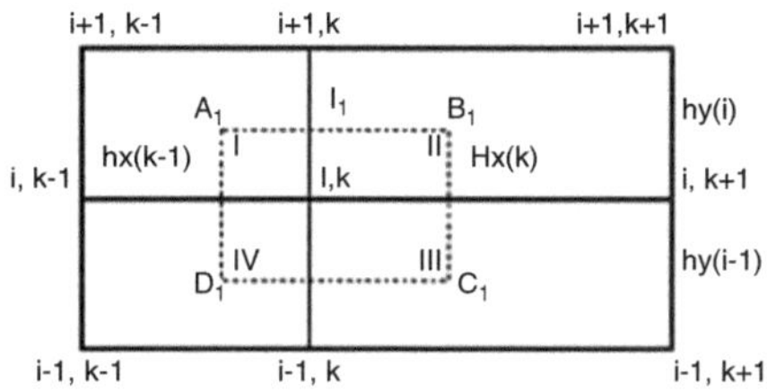

Fig. 2.5 Internal node of the network: I-IV-semi-regions

$$\lim_{(x^2+y^2+z^2)\to\infty} \phi(x, y, z) < \infty. \tag{2.17}$$

It should be noted that the solution of the elliptic type equations in the system of Eqs. 2.14 and 2.15 should be brought to the system of algebraic equations.

The finite element method is the theoretical basis of almost all methods applied recently for calculating the electric field distribution [8]. In addition to being universal, this method is also an effective and widespread method for solving field problems. For the application of this method, initial, boundary, and initial-boundary conditions must be known.

It is appropriate to replace differential equations with integral equations for the development of numerical algorithms in the case of nonlinearity of the environment. Therefore, in the case of Eqs. 2.14 and 2.15, the system of differential equations should be replaced by the system of integral equations. According to ϕ scalar electric potentials, the finite difference scheme in the region Q is constructed according to Figs. 2.4 and 2.5.

To solve the system of differential Eqs. 2.14 and 2.15, we choose the following initial and boundary conditions for the electric field intensity vector $\vec{E}$ and the scalar electric potential ϕ:

$$\begin{gathered}
\frac{\partial \phi}{\partial y} = \alpha_1(x) 1\text{–}4 \text{ in the area;} \\
\frac{\partial \phi}{\partial x} = 0 \ 1\text{–}2 \text{ in the area;} \\
\phi = \chi \ 2\text{–}3 \text{ in the area;} \\
\phi = \alpha_2(x) \ 3\text{–}4 \text{ in the area,}
\end{gathered} \tag{2.18}$$

where $\alpha_1(x)$, $\alpha_2(x)$ are the given functions, $\chi = const$ is a known number.

The discontinuity condition is met at the inner boundaries of the environment. To simplify the calculations, let's consider the two-dimensional distribution of the electric field inside a microwave rectangular waveguide.

Using the following equations according to the complete current law

$$\begin{cases} E_x = -\frac{\partial \phi}{\partial y}\varepsilon(E) = -\frac{\partial \phi}{\partial y}\left(1{,}8 + 1{,}066 \cdot 10^{-5} E\right), \\ E_y = -\varepsilon\frac{\partial \phi}{\partial x}(E) = -\frac{\partial \phi}{\partial x}\left(1{,}8 + 1{,}066 \cdot 10^{-5} E\right), \end{cases} \tag{2.19}$$

for the electric field in a two-dimensional system, we get integral equations in the following form:

$$\int_l \left(1{,}8 + 1{,}066 \cdot 10^{-5} E\right)\frac{\partial \phi}{\partial x}dx + \int_l \left(1{,}8 + 1{,}066 \cdot 10^{-5} E\right)\frac{\partial \phi}{\partial x}dy = \frac{1}{\varepsilon_0}\iint_Q \rho dxdy. \tag{2.20}$$

Step $hx(\kappa)$, $k = 1,2,\ldots,n_1-1$, $hy(i)$, $i = 1, 2,\ldots,n_2-1$ (n_1– nodes on the OX axis) in Q of the calculation area (Fig. 2.4) number, n_2 – the number of nodes on the OY axis), let's build a non-periodic network and denote it by $\overline{Q}$. Within this, let's look at the arbitrary $\overline{Q}$-(i, k) (Fig. 2.5) grid. Let's assume that node (i, k) is located in region Q_1. This region is surrounded by $l_1(A_1B_1C_1D_1)$ contours. Since the density of charges and the intensity of the electric field within the slots of the network are unchanged [9]. Let's calculate the integral shown in Eq. 2.20 on the $l_1(A_1B_1C_1D_1)$ contours of the region Q_1.

$$\phi_{i,k} = \frac{I + \gamma_1\phi_{i-1,k} + \gamma_2\phi_{i,k+1} + \gamma_3\phi_{i,k=1} + \gamma_4\phi_{i+1,\kappa}}{\gamma_1 + \gamma_2 + \gamma_3 + \gamma_4}, \tag{2.21}$$

$$\begin{cases} \gamma_1 = \frac{\left(1,8 + 1,066 \cdot 10^{-5} E_{i,k-1}\right)hx\left(1,8 + 1,066 \cdot 10^{-5} E_{i,k}\right)hx(k)}{2hy(i)}, \\ \gamma_2 = \frac{\left(1,8 + 1,066 \cdot 10^{-5} E_{i-1,k}\right)hy(i-1) + \left(1,8 + 1,066 \cdot 10^{-5} E_{i,k}\right)hx(i)}{2h(x)}, \\ \gamma_3 = \frac{\left(1,8 + 1,066 \cdot 10^{-5} E_{i-1,k-1}\right)hy(i-1) + \left(1,8 + 1,066 \cdot 10^{-5} E_{i,k-1}\right)hy(i)}{2hx(k-1)}, \\ \gamma_4 = \frac{\left(1,8 + 1,066 \cdot 10^{-5} E_{i-1,k-1}\right)hx(k-1) + \left(1,8 + 1,066 \cdot 10^{-5} E_{i-1,k}\right)hx(k)}{2hy(i-1)}, \end{cases} \tag{2.22}$$

$$I = I_I hx(k-1)hy(i)/4 + I_{II} hx(k)hy(i)/4+ \\ +I_{III} hx(k)hy(i-1)/4 + I_{IV} hx(k-1)hy(i-1)/4. \tag{2.23}$$

Accordingly, the electric field intensity E at point $\chi_2\left(y_i + \frac{hy(i)}{2}, x_k + \frac{hx(k)}{2}\right)$ is defined as:

$$E_x = \frac{\phi_{i,k+1} - \phi_{i+1,k+1} + \phi_{i,k} - \phi_{i+1,k}}{2hy(i)}, \tag{2.24}$$

$$E_y = \frac{\phi_{i+1,k+1} - \phi_{i,k} + \phi_{i,k+1} - \phi_{i+1,k}}{2hx(k)}. \tag{2.25}$$

According to the formula (2.21), by finding the value of the potential at any internal point of the network, the value of its neighboring potentials can also be determined. Different boundary conditions should be given according to (2.18) for each oblate of the computational domain. Difference scheme and equations of this scheme should be established for each area. $\phi = \alpha_2(x)$ Dirichlet conditions are given in region 3–4. In this case, the potential of the boundary node is set once as set $\alpha_2(y)$ at point y_{i,n_1}. In the region 2–3, the value of the potential remains unchanged and is equal to χ. $\partial\phi/\partial x = 0$ symmetry conditions are given in the region 1–2. Let's choose contour $l_4(A_4B_4C_4D_4)$ to go beyond the boundary conditions. Calculating (2.20) on the contour $l_4(A_4B_4C_4D_4)$ along the region Q_4, we get

$$\phi_{i,1} = \frac{\gamma_1\phi_{i-1,1} + \gamma_2\phi_{i,2} + \gamma_4\phi_{i+1,1}}{\gamma_1 + \gamma_2 + \gamma_3}. \tag{2.26}$$

Let's look at the given case under the following condition:

$$\left.\frac{\partial x}{\partial y}\right|_{1-4} = \alpha_1(x).$$

Where, the integration contour should be selected so that its lower part passes through the region $l_3(A_3B_3C_3D_3)$ (Fig. 2.6b).

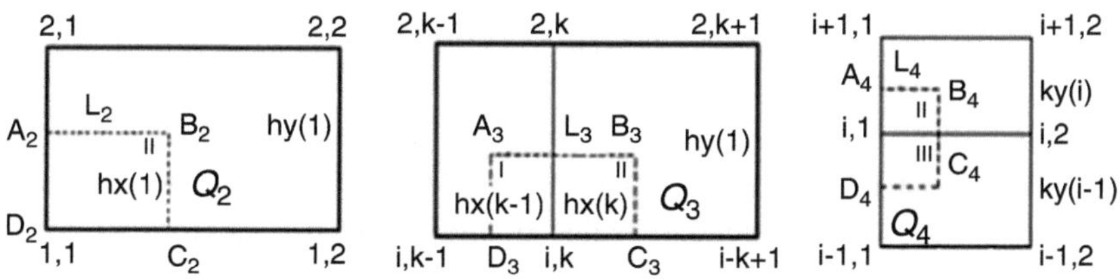

Fig. 2.6 Border nodes of the network

Let's carry out integration along the contour of $l_3(A_3B_3C_3D_3)$

$$\phi_{1,k} = \frac{\gamma_2\phi_{2,k} + \gamma_1\phi_{1,k+1} + \gamma_4\phi_{3,k}}{\gamma_1 + \gamma_3 + \gamma_4}, \tag{2.27}$$

where

$$\gamma = \int_{x_k + \frac{hx(k-1)}{2}}^{x_K + \frac{hx(k)}{2}} \alpha_1(x)dx, \gamma_1 = 0, \gamma_2 = \frac{hy(1)}{2hx(k)},$$

$$\gamma_3 = \frac{hy(1)}{2hx(k-1)}, \gamma_4 = \frac{hx(k-1) + hx(x)}{2hy(1)}.$$

For point 1 (Fig. 2.6)

$$\phi_{1,1} = \frac{\gamma_2\phi_{1,2} + \gamma_4\phi_{2,1} - \gamma}{\gamma_2 + \gamma_4}. \tag{2.28}$$

Where are $l_2(A_2B_2C_2D_2)$ integration contours (Fig. 2.6a):

$$\gamma_1 = 0, \gamma_2 = \frac{hy(1)}{2hx(1)}, \gamma_3 = 0, \gamma_4 = \frac{hx(1)}{2hy(1)}, \gamma = \int_{x_1}^{x_1 + \frac{hx(1)}{2}} \alpha_1(x)dx,$$

At point 4 (Fig. 2.6) the differential equation is as follows:

$$\phi_{1,n_1} = \frac{\gamma_1\phi_{2,n_1} + \gamma_3\phi_{2,n_1-1} - \gamma}{\gamma_1 + \gamma_3}. \tag{2.29}$$

Where l_3 integration contours are symmetrical to the contour shown in Fig. 2.6, *a*:

$$\gamma_1 = \frac{hx(n_1 - 1)}{2hy(1)}, \gamma_2 = 0, \gamma_3 = \frac{hy(1)}{2hx(n_1 - 1)}, \gamma_4 = 0, \gamma = \int_{x_{h1} \frac{hx(n_1-1)}{2}}^{xn_1} \alpha_1(x)dx.$$

Equations 2.24 to 2.27 and 2.29 form the system of linear equations of the difference scheme of the boundary value problem, and it is appropriate to solve this system by iteration methods, by the successive upper relaxation method (Yang's method). So, this method is very simple and easy to implement. Based on the above mathematical algorithm, a block diagram was developed (Fig. 2.7). Based on this

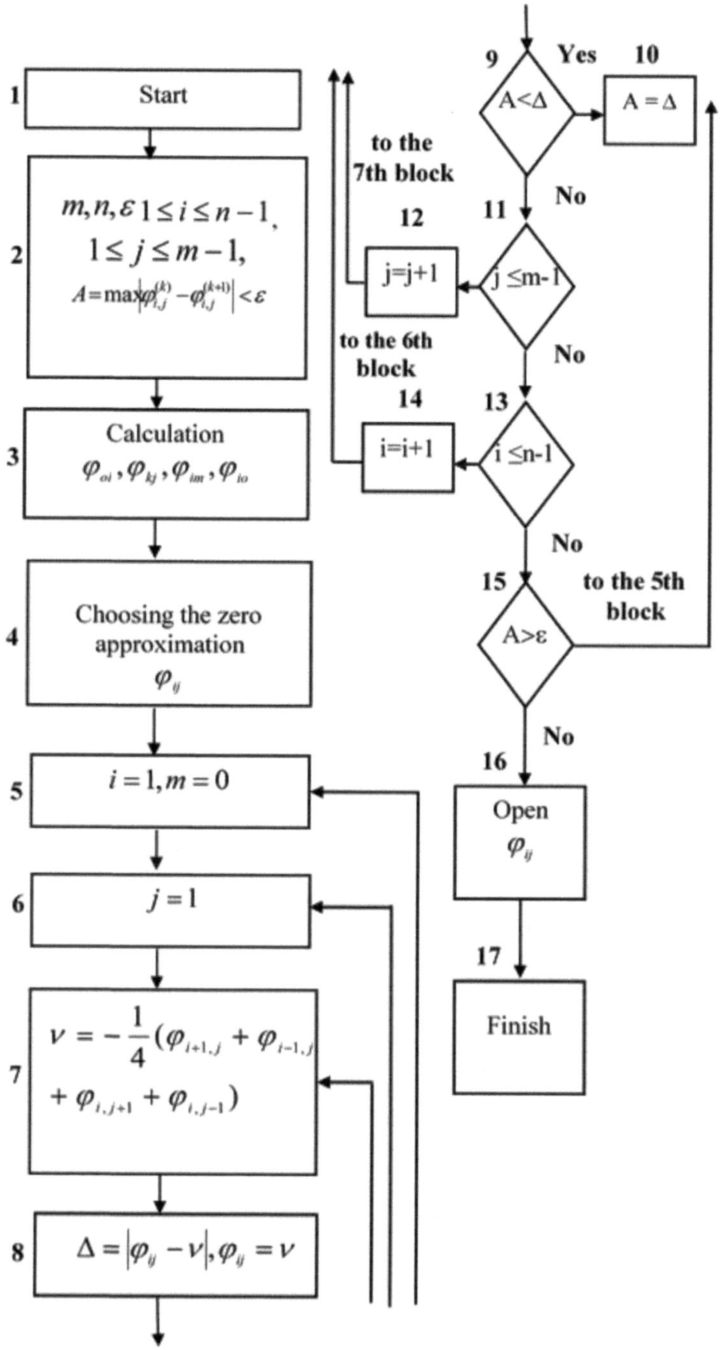

Fig. 2.7 Block diagram of the finite difference method for calculating the electromagnetic field of a microwave rectangular waveguide

Table 2.1 Values of the electric field intensities of a microwave rectangular waveguide with a nonlinear medium operating in the frequency range of 4.9–7.05 *GHz* for *E*-type and *H*-type waves determined by the finite difference method

	For *E*-type wave		For *H*-type wave	
No. of elementary fields	E_x, (10^5) *V/m*	E_y, (10^5) *V/m*	E_x, (10^5) *V/m*	E_y, (10^5) *V/m*
1	2.18	4.45	4.45	2.45
2	3.48	5.24	1.6	2.59
3	4.07	5.28	1.43	3.39
4	4.47	5.29	5.42	4.13
5	4.44	5.36	5.51	3.44
6	4.43	6.07	4.78	3.37
7	4.43	6.42	2.11	2.42
8	2.83	4.75	1.37	1.40
9	3.85	4.27	4.21	1.19
10	4.15	5.47	5.21	1.23
11	4.19	6.03	6.52	1.60
12	3.16	6.25	4.12	1.54
13	2.67	6.26	4.11	1.79
14	2.55	6.29	7.31	2.19
15	2.43	5.59	7.52	3.61
16	2.27	5.14	1.49	1.57
17	2.91	5.12	1.42	1.19
18	3.16	1.20	2.51	4.1
19	4.37	4.37	3.19	5.49
20	5.24	4.78	4.42	6.49

block diagram, calculations were made in the C++ program and the electromagnetic field intensities of the microwave rectangular waveguide with nonlinear medium operating in the *E*-type and *H*-type waves in the frequency range of 4.9–7.05 *GHz* were determined. The obtained numerical values are given in Tables 2.1 and 2.2. Based on the obtained numerical values, the dependencies of the length of the microwave waveguide operating in the frequency range of 4.9–7.05 *GHz* on *E*-type and *H*-type waves with a nonlinear medium and its electric field intensities were established (Fig. 2.8a, b). These dependencies determine how the electromagnetic field is distributed inside the investigated waveguide and the relationship between the electromagnetic and structural parameters of this device. Figure 2.9 shows the 3*D* model (finite difference method) of the electric field intensity distribution for the *E*-type wave of a rectangular waveguide operating in the frequency range of 4.9–7.05 *GHz*. In Fig. 2.10, the dependencies of the length of the *E*-type (*a*) and *H*-type (*b*) waves on the frequency range of 4,9–7,05 *GHz* of the nonlinear medium microwave waveguide on the intensities of its magnetic field, Fig. 2.11 shows the 3*D* model (finite difference method) of the distribution of the magnetic field intensity for the *E*-type wave of the rectangular waveguide operating in the frequency range of 4.9–7.05 *GHz*.

Table 2.2 Values of the magnetic field intensities of the microwave rectangular waveguide with nonlinear medium operating in the frequency range of 4.9–7.05 *GHz* for *E*-type and *H*-type waves determined by the finite difference method

	For *E*-type wave		For *H*-type wave	
Elemen-tary fields	H_x, *A/m*	H_y, *A/m*	H_x, *A/m*	H_y, *A/m*
1	51.86	24.2	53.6	27.94
2	67.11	44.25	24.5	36.3
3	43.56	25.43	36.57	76.23
4	64.16	34.98	44.89	46.98
5	11.34	86.78	65.79	63.39
6	53.12	44.9	34.79	65.15
7	55.56	39.19	74.76	22.37
8	60.47	33.18	24.26	33.71
9	62.12	25.17	11.98	38.18
10	51.45	43.45	33.76	11.56
11	51.9	44.86	35.69	32.96
12	42.45	46.58	73.72	34.48
13	35.19	53.56	35.43	20.11
14	56.61	67.5	67.31	40.81
15	48.13	31.34	34.53	43.12
16	60.16	42.95	33.99	30.15
17	80.78	61.89	64.44	10.35
18	62.34	43.56	76.45	38.48
19	60.45	43.56	41.25	30.05
20	50.23	26.21	40.38	22.38

2.4 Mathematical Models of a Microwave Circular Waveguide in a Nonlinear Medium

Now let's look at the issue of obtaining new mathematical models of the air-filled microwave circular waveguide, taking into account the nonlinearity of the medium. Let's use the cylindrical coordinate system for this. So, let's place the origin of the cylindrical coordinate system in the center of the circular waveguide, and take the axes in the direction of the sides of the circular waveguide, as shown in Fig. 2.12.

In the r, θ, z cylindrical coordinate system, Maxwell's first equation can be written as follows [10]:

$$\frac{\vec{r}_0}{r}\left(\frac{\partial H_z}{\partial \theta} - r\frac{\partial H_0}{\partial z}\right) + \vec{\theta}_0\left(\frac{\partial H_r}{\partial z} - \frac{\partial H_z}{\partial r}\right) + \frac{\vec{z}_0}{r}\left(\frac{\partial (rH_\theta)}{\partial r} - \frac{\partial Hr}{\partial \theta}\right) = \vec{j} + \frac{\partial \vec{D}}{\partial t}. \tag{2.30}$$

Equation 2.30 in vectorial form is equivalent to the following three scalar equations in the cylindrical coordinate system r, θ, z.

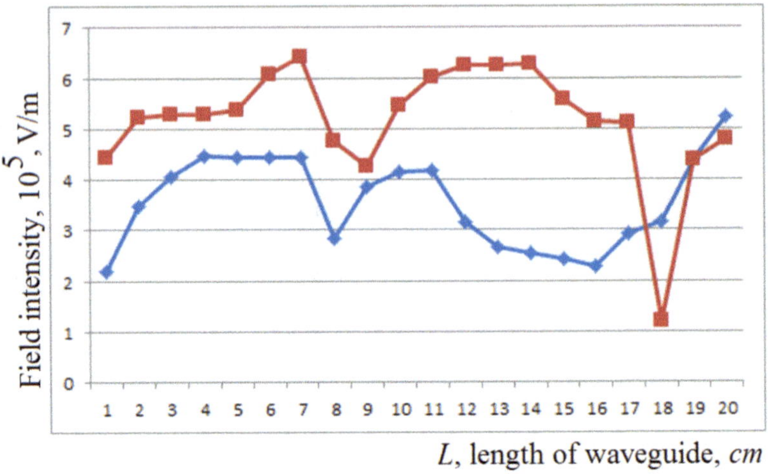

a)

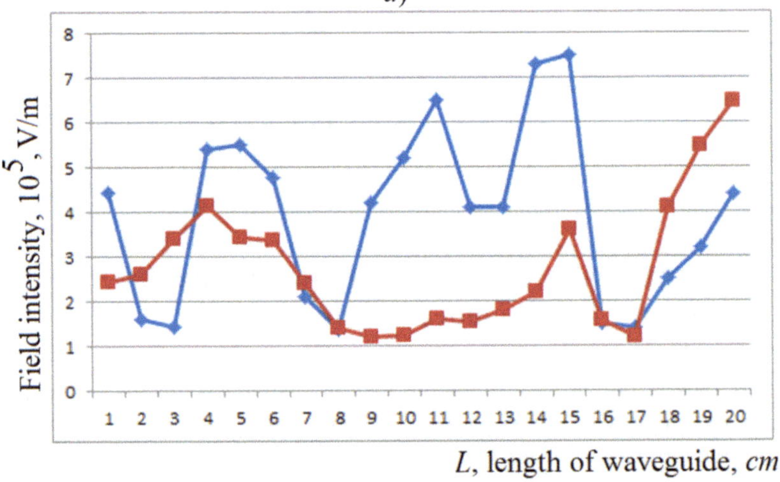

b)

Fig. 2.8 Dependencies of the length of the *E*-type (**a**) and *H*-type (**b**) waveguide in the frequency range of 4.9–7.05 *GHz* with a nonlinear medium microwave on the electric field intensities: E_x – blue color, E_y – red color

$$\begin{cases} \dfrac{\partial H_z}{\partial \theta} - r\dfrac{\partial H_\theta}{\partial z} = j_r + \dfrac{\partial D_r}{\partial t}, \\ \dfrac{\partial H_r}{\partial z} - \dfrac{\partial H_z}{\partial x} = j_\theta + \dfrac{\partial D_\theta}{\partial t}, \\ \dfrac{\partial (rH_\theta)}{\partial r} - \dfrac{\partial H_r}{\partial \theta} = j_z + \dfrac{\partial D_z}{\partial t}. \end{cases} \tag{2.31}$$

Similarly, Maxwell's second equation can be written in scalar form in the cylindrical coordinate system r, θ, z as follows [11]:

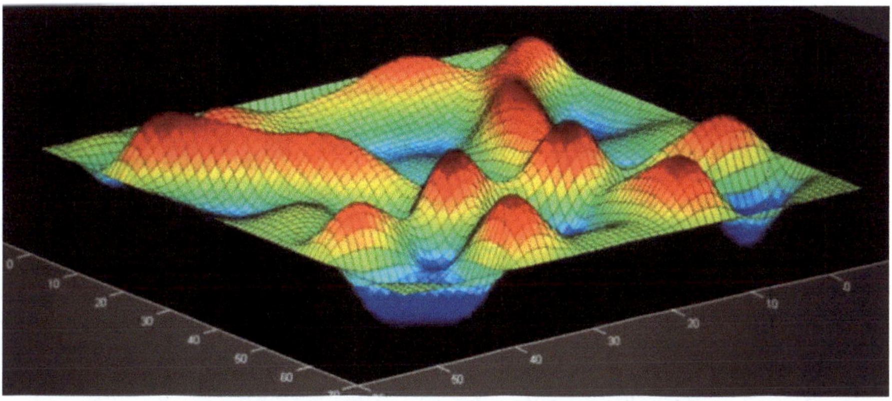

Fig. 2.9 3*D* model of electric field distribution for *E*-type wave of a microwave rectangular waveguide operating in the frequency range of 4.9–7.05 *GHz* (finite difference method)

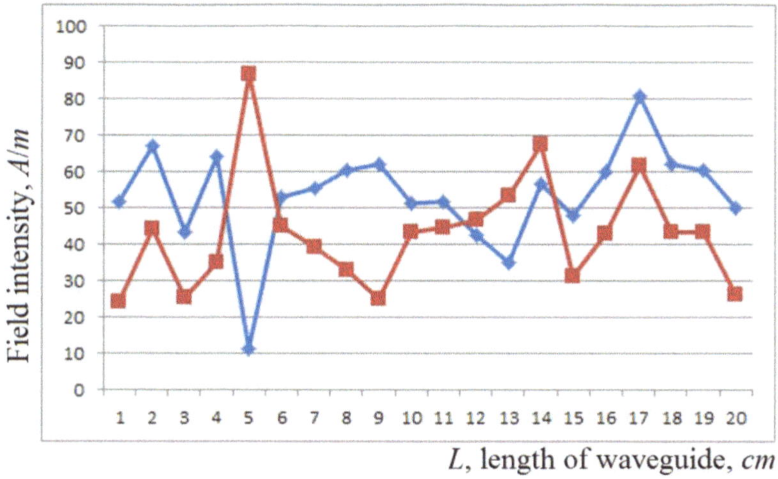

a)

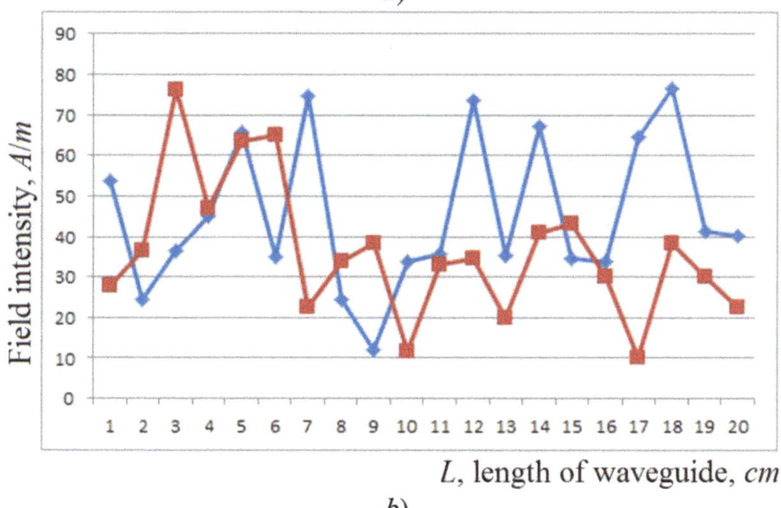

b)

Fig. 2.10 Dependencies of the length of the *E*-type (*a*) and *H*-type (*b*) waveguide in the frequency range of 4.9–7.05 *GHz* with a nonlinear medium on the magnetic field intensities: H_x - blue color. H_y - red color

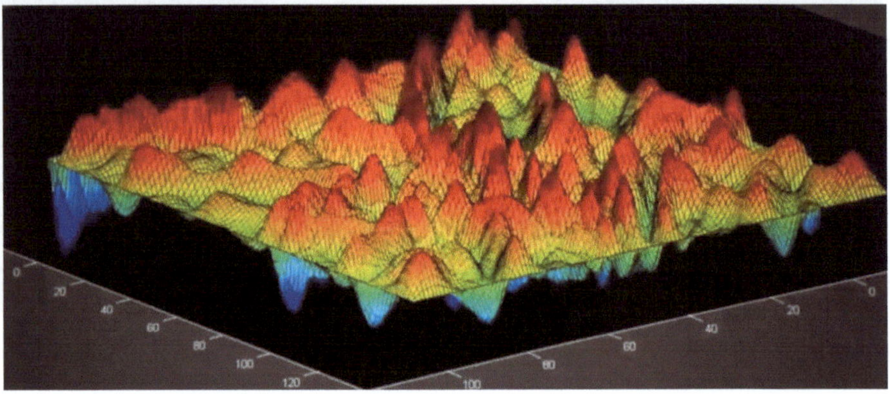

Fig. 2.11 Operating in the frequency range of 4.9–7.05 *GHz* 3*D* model of magnetic field distribution for *E*-type wave of microwave rectangular waveguide (finite difference method)

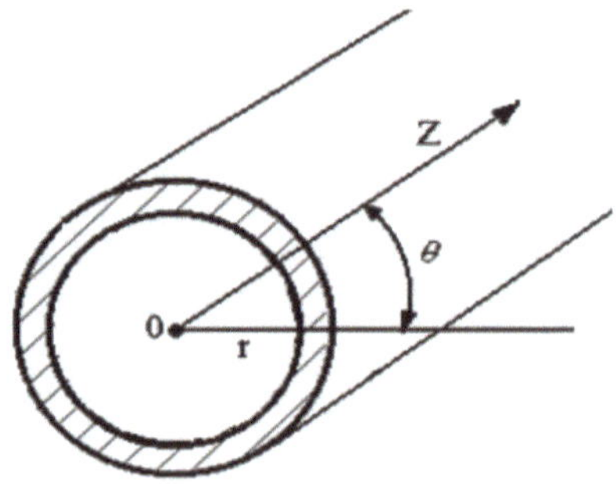

Fig. 2.12 Microwave circular waveguide in cylindrical coordinate system

$$\begin{cases} \dfrac{\partial E_z}{\partial \theta} - r\dfrac{\partial E_\theta}{\partial z} = -\dfrac{\partial B_r}{\partial t}, \\ \dfrac{\partial E_r}{\partial z} - \dfrac{\partial E_z}{\partial r} = -\dfrac{\partial B_\theta}{\partial t}, \\ \dfrac{\partial (rE_\theta)}{\partial r} - \dfrac{\partial E_r}{\partial \theta} = -\dfrac{\partial B_z}{\partial t}. \end{cases} \tag{2.32}$$

In a microwave circular waveguide, transverse electric (*TM* or *E*-type) type waves and transverse magnetic (*TE* or *H*-type) type waves can be propagated. For *E*-type waves, there are components of the electromagnetic field E_r, E_θ, E_z, H_r, H_θ, $H_z = 0$, and for *H*-type waves E_r, E_θ, $E_z = 0$, H_r, H_θ, H_z. So let's look at both the above cases:

(a) For E-type waves, Eqs. 2.31 and 2.32 have the following form:

$$\begin{cases} -r\dfrac{\partial H_\theta}{\partial z} = j_r + \dfrac{\partial D_r}{\partial t}, \\ \dfrac{\partial H_r}{\partial z} = j_\theta + \dfrac{\partial D_\theta}{\partial t}, \\ \dfrac{\partial (rH_\theta)}{\partial r} - \dfrac{\partial H_r}{\partial \theta} = j_z + \dfrac{\partial D_z}{\partial t}, \\ \dfrac{\partial E_z}{\partial \theta} - r\dfrac{\partial E_\theta}{\partial z} = -\dfrac{\partial B_r}{\partial t}, \\ \dfrac{\partial E_r}{\partial z} - \dfrac{\partial E_z}{\partial r} = -\dfrac{\partial B_\theta}{\partial t}, \\ \dfrac{\partial (rE_\theta)}{\partial r} - \dfrac{\partial E_r}{\partial \theta} = -\dfrac{\partial B_z}{\partial t}. \end{cases} \tag{2.33}$$

$$\begin{cases} -r\dfrac{\partial H_\theta}{\partial z} = j_r + \dfrac{\partial D_r}{\partial t}, \\ \dfrac{\partial H_r}{\partial z} = j_\theta + \dfrac{\partial D_\theta}{\partial t}, \\ \dfrac{\partial (rH_\theta)}{\partial r} - \dfrac{\partial H_r}{\partial \theta} = j_z + \dfrac{\partial D_z}{\partial t}, \\ \dfrac{\partial E_z}{\partial \theta} - r\dfrac{\partial E_\theta}{\partial z} = -\dfrac{\partial B_r}{\partial t}, \\ \dfrac{\partial E_r}{\partial z} - \dfrac{\partial E_z}{\partial r} = -\dfrac{\partial B_\theta}{\partial t}, \\ \dfrac{\partial (rE_\theta)}{\partial r} - \dfrac{\partial E_r}{\partial \theta} = -\dfrac{\partial B_z}{\partial t}. \end{cases} \tag{2.34}$$

$\vec{j} = \sigma\vec{E}, \vec{D} = \varepsilon_a\vec{E}$ [12] and $\vec{B} = \mu_a\vec{H}$ [13] in (2.33), we get:

$$\begin{cases} -r\dfrac{\partial H_\theta}{\partial z} = \sigma E_r + \left(\varepsilon_a + E_r\dfrac{\partial \varepsilon_a}{\partial E_r}\right)\dfrac{\partial E_r}{\partial t}, \\ \dfrac{\partial H_r}{\partial z} = \sigma E_\theta + \left(\varepsilon_a + E_\theta\dfrac{\partial \varepsilon_a}{\partial E_\theta}\right)\dfrac{\partial E_\theta}{\partial t}, \\ \dfrac{\partial (rH_\theta)}{\partial r} - \dfrac{\partial H_r}{\partial \theta} = \sigma E_z + \left(\varepsilon_a + E_z\dfrac{\partial \varepsilon_a}{\partial E_z}\right)\dfrac{\partial E_z}{\partial t}, \\ \dfrac{\partial E_z}{\partial \theta} - r\dfrac{\partial E_\theta}{\partial z} = -\mu_a\dfrac{\partial H_r}{\partial t}, \\ \dfrac{\partial E_r}{\partial z} - \dfrac{\partial E_z}{\partial r} = -\mu_a\dfrac{\partial H_\theta}{\partial t}, \\ \dfrac{\partial (rE_\theta)}{\partial r} - \dfrac{\partial E_r}{\partial \theta} = -\mu_a\dfrac{\partial H_z}{\partial t} = 0. \end{cases} \tag{2.35}$$

(b) Since $E_z = 0$ in the case of H-type waves, the system of Eqs. 2.31 and 2.32 has the following form:

$$\begin{cases} \dfrac{\partial H_z}{\partial \theta} - r\dfrac{\partial H_\theta}{\partial z} = \sigma E_r + \left(\varepsilon_a + E_r\dfrac{\partial \varepsilon_a}{\partial E_r}\right)\dfrac{\partial E_r}{\partial t}, \\ \dfrac{\partial H_r}{\partial z} - \dfrac{\partial H_z}{\partial r} = \sigma E_\theta + \left(\varepsilon_a + E_\theta\dfrac{\partial \varepsilon_a}{\partial E_\theta}\right)\dfrac{\partial E_\theta}{\partial t}, \\ \dfrac{\partial (rH_\theta)}{\partial r} - \dfrac{\partial H_r}{\partial \theta} = \sigma E_z + \left(\varepsilon_a + E_z\dfrac{\partial \varepsilon_a}{\partial E_z}\right)\dfrac{\partial E_z}{\partial t} = 0, \\ r\dfrac{\partial E_\theta}{\partial z} = \mu_a\dfrac{\partial H_r}{\partial t}, \\ \dfrac{\partial E_r}{\partial z} = -\mu_a\dfrac{\partial H_\theta}{\partial t}, \\ \dfrac{\partial (rE_\theta)}{\partial r} - \dfrac{\partial E_r}{\partial \theta} = -\mu_a\dfrac{\partial H_z}{\partial t}. \end{cases} \tag{2.36}$$

Thus, for an E-type wave in the r, θ, z cylindrical coordinate system, we get the following system of differential equations for a microwave circular waveguide:

$$\begin{cases} \dfrac{\sigma}{r}\dfrac{\partial E_r}{\partial t} + \left(\varepsilon_a + E_r\dfrac{\partial \varepsilon_a}{\partial E_r}\right)\dfrac{1}{r}\dfrac{\partial^2 E_r}{\partial \partial t^2} = \dfrac{1}{\mu_a}\left(\dfrac{\partial^2 E_r}{\partial z^2} - \dfrac{\partial^2 E_z}{\partial r \partial z}\right), \\ \sigma\dfrac{\partial E_\theta}{\partial t} + \left(\varepsilon_a + E_\theta\dfrac{\partial \varepsilon_a}{\partial E_\theta}\right)\dfrac{\partial^2 E_\theta}{\partial t^2} = \dfrac{1}{\mu_a}\left(r\dfrac{\partial^2 E_\theta}{\partial z^2} - \dfrac{\partial^2 E_z}{\partial \theta \partial z}\right), \\ \sigma\dfrac{\partial E_z}{\partial t} + \left(\varepsilon_a + E_z\dfrac{\partial \varepsilon_a}{\partial E_z}\right)\dfrac{\partial^2 E_z}{\partial t^2} = \dfrac{r}{\mu_a}\left(\dfrac{\partial^2 E_z}{\partial r^2} + \dfrac{\partial^2 E_z}{\partial \theta^2} + r\dfrac{\partial^2 E_\theta}{\partial z \partial \theta} - \dfrac{\partial^2 E_r}{\partial z \partial r}\right), \\ \left(\varepsilon_a + E_\theta\dfrac{\partial \varepsilon_a}{\partial E_\theta}\right)\dfrac{\partial^2 E_\theta}{\partial r \partial t} - \left(\varepsilon_a + E_r\dfrac{\partial \varepsilon_a}{\partial E_r}\right)\dfrac{\partial^2 E_r}{\partial \theta \partial t} = \dfrac{\partial^2 H_r}{\partial r \partial z} + r\dfrac{\partial^2 H_\theta}{\partial z \partial \theta}, \\ \sigma\mu_a\dfrac{\partial H_\theta}{\partial t} - \left(\varepsilon_a + E_r\dfrac{\partial \varepsilon_a}{\partial E_r}\right)\dfrac{\partial^2 E_r}{\partial z \partial t} + \left(\varepsilon_a + E_z\dfrac{\partial \varepsilon_a}{\partial E_z}\right)\dfrac{\partial^2 E_z}{\partial r \partial t} = r\dfrac{\partial^2 H_\theta}{\partial z^2} + \dfrac{\partial^2 (rH_\theta)}{\partial r^2} - \dfrac{\partial^2 H_r}{\partial \theta \partial r}. \end{cases} \tag{2.37}$$

For the H-type wave, we get the following system of differential equations for the microwave circular waveguide in the r, θ, z cylindrical coordinate system:

$$\begin{cases} \frac{\partial^2 H_z}{\partial\theta\partial z} - r\frac{\partial^2 H_\theta}{\partial z^2} + \sigma\mu_a\frac{\partial H_\theta}{\partial t} = \left(\varepsilon_a + E_r\frac{\partial\varepsilon_a}{\partial E_r}\right)\frac{\partial^2 E_r}{\partial z\partial t}, \\ \frac{\partial^2 H_r}{\partial z^2} - \frac{\partial^2 H_z}{\partial r\partial z} - \frac{\sigma\mu_a}{r}\frac{\partial Hr}{\partial t} = \left(\varepsilon_a + E_\theta\frac{\partial\varepsilon_a}{\partial E_\theta}\right)\frac{\partial^2 E_\theta}{\partial z\partial t}, \\ \frac{\partial^2 H_r}{\partial r\partial z} - \frac{\partial^2 H_z}{\partial r^2} - \frac{\partial^2 H_z}{\partial\theta^2} + r\frac{\partial^2 H_\theta}{\partial\theta\partial z} + \sigma\mu_a\frac{\partial H_z}{\partial t} = \\ = \left(\varepsilon_a + E_\theta\frac{\partial\varepsilon_a}{\partial E_\theta}\right)\frac{\partial^2 E_\theta}{\partial r\partial t} - \left(\varepsilon_a + E_r\frac{\partial\varepsilon_a}{\partial E_r}\right)\frac{\partial^2 E_r}{\partial\theta\partial t}, \\ -\frac{1}{\mu_a}\frac{\partial H_z}{\partial\theta}\left(\frac{\partial(rE_\theta)}{\partial r} - \frac{\partial E_r}{\partial\theta}\right) + \frac{r}{\mu_a}\frac{\partial H_\theta \partial E_r}{\partial z^2} = \sigma\frac{\partial E_r}{\partial t} + \left(\varepsilon_a + E_r\frac{\partial\varepsilon_a}{\partial E_r}\right)\frac{\partial^2 E_r}{\partial t^2}, \\ \frac{r}{\mu_a}\frac{\partial H_r \partial H_\theta}{\partial z^2}\frac{\partial^2 H_Z}{\partial r\partial t} = \sigma\frac{\partial E_\theta}{\partial t} + \left(\varepsilon_a + E_\theta\frac{\partial\varepsilon_a}{\partial E_\theta}\right)\frac{\partial^2 E_\theta}{\partial t^2}. \end{cases} \tag{2.38}$$

Thus, if we consider (2.12) in (2.33) and (2.40), the new mathematical models characterizing the electromagnetic field of the air-filled microwave circular waveguide will have the following form:

(a) For an E-type wave (TM-wave) of a microwave circular waveguide:

$$\begin{cases} \frac{\sigma}{r}\frac{\partial E_r}{\partial t} + \left[1{,}8\varepsilon_0 + 1{,}066\cdot 10^{-5}(\varepsilon_0+1)E_r\right]\frac{1}{r}\frac{\partial^2 E_r}{\partial t^2} = -\frac{1}{\mu_a}\left(\frac{\partial^2 E_r}{\partial z^2} - \frac{\partial^2 E_z}{\partial r\partial z}\right), \\ \sigma\frac{\partial E_\theta}{\partial t} + \left[1{,}8\varepsilon_0 + 1{,}066\cdot 10^{-5}(\varepsilon_0+1)E_\theta\right]\frac{\partial^2 E_\theta}{\partial t^2} = \frac{1}{\mu_a}\left(r\frac{\partial^2 E_\theta}{\partial z^2} - \frac{\partial^2 E_z}{\theta y\partial z}\right), \\ \sigma\frac{\partial E_z}{\partial t} + \left[1{,}8\varepsilon_0 + 1{,}066\cdot 10^{-5}(\varepsilon_0+1)E_z\right]\frac{\partial^2 E_z}{\partial t^2} = \frac{r}{\mu_a}\left(\frac{\partial^2 E_z}{\partial\partial r^2} + \frac{\partial^2 E_z}{\partial\theta^2} + r\frac{\partial^2 E_\theta}{\partial z\partial\theta^2} - \frac{\partial^2 E_r}{\partial z\partial r}\right), \\ \left[1{,}8\varepsilon_0 + 1{,}066\cdot 10^{-5}(\varepsilon_0+1)E_\theta\right]\frac{\partial^2 E_\theta}{\partial r\partial t} - \left[1{,}8\varepsilon_0 + 1{,}066\cdot 10^{-5}(\varepsilon_0+1)E_r\right]\frac{\partial^2 E_r}{\partial\theta\partial t} = \frac{\partial^2 H_r}{\partial r\partial z} + \frac{\partial^2 H_\theta}{\partial z\partial\theta}, \\ \sigma\mu_a\frac{\partial H_\theta}{\partial t} - \left[1{,}8\varepsilon_0 + 1{,}066\cdot 10^{-5}(\varepsilon_0+1)E_r\right]\frac{\partial^2 E_r}{\partial z\partial t} + \left[1{,}8\varepsilon_0 + 1{,}066\cdot 10^{-5}(\varepsilon_0+1)E_z\right]\frac{\partial^2 E_z}{\partial r\partial t} = \\ = r\frac{\partial^2 H_\theta}{\partial z^2} + \frac{\partial^2(rH_\theta)}{\partial r^2} - \frac{\partial^2 H_r}{\partial\theta\partial r}. \end{cases} \tag{2.39}$$

(b) For the *H*-wave (*TE*-wave) of a microwave circular waveguide:

$$
\begin{cases}
\frac{\partial^2 H_z}{\partial\theta\partial z} - r\frac{\partial^2 H_\theta}{\partial z^2} + \sigma\mu_a\frac{\partial H_\theta}{\partial t} = \left[1{,}8\varepsilon_0 + 1{,}066\cdot 10^{-5}(\varepsilon_0+1)E_r\right]\frac{\partial^2 E_r}{\partial z\partial t},\\
\frac{\partial^2 H_r}{\partial z^2} - \frac{\partial^2 H_z}{\partial r\partial z} - \frac{\sigma\mu_a}{r}\frac{\partial Hr}{\partial t} = \left[1{,}8\varepsilon_0 + 1{,}066\cdot 10^{-5}(\varepsilon_0+1)E_\theta\right]\frac{\partial^2 E_\theta}{\partial z\partial t},\\
\frac{\partial^2 H_r}{\partial r\partial z} - \frac{\partial^2 H_z}{\partial r^2} - \frac{\partial^2 H_z}{\partial\theta^2} + r\frac{\partial^2 H_\theta}{\partial\theta\partial z} + \sigma\mu_a\frac{\partial H_z}{\partial t} =\\
= \left[1{,}8\varepsilon_0 + 1{,}066\cdot 10^{-5}(\varepsilon_0+1)E_\theta\right]\frac{\partial^2 E_\theta}{\partial r\partial t} - \left[1{,}8\varepsilon_0 + 1{,}066\cdot 10^{-5}(\varepsilon_0+1)E_r\right]\frac{\partial^2 E_r}{\partial\theta\partial t},\\
-\frac{1}{\mu_a}\frac{\partial H_z}{\partial\theta}\left(\frac{\partial(rE_\theta)}{\partial r} - \frac{\partial E_r}{\partial\theta}\right) + \frac{r}{\mu_a}\frac{\partial H_\theta \partial E_r}{\partial z^2} = \sigma\frac{\partial E_r}{\partial t} + \left[1{,}8\varepsilon_0 + 1{,}066\cdot 10^{-5}(\varepsilon_0+1)E_r\right]\frac{\partial^2 E_r}{\partial t^2},\\
\frac{r}{\mu_a}\frac{\partial H_r \partial H_\theta}{\partial z^2}\frac{\partial^2 H_Z}{\partial r\partial t} = \sigma\frac{\partial E_\theta}{\partial t} + \left[1{,}8\varepsilon_0 + 1{,}066\cdot 10^{-5}(\varepsilon_0+1)E_\theta\right]\frac{\partial^2 E_\theta}{\partial t^2}.
\end{cases}
\tag{2.40}
$$

As you can see, these mathematical models are a system of differential equations of the second order with a special formulation. It is impossible to solve these equations by analytical methods. Therefore, numerical methods should be applied to solve these equations. For this, let's look at the calculation of the electromagnetic field of the microwave circular waveguide by the finite difference method, analogously to Sect. 2.3.

Based on the mathematical algorithm and block diagram shown in Fig. 2.7, calculations were made in the C++ program and the electromagnetic field intensities of the microwave circular waveguide with a nonlinear medium operating at a frequency of 9 *GHz* in *E*-type and *H*-type waves were determined. Based on the obtained numerical values, the dependencies of the length of the circular waveguide with a microwave operating at a frequency of 9 *GHz* in *E*-type and *H*-type waves were established (Fig. 2.12a, b) (Table 2.3). These dependencies determine how the electromagnetic field is distributed inside the investigated waveguide and the relationship between the electromagnetic and structural parameters of this device. Figure 2.13 shows the 3*D* model (finite difference method) of the electric field intensity distribution for the *E*-type wave of a circular waveguide operating at a frequency of 9 *GHz*. In Fig. 2.14, the dependencies of the length of the circular waveguide with a microwave operating at a frequency of 9 *GHz* in *E*-type (*a*) and *H*-type (*b*) waves (Table 2.4), Figs. 2.15 and 2.16 shows the 3*D* model (finite difference method) of the frequency distribution of the magnetic field for the *E*-type wave of a circular waveguide operating at a frequency of 9 *GHz*.

Table 2.3 Values of the electric field intensities of a microwave circular waveguide with a nonlinear medium operating at a frequency of 9 *GHz* for *E*-type and *H*-type waves determined by the finite difference method

	For *E*-type wave		For *H*-type wave	
Elemen-tary fields	E_r, (10^5) *V/m*	E_z, (10^5) *V/m*	E_r, (10^5) *V/m*	E_z, (10^5) *V/m*
1	10.5	12.4	44.7	80.6
2	12.6	19.6	42.8	85.7
3	13.8	25.7	37.6	90.8
4	75.6	72.8	49.8	27.6
5	34.5	10.7	58.7	19.5
6	21.8	10.4	93.6	20.3
7	14.9	18.13	27.6	25.7
8	12.6	17.4	33.5	28.2
9	14.8	25.6	42.6	34.6
10	16.5	14.7	45.7	10.9
11	18.4	9.2	58.2	26.6
12	26.8	8.6	65.1	40.6
13	33.1	16.2	90.9	68.7
14	70.2	18.8	33.6	76.1
15	11.3	24.4	28.3	80.0
16	10.9	32.9	26.7	71.2
17	19.9	45.3	16.5	39.5
18	22.6	57.6	45.4	35.0
19	39.5	23.5	49.8	19.6
20	87.1	12.9	68.2	13.7

2.5 Calculation of the Electromagnetic Field of a Microwave Rectangular Waveguide in a Nonlinear Medium by the Finite Element Method

The finite element method is one of the most effective methods for solving field problems, as is the finite difference method. Therefore, since the equations in the system of Eqs. 2.14 and 2.15 are elliptic type equations, let's look at their solution by applying the finite element method. The mathematical model of the electric field of a microwave rectangular waveguide in a nonlinear medium is as follows:

$$\frac{\partial}{\partial x}\left[\left(1,8+1,066\cdot 10^{-5}E\right)\frac{\partial \phi}{\partial x}\right]+\frac{\partial}{\partial y}\left[\left(1,8+1,066\cdot 10^{-5}E\right)\frac{\partial \phi}{\partial y}\right]+ \\ +\frac{\partial}{\partial z}\left[\left(1,8+1,066\cdot 10^{-5}E\right)\frac{\partial \phi}{\partial z}\right]=-\frac{\rho}{\varepsilon_0}, \tag{2.41}$$

where E is the intensity of the electric field, *V/m*, φ is the scalar electric potential, ρ is the volume charge density, $\varepsilon_0 = 8.85\cdot 10^{-12}$ *F/m* is the electric constant.

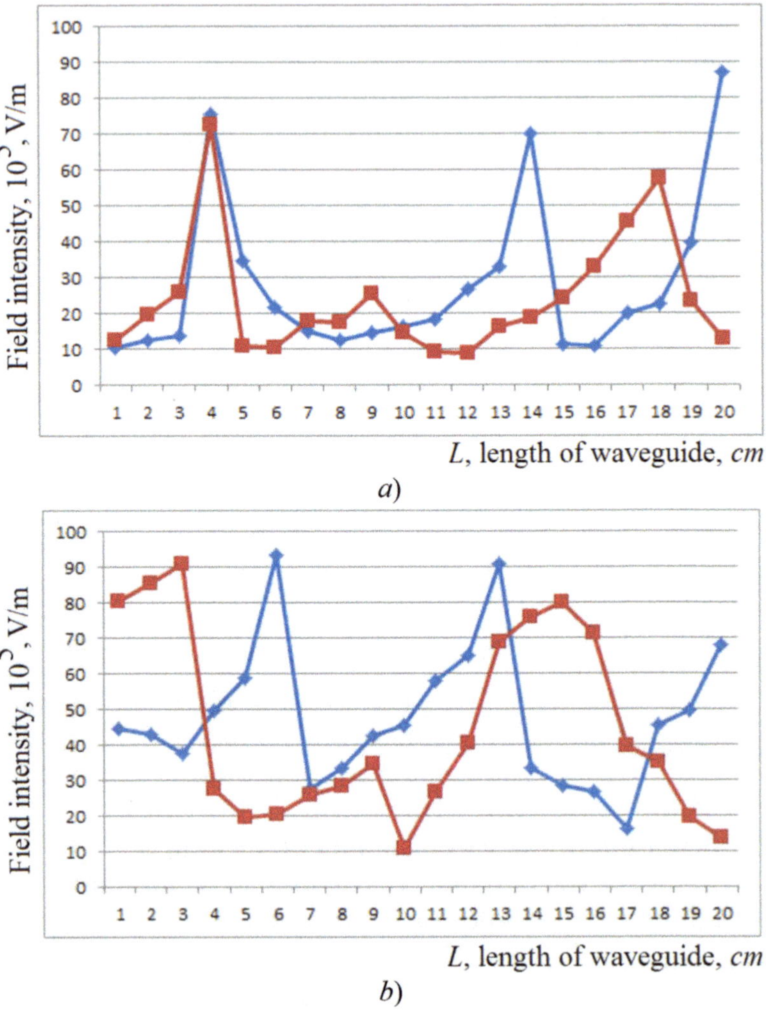

Fig. 2.13 Dependencies of the length of the *E*-type (**a**) and *H*-type (**b**) nonlinear medium microwave waveguide operating at a frequency of 9 *GHz* on its electric field intensities: E_r – blue color, E_z – red color

As can be seen from Eq. 2.41, this equation is an elliptic-type differential equation with special formulation.

The finite element method is one of the most powerful mathematical tools in use today. Using this method, the calculation area of the microwave rectangular waveguide is divided into finite elementary elements and the field intensity is calculated in each element. It should be noted that the division of the computational domain into finite elementary elements is the first stage of solving the problem under consideration. Failure to divide the computational domain into finite elemental exact

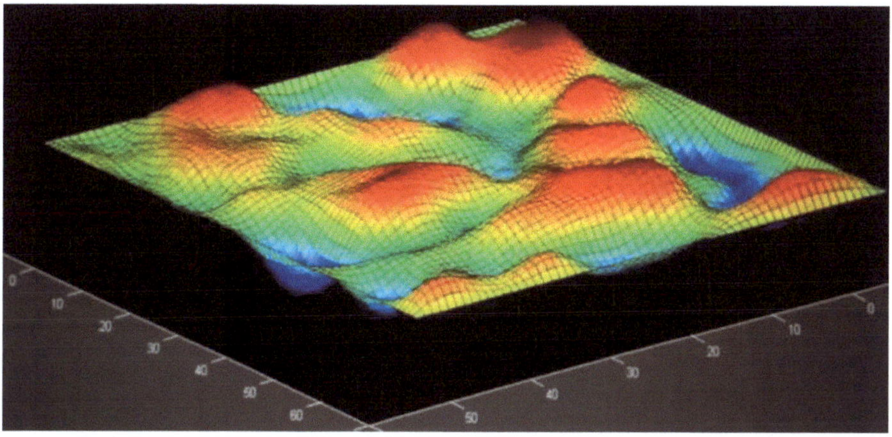

Fig. 2.14 3*D* model of the distribution of electric field intensities of a nonlinear medium microwave waveguide operating at a frequency of 9 *GHz* in *E*-type (**a**) and *H*-type (**b**) waves (finite difference method)

Table 2.4 Values of the magnetic field intensities of a microwave circular waveguide with a nonlinear medium operating at a frequency of 9 *GHz* for *E*-type and *H*-type waves determined by the finite difference method

Eleme-ntary fields	For *E*-type wave		For *H*-type wave	
	H_r, *A/m*	H_z, *A/m*	H_r, *A/m*	H_z, *A/m*
1	50.14	14.13	33.12	29.99
2	53.13	38.12	29.24	39.58
3	52.34	30.59	36.56	86.85
4	54.91	29.35	50.14	35.45
5	16.43	77.06	60.11	74.87
6	60.14	53.86	29.18	74.78
7	41.17	29.43	72.82	34.55
8	57.17	40.03	30.48	42.57
9	52.23	32.86	18.23	43.83
10	60.74	40.01	41.41	18.37
11	45.19	56.78	24.36	47.81
12	14.61	51.07	74.49	25.76
13	39.79	54.89	14.42	29.51
14	39.84	57.18	69.75	39.16
15	51.95	29.27	43.88	53.75
16	49.34	41.73	25.46	34.15
17	65.84	63.89	68.56	17.41
18	52.76	47.43	63.46	37.31
19	58.78	47.89	53.56	18.75
20	33.31	30.92	44.79	30.58

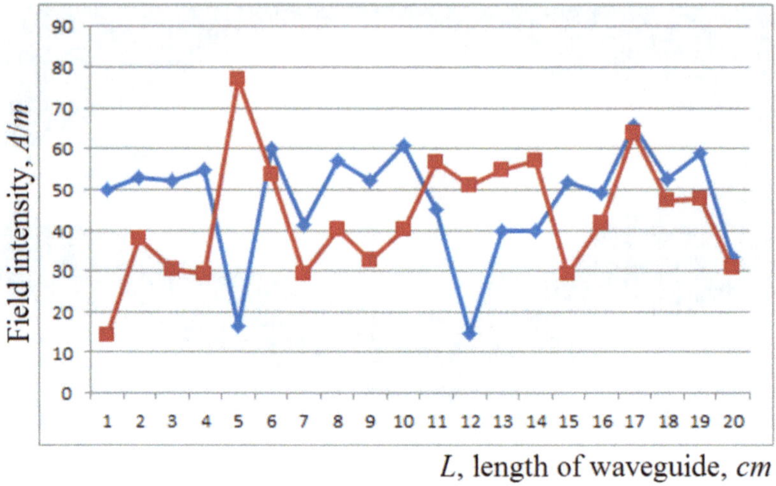

a)

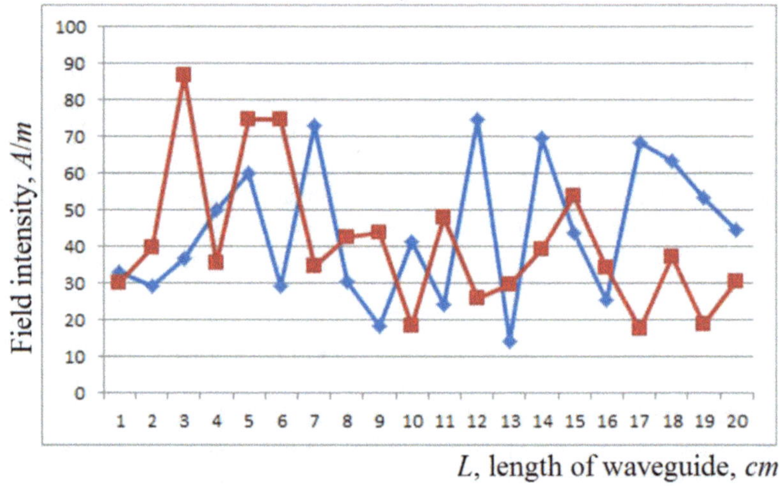

b)

Fig. 2.15 Dependencies of the length of a microwave circular waveguide with a nonlinear medium operating at a frequency of 9 *GHz* in *E*-type (**a**) and *H*-type (**b**) waves on the intensity of its magnetic field: H_r – blue color, H_z – red color

elements leads to imprecise results, regardless of the exact performance of subsequent operations. When dividing the computational domain into finite elementary elements, it is appropriate to consider two main factors [14]:

- Finite element elements should be soft to increase the accuracy of calculations.
- When dividing the computational area into finite elementary elements, it is necessary to work to use fewer elements or nodes.

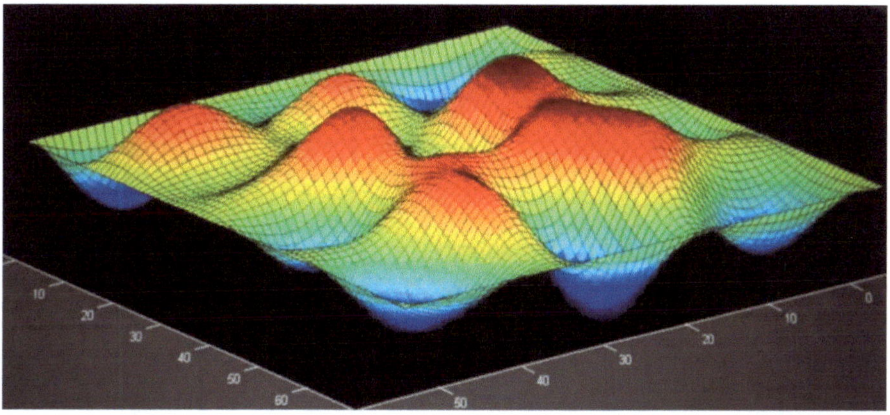

Fig. 2.16 3*D* model of magnetic field intensity distribution (finite difference method) of a microwave circular waveguide with nonlinear medium operating at 9 *GHz* in *E*-type (**a**) and *H*-type (**b**) waves

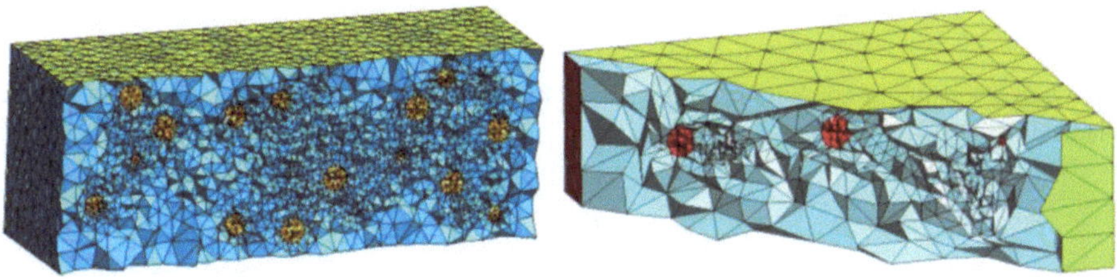

Fig. 2.17 3*D* view of the computational domain of a microwave rectangular waveguide divided into finite elements

Figure 2.17 shows a three-dimensional view of the division of the calculation area of the microwave rectangular waveguide into finite elements, and Fig. 2.18 shows the division of the calculation area of the microwave rectangular waveguide into 20 finite elements.

When dividing the calculation area of the microwave rectangular waveguide into 20 finite elements, the number of the elements at the nodes should be numbered counterclockwise. In this case, the number of borders of elements is 11, the number of nodes is 17, and the number of elements is equal to 20. Now, in order to simplify the calculations, let's simplify Eq. 2.41 and make the following substitution:

$$\varphi = 1,8 + 1,066 \cdot 10^{-5} \cdot E \tag{2.42}$$

Considering (2.42) in (2.41), we get

$$\frac{\partial^2 \phi}{\partial x^2} + \frac{\partial^2 \phi}{\partial y^2} + \frac{\partial^2 \phi}{\partial z^2} + \frac{\rho}{\varepsilon_0} = 0. \tag{2.43}$$

Let's take the first order zero boundary condition as the boundary condition.

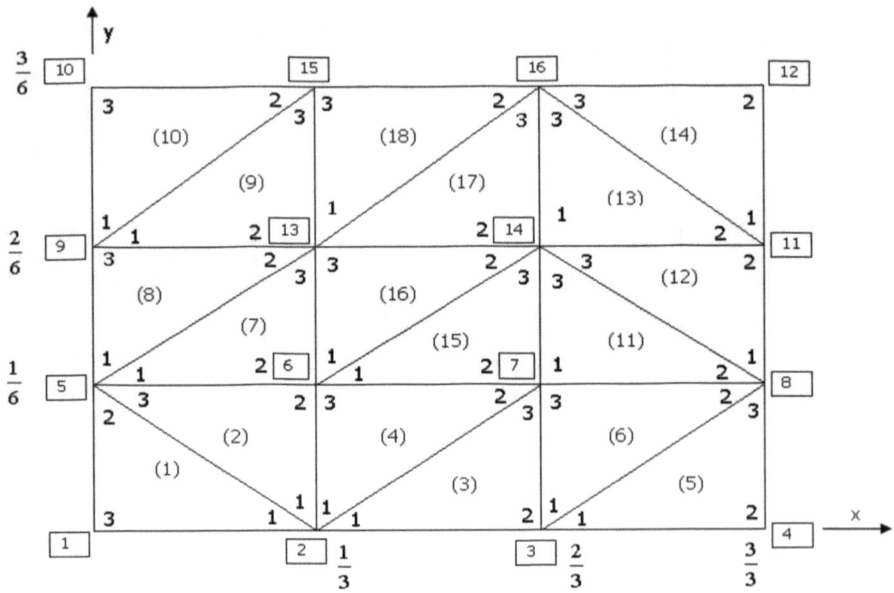

Fig. 2.18 Division of computational domain of microwave rectangular waveguide into 20 finite elements

$$\phi_r = 0.$$

It is known from the calculus of variations [15] that the solution of differential equations in the case of first and second order homogeneous boundary problems corresponds to the energy minimum of the following energy functional:

$$F = \iint_S \left\{ \left[\left(\frac{\partial \varphi}{\partial x} \right)^2 + \left(\frac{\partial \varphi}{\partial y} \right)^2 + \left(\frac{\partial \varphi}{\partial z} \right)^2 \right] + \frac{\rho}{\varepsilon_0} \right\} ds. \tag{2.44}$$

Lu et al. [16] methods of minimizing the functional (2.44) by applying the finite element method are given. To simplify the calculations, let's assume that there are no components of the electric field in the direction of the z axis. That is, $\frac{\partial \phi}{\partial z} = 0$. Then the functional (2.44) will be as follows:

$$F = \iint_S \left\{ \left[\left(\frac{\partial \varphi}{\partial x} \right)^2 + \left(\frac{\partial \varphi}{\partial y} \right)^2 \right] + \frac{\rho}{\varepsilon_0} \right\} ds. \tag{2.45}$$

Now let's look at the methodology of minimizing the functional (2.45) by applying the finite difference method. When dividing the calculation area into triangular elements, the scalar electric potential φ is determined by its vertex values within an arbitrary triangle with vertices i, j, k and becomes a linear function of coordinates.

$$\phi = (\alpha_i + \beta_i x + \gamma_i y)\phi_i + (\alpha_j + \beta_j x \gamma_j y) + \\ +(\alpha_k + \beta_k x + \gamma_k y)\phi_k / 2T_m. \tag{2.46}$$

(2.46), let's adopt the following notation:

$$N_i = \alpha_i + \beta_i x + \gamma_i y,$$
$$N_j = \alpha_j + \beta_j x + \gamma_j y,$$
$$N_k = \alpha_k + \beta_k x + \gamma_k y.$$

Taking these notations into account, (2.46) can be written as follows:

$$\phi = N_i \phi_i + N_j \phi_j + N_k \phi_k. \tag{2.47}$$

If we write (2.47) in matrix form, we get

$$\phi = [N][\phi]_m^{\mathrm{T}}. \tag{2.48}$$

α_i, α_j, α_k, β_i, β_j, β_k, γ_i, γ_j, γ_k and T_m in (2.46) are defined as:

$$\alpha = \frac{y_j x_k - y_k x_j}{2\,\mathrm{T}m}, \quad \beta = \frac{x_j - x_k}{2\,\mathrm{T}m}, \quad \gamma = \frac{y_k - y_j}{2\,\mathrm{T}m};$$
$$\alpha = \frac{y_k x_i - y_i x_k}{2\,\mathrm{T}m}, \quad \beta = \frac{x_k - x_i}{2\,\mathrm{T}m}, \quad \gamma = \frac{y_i - y_k}{2\,\mathrm{T}m};$$
$$\alpha = \frac{y_i x_j - y_j x_i}{2\,\mathrm{T}m}, \quad \beta = \frac{x_i - x_j}{2\,\mathrm{T}m}, \quad \gamma = \frac{y_j - y_i}{2\,\mathrm{T}m};$$
$$\mathrm{T}m = \frac{y_i \beta_i + y_j \beta_j + y_k \beta_k}{2}.$$

The finite element network of the computational domain of the microwave rectangular waveguide consists of $N = 17$ nodes, $M = 18$ triangles. Therefore, when the functional (2.45) is minimized, its derivative is equal to zero:

$$\frac{\partial F}{\partial \phi_n} = 0, (n = 1, 2, \ldots.., N). \tag{2.49}$$

Taking into account (2.49) in (2.45) and (2.46) and performing simple transformations, we get the following system of equations:

$$([\mathrm{T}] + [L])[\phi]^{\mathrm{T}} = [W]^{\mathrm{T}}, \tag{2.50}$$

where $[T]$ and $[L]$ are square matrices of N arrangement, $[\varphi]$ is a column consisting of scalar electric potentials, and $[W]^T$ is a column consisting of volume charge densities at nodes.

The matrix $[T]$ is obtained by combining triangular matrices $[T]_m$, and its elements are calculated using the following formula [17]:

$$\mathrm{T}(n,l) = \frac{\mathrm{T}_m}{\varepsilon_0 \rho_m}\left\{\beta_n\beta_l + \gamma_n\gamma_l + \left(\frac{\rho^2+1}{\rho_m^2} + k^2\right)\varpi\right\}, \tag{2.51}$$

where ω=1/12 when $n \neq l$; ω=1/6 when $n \neq l$.

$$n \neq l, \omega = \frac{1}{12};$$
$$n \neq l, \omega = \frac{1}{6}.$$

The elements of the $[L]$ matrix are found from the following expression [18]:

$$L(n,l) = -\frac{\chi_k}{9\mathrm{T}_k}\left(\sum \mathrm{T}\right)_n \left(\sum \mathrm{T}\right)_l, \tag{2.52}$$

where the number of nodes belonging to n and $l - k$ is the sum of the areas of triangles belonging to $(\Sigma T)_n$ and $(\Sigma T)_l - k$.

It should be noted that the degree of the system (2.50) is equal to $4 \cdot \mathrm{N} = 4 \cdot 17 = 68$, where $\mathrm{N} = 17$ is the number of nodes in the triangular mesh. The values of the scalar electric potential $[\varphi]^T$ at the nodes of the network of triangular elements (2.50) are determined as a result of solving the system of linear equations.

Thus, the values of the components of the scalar electric potential in the m-th triangle are determined as follows:

$$\begin{cases} \phi_x = \beta_i\phi_{xi} + \beta_j\phi_{xj} + \beta_k\phi_{xk}; \\ \phi_y = \gamma_i\phi_{xi} + \gamma_j\phi_{xj} + \gamma_k\phi_{xk}. \end{cases} \tag{2.53}$$

It can be easily determined from the expression (2.42):

$$\begin{cases} E_x = (0,93\phi_x - 1,69) \cdot 10^5; \\ E_y = \left(0,93\phi_y - 1,69\right) \cdot 10^5. \end{cases} \tag{2.54}$$

Considering (2.53) in (2.54), we get

Table 2.5 Values of the electric field intensities of a microwaverectangular waveguide with a nonlinear medium operating in the frequency range of 4.9–7.05 *GHz* for *E*-type and *H*-type waves determined by the finite element method

	For *E*-type wave		For *H*-type wave	
Number of elementary regions	E_x, (10^5) *V/m*	E_y, (10^5)*V/m*	E_x, (10^5) *V/m*	E_y, (10^5) *V/m*
1	2.23	4.43	4.35	2.50
2	3.44	5.25	1.65	2.55
3	4.18	5.30	1.50	3.30
4	4.34	5.33	5.47	4.10
5	4.37	5.42	5.61	3.47
6	4.48	6.24	4.72	3.35
7	4.40	6.45	2.14	2.40
8	2.88	4.72	1.39	1.37
9	3.75	4.25	4.25	1.19
10	4.10	5.42	5.26	1.30
11	4.30	6.09	6.50	1.59
12	3.12	6.22	4.15	1.60
13	2.67	6.21	4.16	1.78
14	2.46	6.21	7.33	2.25
15	2.40	5.54	7.45	3.68
16	2.25	5.14	1.50	1.59
17	2.94	5.14	1.40	1.26
18	3.15	1.15	2.50	4.16
19	4.30	4.39	3.14	5.48
20	5.22	4.76	4.49	6.47

$$\begin{cases} E_x = \left[0,93\left(\beta_i \phi_{xi} + \beta_j \phi_{xj} + \beta_k \phi_{xk}\right) - 1,69\right] \cdot 10^5; \\ E_y = \left[0,93\left(\gamma_i \phi_{xi} + \gamma_j \phi_{xj} + \gamma_k \phi_{xk}\right) - 1,69\right] \cdot 10^5. \end{cases} \tag{2.55}$$

According to the expressions (2.55) for *E*-type and *H*-type waves in the frequency range of 4.9–7.05 *GHz*, the electric field intensities of the microwave rectangular waveguide with structural dimensions $a = 40$ *mm*, $b = 20$ *mm* using the finite element method prices are set. The assigned values are listed in Table 2.5. Calculations were made in the C++ program and the electromagnetic field intensities of the microwave rectangular waveguide with nonlinear medium operating in the frequency range of 4.9–7.05 *GHz* in *E*-type and *H*-type waves were determined. Based on the obtained numerical values, the dependencies of the length of the rectangular waveguide of the microwave operating in the frequency range of 4.9–7.05 *GHz* on *E*-type and *H*-type waves were established (Fig. 2.19a, b). These dependencies determine how the electromagnetic field is distributed inside the investigated waveguide and the relationship between the electromagnetic and structural parameters of this device. Figure 2.20 shows the 3*D* model (finite element method) of the electric field intensity distribution for the *E*-type wave of the microwave rectangular waveguide with nonlinear medium operating in the

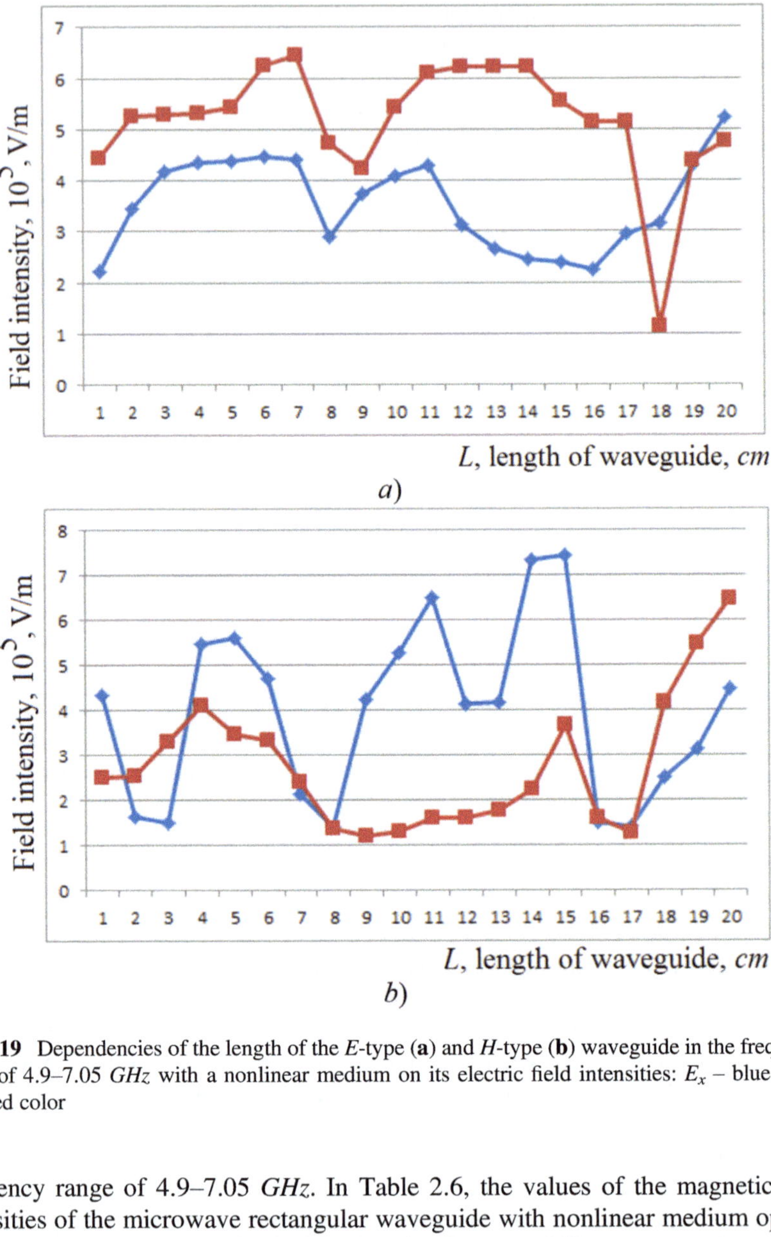

Fig. 2.19 Dependencies of the length of the *E*-type (**a**) and *H*-type (**b**) waveguide in the frequency range of 4.9–7.05 *GHz* with a nonlinear medium on its electric field intensities: E_x – blue color, E_y – red color

frequency range of 4.9–7.05 *GHz*. In Table 2.6, the values of the magnetic field intensities of the microwave rectangular waveguide with nonlinear medium operating in the frequency range of 4.9–7.05 *GHz* for *E*-type and *H*-type waves determined by the finite element method (Fig. 2.21). Dependencies of the length of the *E*-type (*a*) and *H*-type (*b*) waves in the frequency range of 4.9–7.05 *GHz* with a nonlinear medium microwave waveguide on the intensities of its magnetic field, in Fig. 2.22, 3*D* model (finite element method) of the distribution of the magnetic field intensity

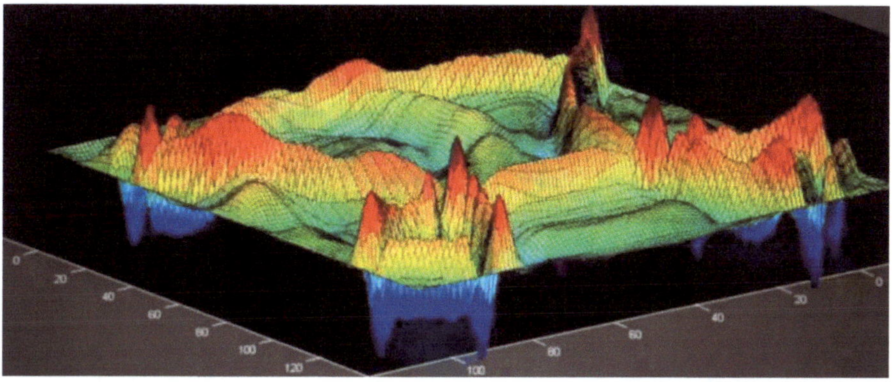

Fig. 2.20 $3D$ model of electric field intensity distribution for E-type wave of microwave rectangular waveguide operating in the frequency range of 4.9–7.05 GHz (finite element method)

Table 2.6 Values of the magnetic field intensities of themicrowave rectangular waveguide with nonlinear medium operating in the frequency range of 4.9–7.05 GHz for E-type and H-type waves determined by the finite element method

	For E-type wave		For H-type wave	
Number of elementary regions	H_x, A/m	H_y, A/m	H_x, A/m	H_y, A/m
1	50.78	24.79	53.67	28.72
2	66.45	42.87	25.78	36.46
3	43.79	25.45	34.87	76.54
4	62.17	34.78	47.34	46.38
5	11.45	83.49	67.81	61.56
6	51.93	44.12	34.13	65.45
7	53.58	39.23	75.76	22.34
8	60.52	33.11	24.18	35.12
9	64.14	25.29	12.36	38.89
10	52.95	43.97	31.57	11.48
11	53.12	46.76	35.34	32.92
12	44.14	47.56	75.65	34.76
13	35.11	51.65	35.93	20.23
14	54.78	65.97	67.14	40.99
15	48.34	30.87	33.12	43.28
16	60.31	42.78	33.85	30.11
17	80.14	63.21	63.45	10.47
18	60.71	45.41	76.54	36.76
19	62.36	45.87	42.32	30.05
20	50.12	26.13	40.59	21.22

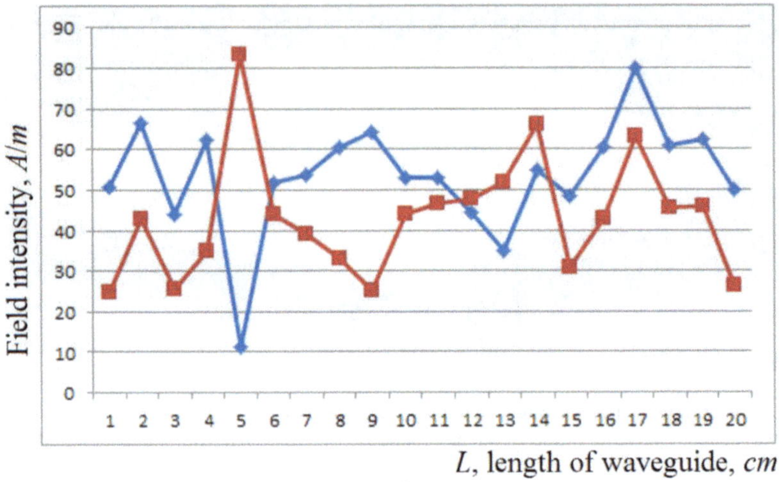

a)

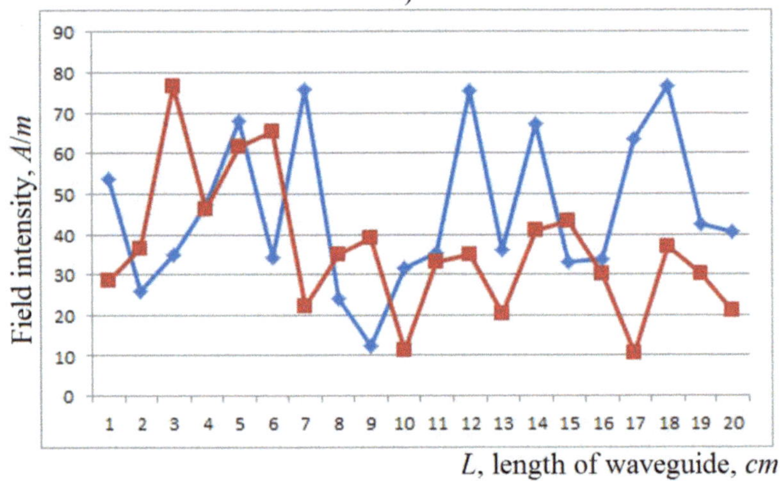

b)

Fig. 2.21 Dependencies of the length of the *E*-type (**a**) and *H*-type (**b**) waveguide in the frequency range of 4.9–7.05 *GHz* with a nonlinear medium on its magnetic field intensities: H_x – blue color, H_y – red color

for an *H*-type wave of a rectangular waveguide operating in the frequency range of 4.9–7.05 *GHz* is shown.

Based on the above mathematical algorithm and block diagram, calculations were made in C++ program and the electromagnetic field intensities of the microwave circular waveguide with nonlinear medium operating at 9 *GHz* frequency in *E*-type and *H*-type waves were determined. Based on the obtained numerical values, the dependencies of the length of the circular waveguide with a microwave operating at a frequency of 9 *GHz* in *E*-type and *H*-type waves were established (Fig. 2.23a, b) (Table 2.7). These dependencies determine how the electromagnetic field is

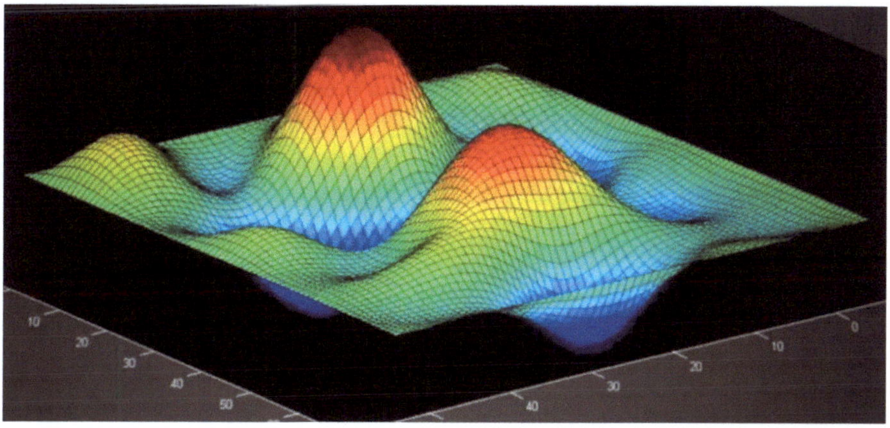

Fig. 2.22 3*D* model of magnetic field intensity distribution for *H*-type wave of a microwave rectangular waveguide operating in the frequency range of 4.9–7.05 *GHz* (finite element method)

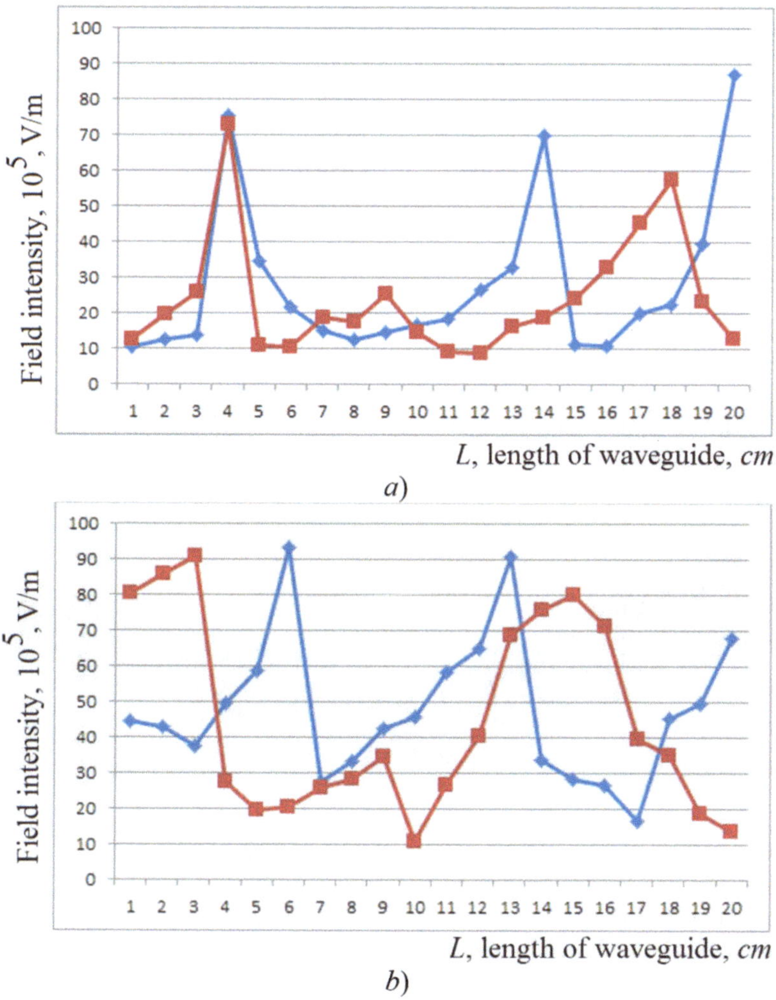

Fig. 2.23 Dependencies of the length of the circular waveguide with a nonlinear medium operating at a frequency of 9 *GHz* in *E*-type (**a**) and *H*-type (**b**) waves and its electric field intensities: E_r – blue color, E_z – red color

Table 2.7 Values of the electric field intensities of a microwave circular waveguide with a nonlinear medium operating at a frequency of 9 *GHz* for *E*-type and *H*-type waves determined by the finite difference method

	For *E*-type wave		For *H*-type wave	
Number of elementary regions	E_r, (10^5) *V/m*	E_z, (10^5) *V/m*	E_r, (10^5) *V/m*	E_z, (10^5) *V/m*
1	10.57	12.44	44.74	80.66
2	12.64	19.63	42.89	85.75
3	13.89	25.75	37.63	90.86
4	75.61	72.87	49.85	27.63
5	34.53	10.73	58.72	19.55
6	21.84	10.45	93.64	20.34
7	14.96	18.66	27.63	25.75
8	12.69	17.49	33.57	28.22
9	14.82	25.65	42.63	34.63
10	16.54	14.73	45.77	10.99
11	18.43	9.25	58.23	26.64
12	26.85	8.62	65.12	40.62
13	33.18	16.21	90.94	68.77
14	70.29	18.83	33.66	76.14
15	11.32	24.42	28.32	80.01
16	10.97	32.95	26.75	71.23
17	19.91	45.39	16.58	39.53
18	22.62	57.66	45.44	35.02
19	39.54	23.54	49.81	18.78
20	87.16	12.97	68.20	13.9

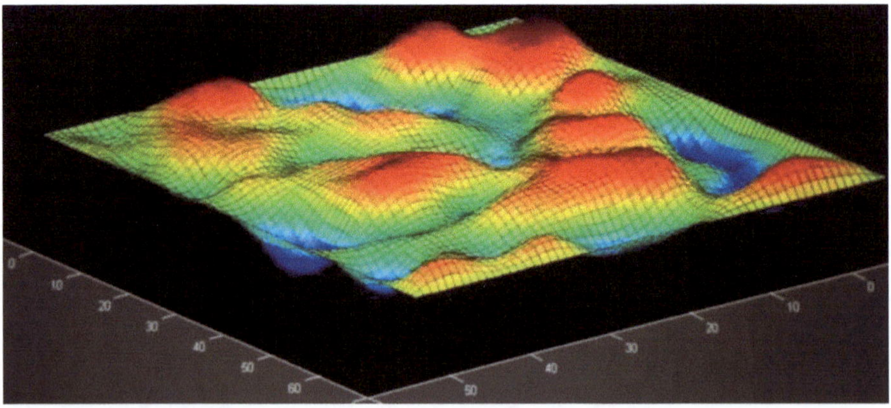

Fig. 2.24 3*D* model of electric field intensity distribution for an *E*-type wave of a microwave circular waveguide operating at 9 *GHz* (finite difference method)

distributed inside the investigated waveguide and the relationship between the electromagnetic and structural parameters of this device. Figure 2.24 shows the 3*D* model (finite element method) of the electric field intensity distribution for the *E*-type wave of a circular waveguide operating at a frequency of 9 *GHz*. Figure 2.25

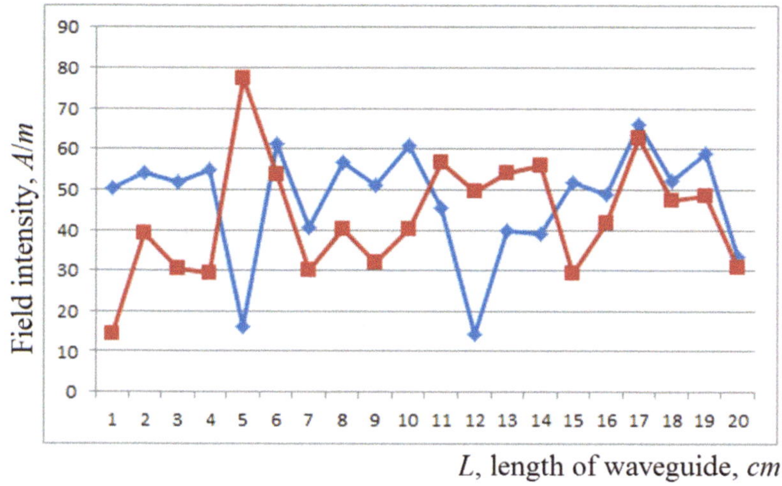

a)

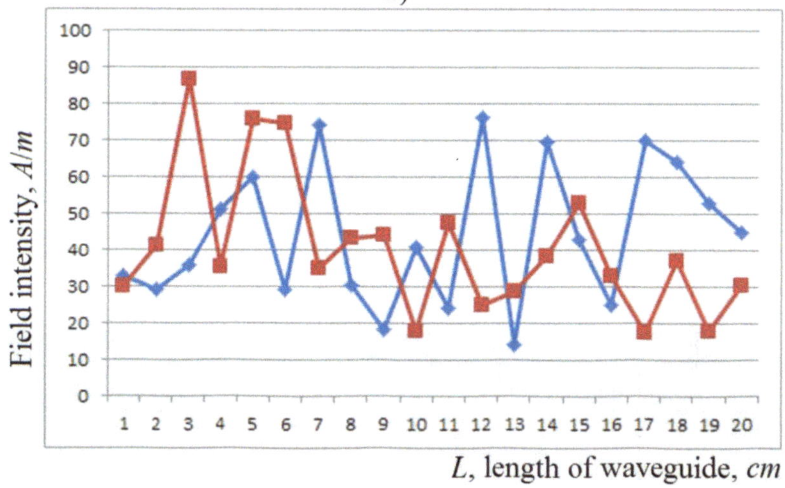

b)

Fig. 2.25 Dependencies of the length of the circular waveguide with a nonlinear medium operating at a frequency of 9 *GHz* in *E*-type (**a**) and *H*-type (**b**) waves and its magnetic field intensities: H_r – blue color, H_z – red color

shows the dependencies of the length of the circular waveguide with a nonlinear environment microwave operating at a frequency of 9 *GHz* in *E*-type (*a*) and *H*-type (*b*) waves (Table 2.8). Figure 2.26 shows the 3*D* model (finite element method) of the frequency distribution of the magnetic field for the *E*-type wave of a circular waveguide operating at a frequency of 9 *GHz*.

Table 2.8 Values determined by finite element method of magnetic field intensities of a microwave circular waveguide with nonlinear medium operating at 9 *GHz* for *E*-type and *H*-type waves

Elemen-tary regions	For *E*-type wave		For *H*-type wave	
	H_r, *A/m*	H_z, *A/m*	H_r, *A/m*	H_z, *A/m*
1	50.39	14.35	32.95	29.98
2	54.15	39.12	29.35	41.17
3	51.75	30.54	35.89	86.73
4	54.78	29.14	51.16	35.69
5	16.33	77.34	60.13	75.78
6	61.16	53.87	29.22	74.68
7	40.69	30.05	74.14	35.21
8	56.76	40.34	30.33	43.32
9	51.11	31.93	18.34	44.13
10	60.87	40.04	40.85	18.01
11	45.31	56.76	24.11	47.78
12	14.21	49.75	76.46	24.94
13	39.72	54.15	14.22	28.97
14	38.94	56.14	69.68	38.35
15	51.86	29.39	43.14	52.89
16	48.81	41.54	25.18	33.03
17	66.25	62.76	70.02	17.36
18	52.13	47.35	64.16	37.17
19	58.79	48.57	52.88	18.12
20	33.46	30.64	44.92	30.43

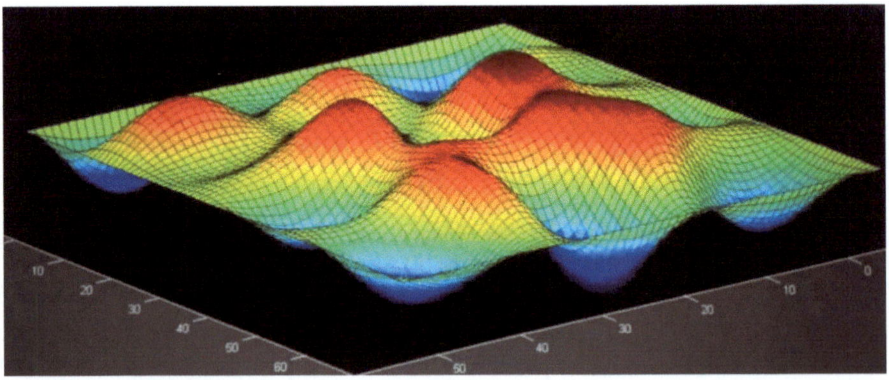

Fig. 2.26 3*D* model of magnetic field intensity distribution for *E*-type wave of microwave circular waveguide operating at 9 *GHz* frequency (finite element method)

2.6 Evaluation of the Error in Calculating the Electromagnetic Field of Microwave Devices Using the Finite Difference Method

Two types of errors can occur in calculating the electromagnetic field of microwave devices by the finite difference method [19]: (1) Errors in the construction of the finite difference scheme when bringing the obtained differential equations to the system of linear eqs. (2) Errors arising during the solution of finite difference equations. To evaluate the error in calculating the electromagnetic field of microwaven devices using the finite difference method

$$\frac{\partial^2 \phi}{\partial x^2}$$

Let's look at the error of the limit. The error of this threshold is sought as follows:

$$\frac{h}{3!}\left[m(1-m^2)\left(\frac{\partial^3 \phi}{\partial x^3}\right)_0\right]-\frac{h^2}{4!}\left[m(1+m^3)\left(\frac{\partial^2 \phi}{\partial x^4}\right)_0\right]+\ldots, \tag{2.56}$$

When $m = 1$, the error of limit $\partial^2\phi/\partial x^2$ according to Eq. 2.56 has the following form:

$$-\frac{2h^2}{4!}\left(\frac{\partial^2 A}{\partial x^4}\right)_0 - \ldots \tag{2.57}$$

As can be seen from (2.57), the error in the form of the finite difference scheme depends on h^2. From this it can be concluded that the error generated in asymmetric networks is more than the error generated in symmetric networks. Therefore, it is more appropriate to divide the computing area into a symmetrical network. Now let's denote the radius of the circle covering the entire calculation area by r, and the eighth degree of the special derivative of the exact solution by E_8. Then the upper bound value of the network error is as follows:

$$\rho_3 = \frac{\mathrm{E}_8 h^6 r^2}{12096}. \tag{2.58}$$

$r = 6$; $h = 0.2$; $E_8 = 1.5 \times 10^5$, the grid error of the finite difference method is $\rho_3 = 0.029\%$.

Let's define the calculation error as follows. If we denote the maximum value of the remainder by Δn, and the radius of the circle covering the calculation area by r, then the error of the finite difference equations does not exceed the following limit:

$$\rho_p = \frac{\Delta n r^2}{4h^2}. \tag{2.59}$$

$r = 6$, $h = 0.2$, $\Delta n = 3.5 \cdot 10^{-4}$, $\rho_p = 0.0787$ % .
Thus, the total error of the finite difference method is as follows:

$$\rho = \rho_3 + \rho_p = 0.029 + 0.0787 = 0.107\%.$$

2.7 Estimating the Error in Calculating the Electromagnetic Field of Microwave Devices Using the Finite Element Method

As mentioned, a certain error is made when calculating the electromagnetic field of microwave devices using the finite element method. At this time, mainly two types of errors occur: (1) discretization error and (2) calculation error.

The discretization error is related to the replacement of the initial state of the problem with a simpler model as a result of discretization. This is also called discretization of the model. The cost of discretization depends on the nature of the function sought, boundary conditions and degree of discretization [20]. When the mesh of finite elements is imprecise and non-smooth, the maximum error of approximation is determined as follows [21]:

$$|\phi - \phi_t| \leq L \cdot k_{\tau+1} h^{\tau+1}, \tag{2.60}$$

where φ is the exact solution, φ_t is the approximation of the solution, L is a constant dependent on τ and independent of t, $k_{\tau\,+\,1}$ is the upper value of the (τ + 1)-th derivative of the exact solution.

The following statement is true for the m-th derivative of the approximation function [22]:

$$\left|\partial^m (\phi - \phi_t)/\partial x_i^m\right| \leq L \cdot h^{\tau+1-m} \cdot k_{\tau+1}, \tag{2.61}$$

where x_i is the i-th coordinate.

For triangular finite elements $\tau = 1$. Then we can write expression (2.60) as follows:

$$|\phi - \phi_t| \leq L k_2 h^2, \tag{2.62}$$

where the error of the $E_i = \pm \frac{\partial \phi}{\partial x_i}$ value of the i-th component of the electric field intensity is determined as follows:

$$|\partial(\phi - \phi_t)/\partial x_i| \leq L \cdot k_2 \cdot h. \tag{2.63}$$

It should be noted that the shape of the triangle also has a significant effect on the approximation error. Studies have shown that more accurate results are obtained if the triangle is of an equilateral configuration. Therefore, the following condition must be fulfilled [23]:

$$\frac{1}{\sin\theta} \leq \frac{h^2}{2S_{\Delta m}} \leq \frac{2}{\sin\theta}, \tag{2.64}$$

where θ is the small angle of the triangle, $S_{\Delta m}$ is the area of the triangle.

The calculation error that occurs during the calculation of the electromagnetic field of microwave devices using the finite element method is determined as follows [24]:

$$\varepsilon_B = 10^{-(5+2)} C_{n(k)} N, \tag{2.65}$$

where $C_{n(k)}$ is the number associated with the matrix of the system of linear algebraic equations, N is the number of elements.

To determine $C_{n(k)}$, the following condition must be satisfied [25]:

$$C_{n(k)} \leq C(h_{max}/h_{min})N, \tag{2.66}$$

where C is a constant, h_{max}/h_{min} are the smallest and largest distances between the nodes of the finite element mesh.

$C = 0.83$, $h_{max} = 3\ mm$, $h_{min} = 1\ mm$, $N = 20$, $C_{n(k)} \leq 0.83 \cdot 3 \cdot 20 = 50$. $S = 1$, $\varepsilon_B = 10^{-3} \cdot 50 \cdot 20 = 1\%$.

Now let's evaluate the calculation error that occurs during the calculation of the electromagnetic field of microwave devices using the finite element method. Given the expression (2.63), the following expression is true for the absolute error of the electric field intensity:

$$\Delta = |E - E_t| \leq L^1 h, \tag{2.67}$$

where E is the true value of the modulus of the electric field intensity, E_t is the approximate solution determined using the finite element method, $L^1 = L \cdot K_2$ is a constant.

Let's assume that there are two solutions to the problem in networks with steps h_1 and h_2 ($h_1 > h_2$).

Then

$$\Delta_1 = |E – E_{t1}| \leq L^1 h_1, \tag{2.68}$$

$$\Delta_2 = |E – E_{t2}| \leq L^1 h_2. \tag{2.69}$$

Taking into account the maximum error, it is possible to pass from inequalities (2.68) and (2.69) to the following equations:

$$\Delta_1 = |E – E_{t1}| \leq L^1 h_1, \tag{2.70}$$

$$\Delta_2 = |E – E_{t2}| \leq L^1 h_2. \tag{2.71}$$

If we separate expressions (2.70) and (2.71), we get

$$|E_{t2} – E_{t1}| = L^1 (h_1 – h_2) = L^1 h_1 \left(1 – \frac{h_2}{h_1}\right). \tag{2.72}$$

Make the following substitutions:

$$\Delta_r = |E_{t2} – E_{t12}|; r = \frac{h_2}{h_1},$$

$$\Delta_r = \Delta_1 \left(1 - \frac{1}{r}\right). \tag{2.73}$$

The algorithm for determining the sampling error is given in the block diagram shown in Fig. 2.27.

Thus, we can determine the calculation error for finite element 1:

$$\Delta_1 = \frac{\Delta_r}{1 - \frac{1}{r}}. \tag{2.74}$$

It should be noted that the calculation error that occurs during the calculation of the electromagnetic field of microwave devices has its smallest value $r = 2$ [26] when he gets it.

Therefore, (2.74) takes the following form:

$$\Delta_1 = 2(\Delta_r). \tag{2.75}$$

The expression (2.75) can be written in the general case as follows:

$$\Delta_n = 2(\Delta_{rn}). \tag{2.76}$$

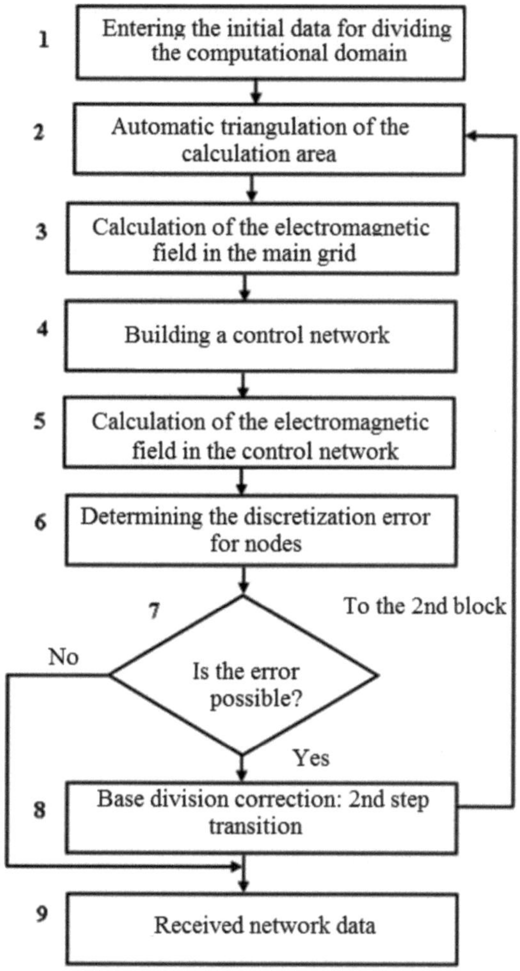

Fig. 2.27 Block diagram of sampling error determination

According to the algorithm shown in Fig. 2.27 and the expression (2.76), the numerical values of the discretization error during the determination of the intensity of the electric field of the microwave rectangular waveguide were determined. These values are listed in Table 2.9.

As can be seen from the values given in Table 2.9, the sampling error can be omitted during the determination of the electric field intensity of the microwave rectangular waveguide.

Table 2.9 Numerical values of discretization error during determination of electric field intensity of microwave rectangular waveguide

The number of finite elements	For *E*-type wave 2(Δ*r*)		For *H*-type wave 2(Δ*r*)	
	Ex	*Ey*	*Ex*	*Ey*
1	0.06	0.04	0.06	0.04
2	0.04	0.1	0.06	0.04
3	0.12	0.1	0.06	0.1
4	0.1	0.06	0.08	0.1
5	0.06	0.06	0.08	0.02
6	0.08	0.04	0.06	−0.02
7	0.04	0	0.04	0.06
8	0.12	0.1	0.08	0.04
9	0.08	0.14	0.06	0.08
10	0.04	0.04	0.02	0.12
11	0.08	0.04	0.12	0.08
12	0.04	0.08	0.06	0.12
13	0.08	0.06	0.04	0.02
14	0.08	0	0.1	0.1
15	0.06	0.04	0.04	0.06
16	0.1	0.06	0.04	0.1
17	0.04	0.04	0.06	0.08
18	0.08	0.04	0.06	0.06
19	0.02	0.06	0.04	0.1
20	0.08	0.08	0.08	0.1

Fig. 2.28 Air-filled microwave metal circular waveguide

2.8 Modeling the Spatial Structure of the Electromagnetic Field in a Microwave Circular Waveguide

A circular waveguide (Fig. 2.28) is widely used to transmit information over long distances with low losses.

There are a number of existing studies in the world about the transmission of information over long distances with circular waveguides [27]. Using the H_{11} mode for long-distance transmission of information with circular waveguides has several advantages. This is also related to the features of the H_{11} mode. Thus, in a circular waveguide of a certain diameter, the attenuation decreases with the increase of the working frequency range. For example, in a copper waveguide with a diameter of 60 *mm*, the attenuation of the H_{11} wave mode at a frequency of 40 *GHz* is 3 dB/km. Therefore, a large amount of information can be transmitted through such transmitters by regenerating every 30 *km* (if the signal is subject to attenuation up to 90 dB).

It should be noted that some specific issues of the H_{11} wave mode in a circular waveguide were discussed by Na et al. [28] reviewed and resolved by. At this time, the issues with communication lines, including the creation of waveguide sections and communication lines based on them, were resolved.

However, these authors did not take into account the nonlinearity of the environment when conducting research.

As can be seen from Fig. 2.28, when microwave metal circular waveguides filled with air are manufactured and operated, a layer of a certain thickness is applied to its inner surface. The function of this layer is to minimize fading. If the cross-sectional area of an air-filled microwave circular waveguide is greater than the wavelength, then the propagation of electromagnetic waves begins to propagate by reflection from its inner walls. In such a transmitter, the working wave mode will be the hybrid EH_{11} mode. The attenuation of this working mode will be a non-monotonic function of wavelength. Air-filled metal circular waveguides can be used in other ranges than the microwave range. Submillimeter, infrared wave ranges can be shown as an example. In these ranges, circular waveguides can be used in both transmitting and receiving tracts of information.

The following expression is fulfilled for the electric field of a microwave metal circular waveguide filled with air [29]:

$$\frac{1}{r^2}\frac{\partial}{\partial r}\left(r^2\frac{\partial U}{\partial r}\right)+\frac{1}{r^2\sin\theta\ \partial\theta}\left(\sin\theta\frac{\partial\phi}{\partial\theta}\right)+\frac{1}{r^2}\frac{1}{\sin^2\theta}\frac{\partial^2 U}{\partial\alpha^2}+k^2U=0. \tag{2.77}$$

Considering the boundary conditions, we get

$$\frac{1}{r}\frac{\partial}{\partial r}\left(r\frac{\partial U}{\partial r}\right)+\frac{1}{r^2}\frac{\partial^2 U}{\partial\alpha^2}+k^2U=0. \tag{2.78}$$

If we approximate expression (2.77) by the following finite difference scheme, we get

$$\begin{aligned}&\frac{\dot{E}_{i-1,j}+\dot{E}_{i+1,j}-2\dot{E}_{i,j}}{h_r^2}+\frac{1}{r_{i,j}^2}\frac{\dot{E}_{i,j-1}+\dot{E}_{i,j+1}-2\dot{E}_{i,j}}{h^2}+\\&+\frac{1}{r_{i,j}}\frac{\dot{E}_{i+1,j}-\dot{E}_{i-1,j}}{2h_r}+\tilde{\chi}^2\dot{E}=0,\end{aligned} \tag{2.79}$$

The markings here are shown in Fig. 2.29.

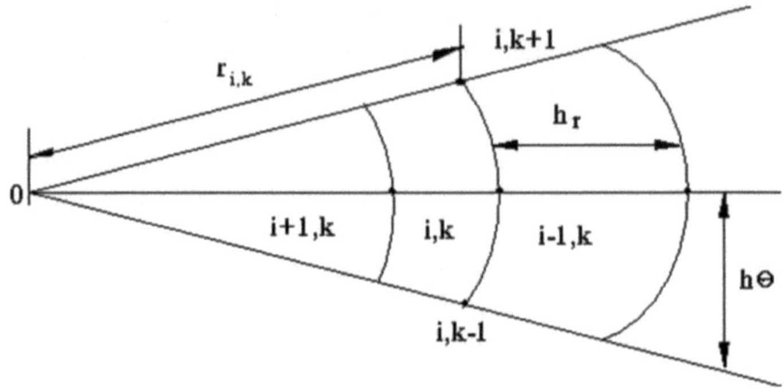

Fig. 2.29 Finite difference scheme of the electric field of a microwave circular waveguide in a cylindrical coordinate system

According to expression (2.79), the parameter $\dot{E}$ can be determined according to the values at the neighboring four nodes at node i, j:

$$\dot{E}_{i,j} = \frac{\left(1-\frac{h_r}{2r_{i,j}}\right)\dot{E}_{i-1,j} + \left(1+\frac{h_r}{2r_{i,j}}\right)\dot{E}_{i+1,j} + \frac{1}{r_{i,j}^2}\left(\frac{hr}{h\theta}\right)^2\left(\dot{E}_{i,j-1} + \dot{E}_{i,j+-1}\right)}{2\left[1+\frac{1}{r_{i,j}^2}\left(\frac{hr}{h\theta}\right)^2\right] - \tilde{\chi}^2 h_r^2}. \quad (2.80)$$

Since $\dot{E}_{i,j}$ is a complex number, we can write in expression (2.80) by dividing it into real and imaginary parts:

$$\begin{cases} \dot{E}_{i,j} = U_{i,j} + jv_{i,j}, \\ U_{i,j} = \dfrac{2A\left[1+\dfrac{1}{r_{i,j}^2}\left(\dfrac{hr}{h\theta}\right)^2\right] + \beta\tilde{\chi}^2 h_r^2}{2\left[1+\dfrac{1}{r_{i,j}^2}\left(\dfrac{hr}{h\theta}\right)^2\right] + \tilde{\chi}^4 h_r^4} \\ v_{i,j} = \dfrac{2\beta\left[1+\dfrac{1}{r_{i,j}^2}\left(\dfrac{hr}{h\theta}\right)^2\right] - A\tilde{\chi}^2 h_r^2}{2\left[1+\dfrac{1}{r_{i,j}^2}\left(\dfrac{hr}{h\theta}\right)^2\right] + \tilde{\chi}^4 h_r^4}, \end{cases} \quad (2.81)$$

where

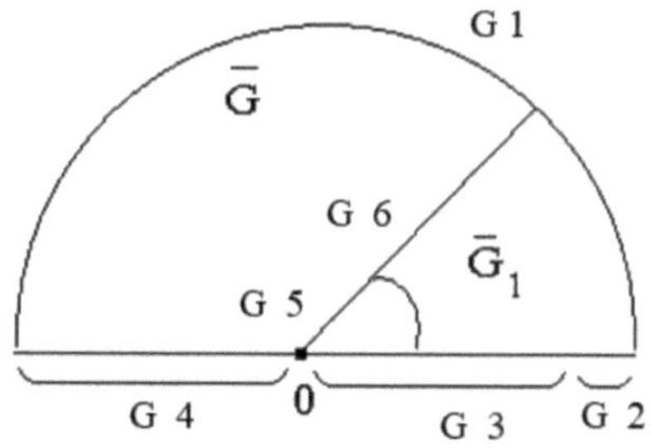

Fig. 2.30 $\overline{G}$ areas where the price of ΔE is calculated

$$A=\left(1-\frac{hr}{2r_{i,j}}\right)U_{i-1,j}+\left(1+\frac{hr}{2r_{i,j}}\right)U_{i+1,j}+\frac{1}{r_{i,j}^2}\left(\frac{hr}{h\theta}\right)^2\left(U_{i,j-1}+U_{i,j+1}\right),$$

$$B=\left(1-\frac{hr}{2r_{i,j}}\right)v_{i-1,j}+\left(1+\frac{hr}{2r_{i,j}}\right)v_{i+1,j}+\frac{1}{r_{i,j}^2}\left(\frac{hr}{h\theta}\right)^2\left(v_{i,j-1}+v_{i,j+1}\right).$$

As you can see, in the cylindrical coordinate system, the node $i = 0, j = 0$ is of special importance. In this node, the value of function $\dot{E}$ is set as follows:

$$\dot{E}_{0,0}=\frac{2\left(\dot{E}_{1,0}+\dot{E}_{1,N}+2\sum\limits_{m-1}^{N-1}\dot{E}_{1,m}\right)}{N\left(4-\widetilde{\chi}^2h_r^2\right)}, \tag{2.82}$$

where $N - \theta$ is the number of nodes in the coordinate.

The expression (2.82) is worked out for $\overline{G}$ semicircles (Fig. 2.30). So, according to the symmetry problem, it can be found for the whole circle.

We can write expression (2.82) by dividing it into real and imaginary parts as in expression (2.80):

$$U_{0,0}=\frac{8\left(U_{1,0}+U_{1,N}+2\sum\limits_{m-1}^{N-1}U_{1,m}\right)+2\left(v_{1,0}+v_{1,N}+2\sum\limits_{m-1}^{N-1}v_{1,m}\right)\widetilde{K}^4h_r^2}{N\left(16+h_r^2\widetilde{K}^4\right)},$$

$$v_0=\frac{8\left(v_{1,0}+v_{1,N}+2\sum\limits_{m-1}^{N-1}v_{1,m}\right)+2\left(U_{1,0}+U_{1,N}+2\sum\limits_{m-1}^{N-1}U_{1,m}\right)}{N\left(16+h_r^2\widetilde{K}^4\right)}. \tag{2.83}$$

To increase the accuracy of the performed calculations, we make the following substitution in (2.79):

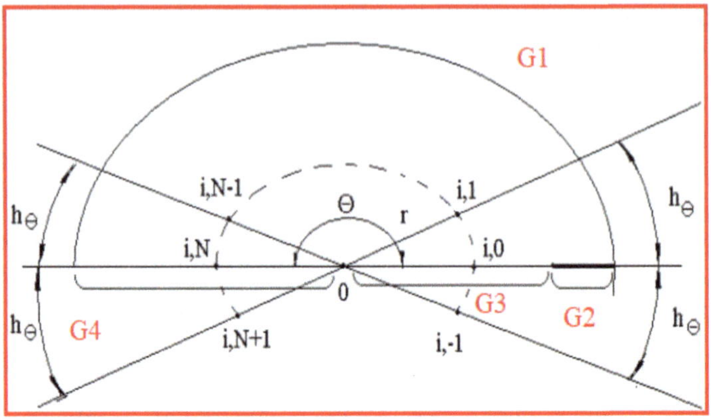

Fig. 2.31 On the provision of boundary conditions at the boundaries of Q_3 and Q_4

$$\Delta\dot{E}_{i,j} = \dot{H}_{i,j} - \dot{H}^{0}_{i,j},$$

where $\dot{E}^{0}_{i,j}$ is the intensity of the electric field at the grid points i, j.

Thus,$\Delta\dot{E}_{i,j}$ determines the variation of the network function. The inclusion of the difference function $\Delta\dot{E}_{i,j}$increases the accuracy of determining the parameters of the electric field at the corresponding values of the functions $\Delta\dot{E}_{i,j}$ and $\dot{E}^{0}_{i,j}$. $\Delta\dot{E}_{i,j}$ and $\dot{E}^{0}_{i,j}$ are close numbers. Substitution of variables does not disturb the appearance of Eq. 2.79. Thus, the function $\dot{E}^{0}_{i,j}$ satisfies Eq. 2.79. However, since the intensity of the electric field of the outer surface of the microwave circular waveguide is constant and does not depend on the shape of the region $\overline{G}$, the boundary conditions will change as follows:

$$\Delta\dot{E}_{i,j}\big|_{Q_1} = 0, \tag{2.84}$$

where Q_1 is a semicircle (Fig. 2.31).

$\Delta\dot{E}_{i,j} = 1$ at the grid points located in Q_2. At those points, the value of $\dot{E}^{0}_{i,j}$ is determined as follows:

$$\Delta\dot{E}^{0}_{i,j}\Big|_{Q_2} = \frac{J_0(\widetilde{K}rj)}{J_0(\widetilde{K})}.$$

Properly,

$$\Delta\dot{H}^{0}_{i,j}\Big|_{Q_2} = 1 - \frac{J_0(\widetilde{K}rj)}{J_0(\widetilde{K})}. \tag{2.85}$$

The following condition is met according to symmetry on the boundaries Q_3, Q_4 of the region G:

$$\frac{\partial \dot{E}}{\partial \theta}|_{Q_3,Q_4} = 0. \tag{2.86}$$

(2.86) to establish the difference approximation of the boundary condition, let's add a node not included in the boundaries of Q_3, Q_4 to region $\overline{G}$. Then for the difference approximation of the boundary condition (2.86) we get:

1. For the Q_3 boundary

$$\frac{\dot{E}_{i,1} - \dot{E}_{i-1}}{2h\theta} = 0, \tag{2.87}$$

2. For the Q_4 boundary

$$\frac{\dot{E}_{i,N+1} - \dot{E}_{i,N-1}}{2h\theta} = 0, \tag{2.88}$$

From expressions (2.87) and (2.88) we get

$$\begin{cases} \dot{E}_{i,1} = \dot{E}_{i-1}, \\ \dot{E}_{i,N+1} = \dot{E}_{i,N-1} \end{cases} \tag{2.89}$$

(2.89) boundary conditions also hold for the following functions:

$$\begin{cases} \Delta\dot{E}_{i,1} = \Delta\dot{E}_{i-1}; \\ \Delta\dot{E}_{i,N+1} = \Delta\dot{E}_{i,N-1}. \end{cases} \tag{2.90}$$

After setting the value $\Delta\dot{E}_{i,j}$ in all nodes, if we consider the occurrence of $S = \pi$, we get

$$\Delta\dot{E} = j\frac{2\varepsilon}{\pi}\sum_{i=1}^{M}\sum_{j=1}^{N}\Delta\dot{E}_{i,j}r_i h_{ri} h_{\theta i}, \tag{2.91}$$

where M is the number on the radius.

To determine the values of$\Delta\dot{E}_{i,j}$, an iteration scheme using the upper relaxation method (Yang's method) was used [30]. Here, the value of the function $\Delta\dot{E}_{i,j}^{(n+1)}$ after the next iteration correlates the preceding $\Delta\dot{E}_{i,j}^{(n)}$ iterations as follows:

$$\Delta\dot{E}_{i,j}^{(n+1)} = \Delta\dot{E}_{i,j}^{(n)} + \beta\left[\Delta\dot{E}_{i,j}^{(n+0,5)} - \Delta\dot{E}_{i,j}^{(n)}\right], \tag{2.92}$$

where $\Delta\dot{E}_{i,j}^{(n+0.5)}$ is the value of the function calculated at the $(n + 1)$-th iteration when β=1.

The value of the upper relaxation parameter lies in the interval 1<β<2 and is related to the accumulation of the iteration process. In case viewed β=1.4. Experience shows that despite the application of the upper relaxation method, the iteration process accumulates late. Therefore, it is appropriate to include the following function.

$$y^{(n+1)} = y^{(n)} + \frac{1}{1-q}\left[y^{(n)} - y^{(n-1)}\right], \tag{2.93}$$

where

$$q = \frac{y^{(n)} - y^{(n-1)}}{y^{(n-1)}y^{(n-2)}}.$$

Insertion of (2.93) replaces a large amount of calculations. So let's insert the following substitution:

$$\delta = \frac{1}{1-q}\left[y^{(n)} - y^{(n-1)}\right],$$

This substitution is introduced after calculations until the q parameter stabilizes. In this case, the correction to limit $\Delta\dot{E}^{(n+1)}$ is as follows:

$$\Delta\dot{E}^{(n+1)} = \Delta E^{(n)} + \frac{1}{1-q}\left[\Delta\dot{E}^{(n)} - \Delta\dot{E}^{(n-1)}\right], \tag{2.94}$$

where q is defined analogously to (2.92).

To solve the problem under consideration, h = 0.05; 0.01; 0.415; 0.500 when $\tilde{\chi}^2$ =1,3,5,9,150 using C++ language according to the above algorithm $\Delta\dot{E}$ for are assigned values. Equal grids were used during calculations. Thus, when using uneven networks, despite the reduction in the number of nodes, it complicates the software. Depending on the value of the step H, the number of nodes of the network is taken differently. For $h > 0.4$, the pitch of the grid along the radius of the circular waveguide is $h_r = 0.0833$, and the pitch of the grid according to the angle is $h_\theta = \pi/60$. At this time, the total number of nodes is equal to 12 × 30 = 360, when $h < 0.01$, and $h_r = 0.01$, $h_\theta = \pi/120$. The solution of the problem was carried out only in the area $\overline{G_1}$, which is limited by the values of $p_i = 0.5$ and $\theta_j = \pi/4$. Accordingly, the total number of nodes is equal to 50 × 30 = 1500. $\Delta\dot{E}_{i,j} = 0$ in $\overline{G}$ regions.

It should be noted that the total number of nodes in the network equal to 1500 almost completely reflects the issue under consideration. Since the real $\Delta U_{i,\,j}$ and imaginary $\Delta U_{i,\,j}$ values of $\Delta \dot{E}_{i,j}$ are used, 3000 slots in memory are required for the network. The value of the accuracy parameter ε of the iteration process is taken equal to 0.01. As mentioned, since the iteration process is accumulated late, their number is χ^2 and depending on the values of h, it is in the range of 150–250. In networks with a total number of nodes greater than 360, it takes approximately 4 min to calculate the value of $\Delta \dot{E}$, and in networks with a total number of nodes greater than 1500, it takes approximately 9 min to calculate the value of $\Delta \dot{E}$. The main error in the calculations is the discretization error. This error is reduced by reducing the grid step. To evaluate this error, a network two times smaller than the main network was built and the calculations were repeated in case of $\chi^2 = 1$, $h = 0.333$. On the Q_2 boundary, the number of nodes during the main step is 4, and the number of nodes during the additional step is 8. A comparison of the values of $\Delta \dot{E}$ calculated in the basic and additional steps shows that the error between them does not exceed 3.3% for the modulus and 40′ for the argument. These are the prices that can be released. The result of numerical modeling is shown in Fig. 2.32.

In the zone where the function changes sharply, the value of the discretization error depends strongly on the step h. In this case, this zone is located near the border of Q_2. Taking this into account, the minimum number of nodes on the boundary of Q_2 was taken equal to 5 in all calculations. It should be noted that the accuracy of the calculations is determined at only one value of the parameter $\chi^2(\chi^2 = 1)$.

The process of evaluating the error χ^2 parameter at other values is carried out similarly. At large values of parameter χ^2, a sharp change of function $\Delta \dot{E}$ occurs depending on the radius and angle.

2.9 Determination of the Main Parameters of the Microwave Rectangular Waveguide Tract in the Case of Environmental Nonlinearity

2.9.1 Determination of Critical Frequency in a Microwave Rectangular Waveguide in Case of Environmental Nonlinearity

In the microwave rectangular waveguide, the critical frequency $f_{\kappa r}^{mn}$ is calculated as follows [31]:

$$f_{\kappa r}^{mn} = \frac{c}{2\sqrt{\varepsilon\mu}}\sqrt{\left(\frac{m}{a}\right)^2 + \left(\frac{n}{b}\right)^2}, \tag{2.95}$$

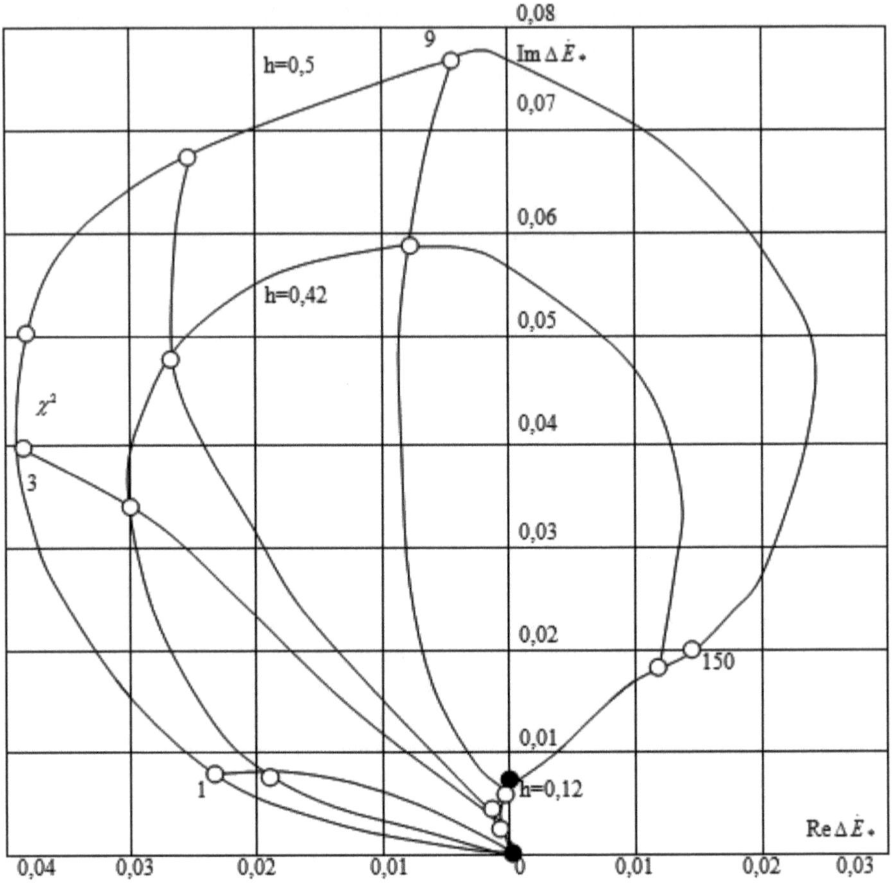

Fig. 2.32 Dependence of the relative intensity of the electric field of a microwave circular waveguide on the step h (in the case of changing χ^2 generalized parameters)

where c is the speed of heat in a vacuum, ε and μ are the electric and magnetic permeabilities of the waveguide dielectric, respectively, and a and b are the dimensions of the large and small wall of the waveguide.

In the expression (2.95), for E-type wave, m and n take positive integer values, while for H-type wave, they can take even zero value. It should be noted that in the air-filled microwave rectangular waveguide, the H_{10} wave is the main wave type. Therefore, $m = 1$ and $n = 0$ for H_{10}-type wave. Then expression (2.95) can be written as follows:

$$f_{\kappa r}^{10} = \frac{c}{2\sqrt{\varepsilon\mu}}. \tag{2.96}$$

As can be seen from the expression (2.96), the critical wavelength for the H_{10}-type wave does not depend on the size of the small wall of the waveguide b.

On the other hand, as we know, $\varepsilon = \varepsilon\ (E)$ [32].

$$\varepsilon = 1,8 + 1,066 \cdot 10^{-5} \cdot E. \tag{2.97}$$

Considering expression (2.97) in (2.96), we get

$$f_{\kappa r}^{10} = \frac{c}{2a\sqrt{\mu\left(1.8 + 1.066 \cdot 10^{-5} \cdot E\right)}}. \tag{2.98}$$

If we make certain transformations in (2.98), we get the following expression between the critical frequency $f_{\kappa r}^{10}$ and the electric field intensity E in the microwave rectangular waveguide in the nonlinear case of the medium:

$$E = \frac{c^2 - 7.2\mu a^2 \left(f_{kr}^{10}\right)^2}{4.264\mu a^2 \left(f_{kr}^{10}\right)^2} \cdot 10^5,\ \left[\frac{V}{m}\right]. \tag{2.99}$$

This expression between the critical frequency $f_{\kappa r}^{10}$ and the electric field intensity E in a microwave rectangular waveguide in the nonlinear case of the environment can be applied very successfully in engineering calculations.

Now let's determine the electric field intensity E based on the expression (2.99). As we know, the studied microwave rectangular waveguide works in the frequency range of 4.9–7.05 *GHz*, so its cross-section dimensions are 40 × 20 *mm*.

It should be noted that in an air-filled microwave rectangular waveguide, the $f \geq f_{\kappa r}^{mn}$ condition must be satisfied to excite a wave at m and n indices at frequency f. Within this condition, we can determine the following: when $f_{\min} = 4.9$ *GHz*, $E_{\min} = 1139 \cdot 10^5$ *V/m*, when $f_{\max} = 7.05$ *GHz*, $E_{\max} = 1423 \cdot 10^5$ *V/m*.

According to the numerical values of $E_{\min}$ and $E_{\max}$ determined based on the expression (2.99), it can be noted that as the working frequency increases in the air-filled microwave rectangular waveguide, the value of the electric field intensity E will also increase. Therefore, by adjusting the frequency of the transmitter, it is possible to control how the value of the electric field intensity E changes.

2.9.2 Determination of the Limit Power Transmitted by a Microwave Rectangular Waveguide Tract in the Case of Ambient Nonlinearity

Let's look at the method of determining the limit power of the H_{10} wave transmitted by a microwave rectangular waveguide tract in the case of environmental nonlinearity. The limiting power of the H_{10} wave transmitted by a microwave rectangular waveguide tract is generally determined as follows:

$$dP_z = \operatorname{Re} \widetilde{P}_z ds, \tag{2.100}$$

where $\widetilde{P}_z$ is the longitudinal component of the Poynting vector.

$$\widetilde{n}_z = \frac{1}{2}\left(z_0, \left[\dot{E}, \overset{*}{H}\right]\right) = -\frac{1}{2}\left(\dot{E}, \left[z_0, \overset{*}{H}\right]\right). \tag{2.101}$$

There is the following relationship between E and H:

$$\dot{H} = \frac{1}{Z_o}\left[z_0, \dot{E}\right]. \tag{2.102}$$

$Z_0 = Z_C^E$ for E-type wave in expression (2.102) $Z_0 = Z_C^H$ for H-type wave.

If we consider the expression (2.102) in (2.101) and at the same time take into account that the longitudinal components of the electric and magnetic fields differ from each other in phase by 90°, we get

$$\widetilde{P}_z = \frac{1}{2}\frac{|E_z|^2}{Z_o}. \tag{2.103}$$

Considering (2.103) in (2.100) and integrating, we get

$$P_Z^E = \frac{1}{2}\frac{1}{Z_o}\int_S |E_Z|^2 dS, \tag{2.104}$$

where dS is the element of the cross-sectional area of the microwave rectangular waveguide (dashed area in Fig. 2.33).

In a microwave rectangular waveguide, power is transmitted through dS.

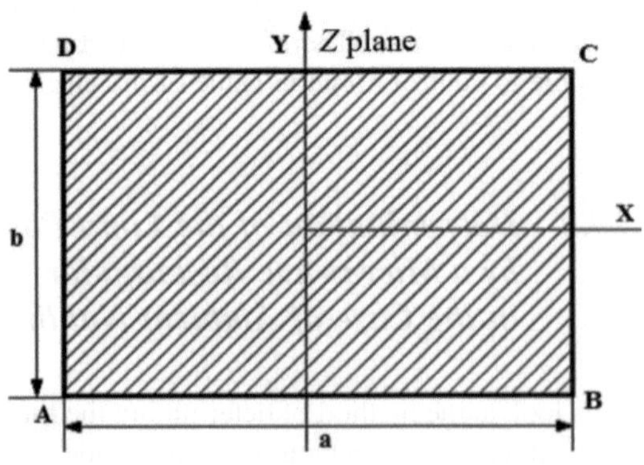

Fig. 2.33 On the determination of transmitted power through dS in a microwave rectangular wave transmitter

Knowing the relationship between electric and magnetic fields, expression (2.104) can be written as follows:

$$P_Z^E = \frac{1}{2}\sqrt{\frac{\varepsilon}{\mu}}\int\limits_S |E_Z|^2 dS, \tag{2.105}$$

where E_Z is the amplitude value of the intensity of the electric field in the center of the waveguide.

If in expression (2.101) we express $\dot{H}$ instead of $\dot{E}$, we get

$$P_Z^E = \frac{Z_o}{2}\int\limits_S |H_Z|^2 dS = \frac{1}{2}\sqrt{\frac{\mu}{\varepsilon}}\int\limits_S |H_Z|^2 dS, \tag{2.106}$$

where H_Z is the amplitude value of the intensity of the electric field at the center of the waveguide.

If we integrate over the area of $a \times b$, we get

$$P_Z = (ab/4)E_Z^2\sqrt{\frac{\varepsilon}{\mu}}. \tag{2.107}$$

That $\mu = 1$ and the components of the electric and magnetic fields from the amplitude values to their true values, i.e., $E_m = \frac{E}{\sqrt{2}}$, $H_m = \frac{H}{\sqrt{2}}$ [33], we get the following two expressions equivalent to (2.107):

$$P_Z^E = (ab/8)E^2\sqrt{\varepsilon}, \tag{2.108}$$

$$P_Z^H = (ab/8)H^2\sqrt{\frac{1}{\varepsilon}}, \tag{2.109}$$

where $E = 0.05{\cdot}10^5$ *V/m* is the true value of the electric field intensity calculated by the finite difference method.

According to this value of E, the relative dielectric strength of the electric field is $\varepsilon = 2$. If we substitute these values of parameters E and ε in (2.108), we get P_Z^E=3.5 *kWt*.

Thus, knowing the value of the field and the cross-sectional area of the waveguide, the power transmitted through a microwave rectangular waveguide tract can be easily determined.

If we take E in (2.108) as E_d – piercing intensity, then the limit power transmitted through a microwave rectangular waveguide tract in pulse mode can be defined as:

$$P_d^E = (ab/8)E_d^2\sqrt{\varepsilon}. \tag{2.110}$$

$E_d = 30$ *kV/cm* for a microwave rectangular waveguide filled with air [34] since

$$P_d^E = 112.5ab\sqrt{\varepsilon}, \tag{2.111}$$

where P_d is *kWt* and a and b are measured in *cm*.

Considering that $a = 4$ *cm*, $b = 2$ *cm* and $\varepsilon = 19.75$, we get from expression (2.111): $P_d^E = 4000kWt$.

It should be noted that the output power (P_b) is several times smaller than the limit power, i.e.,

$$P_b = kP_d, \tag{2.112}$$

where $k = 0.2$–0.3 is the reserve coefficient of electrical strength [35].

Thus, for the power that can be released in a microwave rectangular waveguide operating at a H_{10} wave in pulse mode, we get $P_b = 1200$ *kWt.*

It should be noted that the microwave rectangular waveguide with a cross section of 40 × 20 *mm* works with a limit power of 3580 *kWt* in the frequency range of 4.9–7.05 *GHz* [36]. These values are true if the medium is linear. However, as mentioned above, in real practice, the environment behaves as nonlinear, and in this case, the limit power of the microwave rectangular waveguide with a cross section of 40 × 20 *mm* is $P_d = 4000$ *kWt.*

Therefore, it can be concluded that during the conducted research, the limit power of the microwave rectangular wave-transmitter with a cross section of 40 × 20 *mm* and working in pulse mode in the H_{10} wave is from 3580 to 4000 *kWt*, and the output power is 1074 *kWt* – increased to 1200 *kWt.*

2.9.3 Loss and Extinction in a Microwave Rectangular Waveguide Path in Ambient Nonlinearity

Extinction of electromagnetic waves in the transmitter tract reduces the efficiency of the information transmission system, increases thermal noise, reduces the sensitivity in the receiving part of the telecommunication system, and deteriorates the electrical characteristics of the elements in the transmission line. Extinctions in a microwave rectangular waveguide tract occur in two cases: extinctions occurring in the medium and extinctions occurring on the metal surface of the waveguide tract. In case 1, the attenuations are caused by water droplets on the inner wall of the waveguide. The most common extinction is the extinction that occurs on the metal surface of the waveguide tract. Let's set this fade. For this, let's separate a section of length Δx in the tract of the microwave rectangular waveguide, as shown in Fig. 2.34.

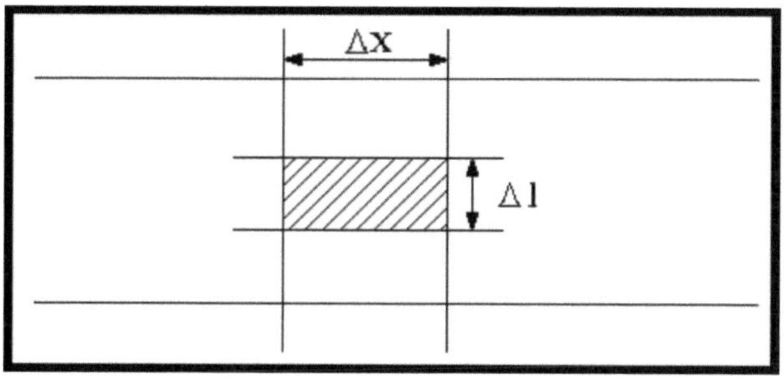

Fig. 2.34 On the determination of extinction on the metal surface of the tract of a microwave rectangular waveguide

The average power of heat loss in a section of length Δx is determined as follows:

$$\Delta P_{n,,m} = \frac{1}{2\sigma\Delta_0}\int_0^{\Delta x} dx \oint_l |H_m|^2 dl = \frac{\Delta x}{2\sigma\Delta_0}\oint_l |H_m|^2 dl, \tag{2.113}$$

ie, loss per unit length

$$P'_{n,m} = \frac{\Delta P_{n,m}}{\Delta_0} = \frac{1}{2\sigma\Delta_0}\oint_l |H_m|^2 dl, \tag{2.114}$$

where H_m is the amplitude value of the magnetic field intensity inside the microwave rectangular waveguide.

The active part of surface resistance

$$R_S = \frac{1}{\sigma\Delta_0} = \sqrt{\frac{\pi f \mu_a}{\sigma}}. \tag{2.115}$$

So for all metals except ferromagnets $\mu_a = \mu_0$. Then

$$R_S = 1,987\sqrt{\frac{f}{\sigma}} \cdot 10^{-3}, Ohm \tag{2.116}$$

where f is the frequency (Hz).

Considering that the material of the wall of the microwave rectangular waveguide is brass with specific permittivity $\sigma = 5{,}65 \cdot 10^7\ Ohm^{-1}\ m^{-1}$, then $R_S = 2{,}21 \cdot 10^{-2}$ Ohm for $f = 7{,}05 \cdot 10^9\ Hz$.

In this case, the expression (2.114) is as follows:

$$P'_{n,m} = 1,105 \cdot 10^{-2}\oint_l |H_m|^2 dl. \tag{2.117}$$

By knowing the wall loss and transmitted power of a microwave rectangular waveguide, the attenuation in the waveguide can be easily determined [37]:

$$\alpha = \frac{P'_{n,m}}{2m}. \tag{2.118}$$

If instead of P and $P'_{n,m}$ in the expression (2.118) we write their numerical values and switch from the amplitude values of electric and magnetic field intensities to their true values, we get

$$\alpha = \frac{4,42 \cdot 10^{-2} \oint\limits_{\partial} H^2 dl}{abE^2\sqrt{\varepsilon}}. \tag{2.119}$$

The values of E and H were determined during the solution of the system of differential Eqs. 2.14 and 2.15. Considering that $E = 0{,}05 \cdot 10^5$ V/m, $H = 0{,}12 \cdot 10^5$ A/m, $\varepsilon = 2$ in (2.119), we get $\alpha = 0{,}0201$ dB/m.

From the expression (2.8), it can be concluded that the attenuation depends on the cross-sectional area of the microwave rectangular waveguide. Thus, as the cross-sectional area increases, the extinction decreases. Therefore, rectangular waveguides for wavelengths of 5 *cm* and less should be made in the form of short sections. For a microwave rectangular waveguide, the attenuation $\alpha = 0{,}0201$ dB/m means that the attenuation is 2,01 *dB* over a length of 100 *m*.

2.9.4 Determination of Standing Wave and Reflection Coefficients in the Tract of a Microwave Rectangular Waveguide in the Case of Environmental Nonlinearity

Standing wave coefficient in the tract of microwave rectangular waveguide means the periodic change of the amplitudes of electric and magnetic field intensities. This, in turn, is related to the interference of incident and reflected waves.

Let's assume that there are a number of irregularities at point M inside the microwave rectangular waveguide (Fig. 2.35a). Then the reflected wave will be directed in the direction of the source of the signal. The amplitude of the reflected wave will be smaller than the amplitude of the main wave. At this time, a standing wave will appear on the left side from point M (Fig. 2.35b).

According to the intensity of the electric field, the standing wave is evaluated by the reflection coefficient Q or the standing wave coefficient K_d.

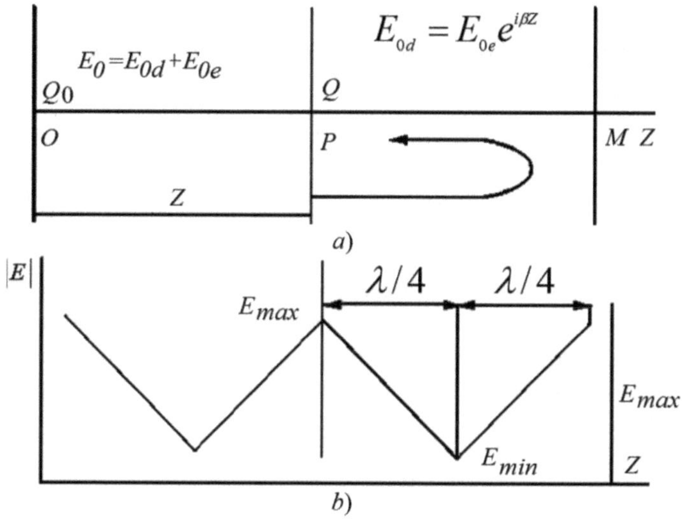

Fig. 2.35 Standing wave in a microwave rectangular waveguide: (**a**) distribution of characteristic points; (**b**) electric field intensity distribution

The reflection coefficient Q is determined by the ratio of the amplitudes of the intensities of the reflected and incident waves:

$$Q = \frac{E_{eks}}{E_{dus}}. \tag{2.120}$$

K_d – the standing wave coefficient (Fig. 2.35b) is determined by the ratio of the maximum value of the electric field intensity of the standing wave to the minimum value:

$$K_d = \frac{E_{\max}}{E_{\min}} = \frac{|E_d| + |E_{eks}|}{|E_d| - |E_{eks}|} = \frac{1+Q}{1-Q}, \tag{2.121}$$

where E_{min}, E_{max} are the maximum and minimum values of the electric field intensity [38].

Considering that $E_{\min} = 7.04 \cdot 10^5$ V/m, $E_{\max} = 7.18 \cdot 10^5$ V/m in (2.121), we get $K_d = 1.02$.

Knowing that $K_d = 1.02$, the reflection coefficient can be determined according to the formula (2.120)

$$Q = \frac{K_d - 1}{K_d + 1} = 0.0099.$$

2.9.5 *Determination of the Quality of a Microwave Rectangular Waveguide in the Case of Ambient Nonlinearity*

It is known that [39] rectangular waveguides for wavelengths of 5 *cm* and less should be made in the form of short sections. So the microwave rectangular waveguide becomes a resonator at this time. The area of this resonator corresponds to the area of a microwave rectangular waveguide.

Let's assume that the volume of the resonator V forms an isolated system. In this case, the energy of the electric field in the volume V of the resonator will be periodically converted to the energy of the magnetic field and vice versa.

If the average energy values of the electric and magnetic fields during a certain period are equal, i.e.,

$$W_{e.or} = W_{m.or}. \tag{2.122}$$

In this case, the process takes place without a source, and the power of external sources becomes active. If the condition $W_{e.or} \neq W_{m.or}$ is satisfied, then the energy of the electric field field in the volume V of the resonator will periodically be converted to the energy of the magnetic field and vice versa only with the presence of sources. At this time, the active power of external sources will be different from zero. If the power of the external sources is only active in the isolated region, then resonance phenomenon will occur. From the above, it can be said that for resonance, the average values of the energies of the electric and magnetic fields must be equal. That is, the equality (2.122) must be satisfied.

The change of the resonance frequency in the resonance model is proportional to the square of the intensity of the electric field [40].

$$\frac{\Delta f}{f} = K_e \frac{E^2}{W_{or}}, \tag{2.123}$$

where W_{or} is the average reserve energy in the resonance model during one period of oscillation, K_e is the coefficient depending on the geometric dimensions, $\frac{\Delta f}{f} = Q_y$ is the loaded quality of the resonance model.

The specific quality of the resonance model is determined as follows [41]:

$$Q_O = 2Q_y. \tag{2.124}$$

Considering expression (2.124) in (2.123), we get for microwave rectangular waveguide

$$Q_0 = 2K_e \frac{E^2}{W_{or}}, \tag{2.125}$$

where E is the true value of the electric field intensity inside the waveguide.

Considering that $K_e = 2\pi$, $W_{or} = 0.7\cdot10^5$ *C*, $E = 0.05\cdot10^5$ *V/m*, then $Q_0 = 4490$.
From the derived expression (2.125) it is seen that the quality of the microwave rectangular waveguide is proportional to the square of the electric field intensity.

2.9.6 Determination of the Reflection Coefficient Phase of a Microwave Rectangular Waveguide Tract in the Case of Ambient Nonlinearity

The phase of the reflection coefficient indicates the change of the wave's phase during its path of 1 *m* length and is determined as follows [42]:

$$\phi = \frac{2\pi}{\lambda_{vv}}, \tag{2.126}$$

where λ_{vv} is the wavelength in the waveguide.

$$\lambda_{vv} = \frac{\lambda_0}{\sqrt{1 - \left(\frac{\lambda_0}{\lambda_{\kappa r}}\right)^2}}, \tag{2.127}$$

where $\lambda_0 = c/f$ is the wavelength in free space, $\lambda_{\kappa r} = 2a$ is the critical wavelength.
Considering the expression (2.127) in (2.126) and also taking into account the expressions λ_0 and λ_{kr}, we get

$$\phi = \frac{2\pi f}{c}\sqrt{1 - \left(\frac{c}{2af}\right)^2}. \tag{2.128}$$

Considering $f = 7.05\cdot10^9$ *Hz*, $c = 3\cdot10^8$ *m/s* və $a = 40\cdot10^{-3}$ *m* in (2.128), we get $\varphi = 39.8\pi\cdot m^{-1}$.

2.9.7 Determination of the Characteristic Resistance of a Microwave Rectangular Waveguide in the Case of Ambient Nonlinearity

When designing microwave telecommunication systems, it is necessary to know the characteristic impedance of the waveguide.
It is known that *E*-type and *H*-type waves can be propagated in a microwave rectangular waveguide. The power transmitted by a microwave rectangular waveguide tract is concentrated between the walls of this waveguide, and the electromagnetic field should be expressed as the sum of *E* and *H* waves. As the distance between

the walls of the waveguide is small, the above-mentioned fact becomes more apparent. One of the most important electrical parameters of a microwave rectangular waveguide operating at H_{10} is its characteristic resistance Z_H.

The characteristic resistance for a transverse electric wave is determined as follows [43]:

$$Z_H = \frac{Z_{CO}}{\sqrt{1 - \left(\frac{\lambda_0}{\lambda_{\kappa r}}\right)^2}}, \tag{2.129}$$

where $Z_{CO} = \sqrt{\frac{\mu_0}{\varepsilon_0}} = 376.7\text{Ohm}$ is the characteristic resistance of a plane wave in vacuum.

Then (2.129) becomes as follows:

$$Z_H = \frac{376,7}{\sqrt{1 - \left(\frac{\lambda_0}{\lambda_{\kappa r}}\right)}}. \tag{2.130}$$

$\lambda_{\kappa r} = 2a$ and $\lambda_0 = c/f$ for a microwave rectangular waveguide.

Considering these expressions in (2.130), we get

$$Z_H = \frac{376.7}{\sqrt{1 - \left(\frac{c}{2af}\right)^2}}. \tag{2.131}$$

If we take into account that $c = 3 \cdot 10^8$ *m/s*, $a = 40$ *mm*, $f = 7.05 \cdot 10^9$ *Hz* in (2.131), we get the dependence of the characteristic resistance for different values of *a* of the large wall of the microwave waveguide (Fig. 2.36).

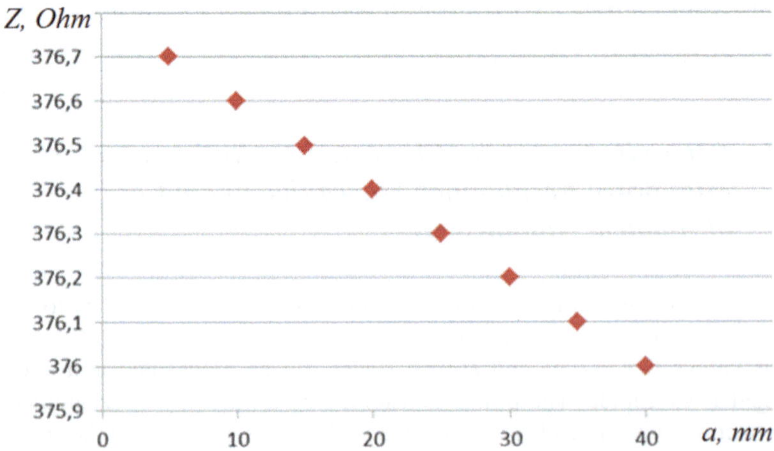

Fig. 2.36 Dependence of characteristic resistance for different values of *a* in microwave rectangular waveguide large wall

Table 2.10 Values of set parameters for a microwave rectangular waveguide tract operating at H_{10} wave

Limit power, P_d, Wt	4000
Releasable power, P_{bur}, Wt	1200
Extinction coefficient, α, dB/m	0.0203
Standing wave coefficient, K_{dd}	1.02
Refractive coefficient, Q	0.0099
Special quality, Q_0	4490
The phase of the reflection coefficient, φ, rad/m	39.8π
Characteristic resistance, Z_H, Ohm	376.7

Thus, the obtained expression (2.131) is very simple and allows to accurately determine the characteristic resistance for different values of the frequency and large wall a of the waveguide.

Thus, the values of the set parameters for the tract of the microwave rectangular waveguide operating at the H_{10} wave are given in Table 2.10.

2.10 Conclusions

1. For the first time, an analytical expression was obtained between the intensity of the electric field and the relative dielectric permittivity in the nonlinear case of the medium. This analytical statement is of great theoretical interest and is the basis for designing telecommunication systems of the microwave waveguide type with different frequency ranges to transmit information over long distances.
2. New mathematical models of the electromagnetic field of a rectangular waveguide operating in the frequency range of 4.9–7.05 *GHz* and a circular waveguide operating at a frequency of 9 *GHz* from microwave telecommunication devices in Cartesian and cylindrical coordinate systems, taking into account the nonlinearity of the environment and the types of waves processed.
3. Effective algorithms for solving new mathematical models of the electromagnetic field of a rectangular waveguide operating in the frequency range of 4.9–7.05 *GHz* and a circular waveguide operating at a frequency of 9 *GHz* from microwave telecommunication devices were proposed, taking into account the nonlinearity of the environment and the types of waves. This allowed to improve the electrical, magnetic, structural and operating parameters and characteristics of microwave rectangular and circular waveguides.
4. Finite element and finite difference methods have been improved taking into account the structural dimensions for modeling the electromagnetic fields of *E*-type (*TH*) and *H*-type (*TE*) waves of microwave rectangular and circular waveguides, and as a result, of 4.9–7.05 *GHz*. The method of minimizing the extinction coefficient of electromagnetic waves in the frequency range and at the frequency of 9 *GHz* has been developed.
5. For *E*-type and *H*-type waves, *3D* models of the distribution of electromagnetic field intensities in the elementary regions of a rectangular waveguide operating in

the frequency range of 4.9–7.05 *GHz* and a circular waveguide operating at a frequency of 9 *GHz* were obtained.

6. Dispersion compositions of different types of electromagnetic waves propagating in rectangular and circular waveguides from microwave telecommunication devices were studied and the effect of field intensities on dispersion dependencies was shown.
7. Mathematical models of the microwave circular waveguide in the cylindrical coordinate system described by special derivative second order differential equations were developed. Taking into account the specific characteristics of the circular waveguide, the algorithm and block diagram of this device were developed using the finite difference method for calculating the intensities of the electric and magnetic fields. According to the developed algorithm and block diagram, the values of the intensities of the electric and magnetic fields in the elementary regions of the microwave circular waveguide were determined. When comparing their values, it was determined that the error was 3.3% according to the module, and 40/ according to the argument. This can also be omitted. The dependence of the intensity of the electric field of the circular waveguide on the *h* parameter was established in case of a change of its generalized parameter according to the values obtained during the numerical modeling.
8. An algorithm and methodology were developed for estimating the discretization error during the calculation of the electromagnetic field of the microwave rectangular waveguide.
9. Limit power, emissive power, extinction coefficient, standing wave of the transmission path of a microwave square wave transmitter in the frequency range of 4.9–7.05 *GHz* for H_{10} mode of *E*-type (*TH*) and *H*-type (T*E*) waves coefficient, reflection coefficient, specific quality, reflection coefficient phase and characteristic resistance in the nonlinear case of the medium, new mathematical expressions were obtained.

References

1. Li, L. (2013). Solution to 3-D electromagnetic problems discretized by a hybrid FEM/MOM method. *Computer Physics Communications, 184*(1), 73–78.
2. Li, M.-Y., et al. (2017). Eight-port orthogonally dual-polarised MIMO antennas using loop structures for 5G smartphone. *IET Microwaves, Antennas & Propagation, 11*(12), 1810–1816.
3. Li, Q., & Yang, T. (2021). Compact UWB half-mode SIW bandpass filter with fully reconfigurable single and dual notched bands. *IEEE Transactions on Microwave Theory and Techniques, 69*, 65–74.
4. Li, S., & Zhu, Z. (2012). A numerical method to determine mode content in circular waveguide based on an integral identity equation. *Applied Mathematical Modelling, 36*(7), 2922–2940.
5. Liang, S., et al. (2017). Sidelobe-level suppression for linear and circular antenna arrays via the cuckoo search-chicken swarm optimisation algorithm. *IET Microwaves, Antennas & Propagation, 11*, 209–218.
6. Liao, X., et al. (2019). A 90° oversized broadband TE 01 -to-TM 11 mode converter for high-power transmission line. *IEEE Transactions on Microwave Theory and Techniques, 67*(12), 4692–4699.

7. Liao, X., et al. (2020). Analysis of the synthesis method for broadband oversized TE 01 -to-TE 11 mode converter. *IEEE Transactions on Microwave Theory and Techniques, 68*(2), 620–627.
8. Lin, Q., et al. (2007). Nonlinear optical phenomena in silicon waveguides: Modeling and applications. *Optics Express, 15*(25), 16604–16644.
9. Lipworth, G. (2016). Phase and magnitude constrained metasurface holography at W-band frequencies. *Optics Express, 24*(17), 19372–19387.
10. Liu, G., et al. (2016). A TE 13 mode converter for high-order mode gyrotron-traveling-wave tubes. *IEEE Transactions on Electron Devices, 63*(1), 486–490.
11. Liu, W. E. I., et al. (2020). Broadband low-profile L-probe fed metasurface antenna with TM leaky wave and TE surface wave resonances. *IEEE Transactions on Antennas and Propagation, 68*(3), 1348–1355.
12. Liu, W. (2022). Larger-than-unity external optical field confinement enabled by metamaterial-assisted comb waveguide for ultrasensitive long-wave infrared gas spectroscopy. *Nano Letters, 22*(15), 6112–6120.
13. Liu, Y., et al. (2019). TCM-based heptaband antenna with small clearance for metal-rimmed mobile phone applications. *IEEE Antennas and Wireless Propagation Letters, 18*(4), 717–721.
14. Liu, Z. Q., & Sun, D. Q. (2019). Transition from rectangular waveguide to empty substrate integrated gap waveguide. *International Journal of Electronics Letters, 55*(11), 654–655.
15. Liu, Z. et al. (2020). *A novel compact diplexer with high isolation based on the mode composite ridged waveguide*. In Proceedings of the International Conference on Microwave and Millimeter Wave Technology (ICMMT) (pp. 1–3).
16. Lonkina, D. V., et al. (2017). Electrodynamic analysis and synthesis of the waveguide bandpass filter on resonant diaphragms with C-shaped aperture. *Radiation and Scattering of Electromagnetic Waves RSEMW*, 434–436.
17. Lu, R., et al. (2021). SIW cavity-fed filtennas for 5G millimeter-wave applications. *IEEE Transactions on Antennas and Propagation, 69*, 5269–5277.
18. Mazur, M., & Mazur, J. (2011). Operation of the phase shifter using complex waves of the circular waveguide with periodical ferrite-dielectric filling. *Journal of Electromagnetic Waves and Applications, 25*(7), 935–947.
19. Medran del Rio, J., et al. (2020). Multilayered balanced dual-band bandpass filter based on magnetically coupled open-loop resonators with intrinsic common-mode rejection. *Applied Sciences, 10*, 3113.
20. Medvedik, M. Y., et al. (2012). Propagation of TM modes in a circular dielectric waveguide filled with a medium with Kerr nonlinearity. *Journal of Communications Technology and Electronics, 57*(4), 363–369.
21. Menachem, Z. (2019). A new technique for the analysis of the physical discontinuity in a hollow rectangular waveguide with dielectric inserts of varying profiles. *Journal of Electromagnetic Waves and Applications, 33*(9), 1145–1162.
22. Meyer, A., et al. (2018). Dispersion-minimized rod and tube dielectric waveguides at W-band and D-band frequencies. *IEEE Microwave and Wireless Components Letters, 28*(7), 555–557.
23. Miao, M. J. (2016). Acceleration of perturbation-based electric field integral equations using fast Fourier transform. *IEEE Transactions on Antennas and Propagation, 64*(10), 4559–4564.
24. Mironov, V. L., & Mironov, S. V. (2017). Two types of Lorentz transformations for massless fields. *Journal of Geometry and Symmetry in Physics, 44*, 83–96.
25. Mitchell, G. A., & Wasykiwskyj, W. (2018). A review of the modal decomposition matrix for calculating the far field of an infinitely flanged rectangular waveguide. *Applied Computational Electromagnetics Society Journal, 33*(12), 1390–1396.
26. Moradpour, H., et al. (2015). Three-particle bell-like inequalities under Lorentz transformations. *Quantum Information Processing, 14*(10), 3913–3931.
27. Muñoz-Enano, J., et al. (2020). Planar microwave resonant sensors: A review and recent developments. *Applied Sciences, 10*, 2615.
28. Na, D.-Y., et al. (2016). Local, explicit, and charge-conserving electromagnetic particle-in-cell algorithm on unstructured grids. *IEEE Transactions on Plasma Science, 44*(8), 1353–1362.

29. Nair, D., & Webb, J. P. (2003). Optimization of microwave devices using 3-D finite elements and the design sensitivity of the frequency response. *IEEE Transactions on Magnetics, 39*(3), 1325–1328.
30. Nakatake, Y., & Watanabe, K. (2011). Numerical analysis of two-dimensional photonic crystal waveguide devices using periodic boundary conditions. *IEICE Transactions on Electronics, E94C*(1), 32–38.
31. Parameswaran, A., & Raghavan, S. (2016). *Analysis of mode of propagation in substrate integrated waveguide using FDTD*. In Proceedings of the international conference on communication systems and networks (pp. 71–75).
32. Parment, F., et al. (2016). Double dielectric slab-loaded air-filled SIW phase shifters for high-performance millimeter-wave integration. *IEEE Transactions on Microwave Theory and Techniques, 64*(9), 2833–2842.
33. Parvathy, A. R., et al. (2018). Circularly polarized split ring resonator loaded slot antenna. *Advanced Electromagnetics, 7*(5), 1–6.
34. Permyakov, V. A., et al. (2019). Analysis of propagation of electromagnetic waves in difficult conditions by the parabolic equation method. *IEEE Transactions on Antennas and Propagation, 67*(4), 2167–2175.
35. Piltyay, S. (2021). Square waveguide polarizer with diagonally located irises for Ka-band antenna systems. *Advanced Electromagnetics, 10*(3), 31–38.
36. Potratz, C., & Glock, H.-W. (2011). Time-domain field and scattering parameter computation in waveguide structures by GPU accelerated discontinuous-Galerkin method. *IEEE Transactions on Microwave Theory and Techniques, 59*(11), 2788–2797.
37. Pradhan, N. C., et al. (2021). Design of compact substrate integrated waveguide based triple- and quad-band power dividers. *IEEE Microwave and Wireless Components Letters, 31*, 365–368.
38. Prazeres, R. (2014). Numerical method for full calculation of the electromagnetic field in a rectangular waveguide within overmoded configuration, using the fast Fourier transform. *European Physical Journal-Applied Physics, 68*(2), 367–376.
39. Qu, Y., et al. (2022). Photo-thermal tuning of graphene oxide coated integrated optical waveguides. *Micromachines, 13*(8), 1194.
40. Quanxin, R. (2022). Linear antenna array with large element spacing for wide-angle beam scanning with suppressed grating lobes. *IEEE Antennas and Wireless Propagation Letters, 21*(6), 1258–1262.
41. Quesada, P. F. D., et al. (2018). Novel spatial domain integral equation formulation for the analysis of rectangular waveguide steps close to arbitrarily shaped dielectric and/or conducting posts. *Radio Science, 53*(4), 406–419.
42. Raheem, N., & Qasem, N. (2020). A compact multi-band notched characteristics UWB microstrip patch antenna with a single sheet of grapheme. *Telecommunication, Computing, Electronics and Control, 18*(4), 1708–1818.
43. Rahimzadeh, R. M., et al. (2020). Residue-pole methods for variability analysis of S-parameters of microwave devices with 3D FEM and mesh deformation. *Journal of Radioengineering, 29*(1), 10–20.

Chapter 3
Modeling of Microwave Waveguide Systems of Complex Structure in Nonlinear Media in a Mobile Computer System

3.1 Modeling of a Microwave Band Rectangular Waveguide Telecommunication Device

The microwave range is now intensively used to transmit information for various purposes. One of the devices operating in this range is the waveguide type telecommunication devices. These devices themselves have different constructions and purposes. They are widely used in telecommunications, television technology, radiolocation, radionavigation, medical technology, radio control, communication technology, antenna technology, and other fields of science and technology. Among the waveguides used for information transmission, the most widespread are the microwave rectangular waveguides. Microwave rectangular waveguides have several advantages. Thus, they have high reliability, stability of parameters and long service life in the process of operation. At the same time, microwave rectangular waveguides are completely isolated from the environment, that is, when transmitting information with such waveguides, radiation of electromagnetic waves (radio waves) to the external environment almost does not occur. Despite the very wide application of microwave rectangular waveguides, there are not enough scientific works that fully reflect the electromagnetic processes propagated in such transmission lines and are of experimental importance. In known works in this field, the medium in which electromagnetic waves (radio waves) propagate is considered linear. Until now, when designing such waveguides, the medium in which electromagnetic waves (radio waves) propagate was considered linear. But in real practice, the medium behaves nonlinearly. At the same time, when modeling such waveguides, complex classical mathematical apparatuses were used, which increase the error of calculations and make them difficult to perform.

Therefore, in order to improve the electrical, magnetic, technical, structural and operational parameters and characteristics of microwave rectangular waveguides, taking into account the nonlinearity of the environment, a great deal of attention has recently been paid to the development of modeling methods that allow the

I. Islamov, *Radio Engineering and Telecommunications Waveguide Systems in the Microwave Range*, https://doi.org/10.1007/978-3-031-37916-1_3

development and algorithmization of new mathematical models that reflect the specific characteristics of these devices. is needed and this issue is of great practical importance.

On the other hand, since rectangular waveguides with microwave range are widely used in production, improving the electrical, magnetic, structural, technical and operational parameters and characteristics of these devices and designing new devices with more optimal constructive dimensions are very important from both a scientific point of view and a production point of view. is relevant.

E-type (*TH*) and *H*-type (*TE*) waves can exist separately in a microwave rectangular waveguide. It should be noted that these waveguides are transmission lines with a special construction that ensure the transmission of electromagnetic waves in one direction without radiating energy to the environment. Transmission lines are divided into two parts: regular transmission lines; irregular lines.

Regular transmitting lines are those lines in which the property of the electromagnetic wave in the direction of their propagation axes remains unchanged or changes according to a periodic law. In the first case, transmission lines are called longitudinal-homogeneous, and in the second case, periodic transmission lines. If the energy of the electromagnetic wave varies along the axis of the transmission line, the transmission lines are called longitudinally inhomogeneous or irregular transmission lines. Transmission lines that channel electromagnetic energy are open and closed. If the electromagnetic energy in the transmission line is unlimited in the transverse direction, it is called an open transmission line. In a closed transmission line, the electromagnetic energy exists only within the closed metal cover. A rectangular waveguide is a closed-type transmission line.

Depending on the material that forms the transmission line, they are metal, dielectric or metal-dielectric. The properties of the transmission lines are strongly dependent on the relationship of their cross section. In this respect, transmission lines are divided into two parts, single-connection or multi-connection. Microwave rectangular waveguide is called closed metal duplex transmission line.

E-type (*TH*) and *H*-type (*TE*) waves in a microwave rectangular waveguide have the following characteristics:

1. *TH* wave. This is also called *E*-type wave. An *E* wave is a wave whose *E* vector has both transverse and parallel components ($E_\perp$ and $E_{||}$). The parallel component of *H* vector is zero ($E_\perp$, $E_{||}$, $H_\perp$, $H_{||} = 0$). The structure and parameters of the *E* wave depend on the frequency, that is, they are dispersed.
2. *TE* wave. This is called an *H*-type wave. An *H* wave is a wave whose $E_\perp$, $E_{||}$ and $H_\perp$ are non-zero. And $E_{||}$ becomes zero ($H_{||}$, $H_\perp$, $E_\perp$, $E_{||} = 0$).

It was mentioned above that *E* and *H* type dispersion waves are excited in a rectangular waveguide. Consider the rectangular waveguide shown in Fig. 3.1*a*, *b*.

It is more convenient to use the *XYZ* Cartesian coordinate system to perform calculations in a microwave rectangular waveguide. Let's look at the *E*-type wave and assume that $E_z \neq 0$, $H_z = 0$.

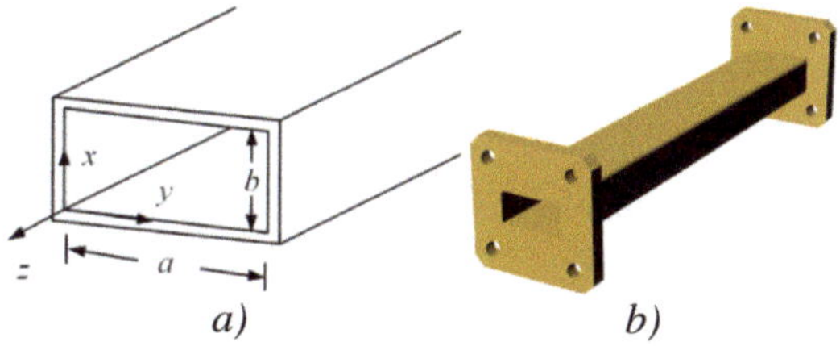

Fig. 3.1 Microwave rectangular waveguide: (*a*) cross-sectional area; (*b*) spatial model

Let's use the following equations to determine the E_z component of the field [1]:

$$\begin{aligned} E_z(x, y, z) = {} & [E_1 \sin(\gamma_{2x} x) + E_2 \cos(\gamma_{2x} x)] \times \\ & \times \left[E_3 \sin\left(\gamma_{2y} y\right) + E_4 \cos\left(\gamma_{2y} y\right)\right] \cdot e^{-\gamma_1 z}, \end{aligned} \tag{3.1}$$

$$\begin{aligned} H_z(x, y, z) = {} & [H_1 \sin(\gamma_{2x} x) + H_2 \cos(\gamma_{2x} x)] \times \\ & \times \left[H_3 \sin\left(\gamma_{2y} y\right) + H_4 \cos\left(\gamma_{2y} y\right)\right] \cdot e^{-\gamma_1 z}. \end{aligned} \tag{3.2}$$

In order to determine the parameters and characteristics of the electromagnetic field of microwave rectangular waveguides, the Leontovich-Shukin boundary condition must first be taken into account. This condition is considered one of the important issues to determine the area when there are metallic bodies with finite dimensions. In order to greatly facilitate the solution of this problem, approximate conditions called the Leontovich-Shukin boundary condition are introduced. Unlike the usual boundary conditions, the Leontovich-Shukin boundary conditions show the relationship between the constituents of their vectors in a medium. The expression of the Leontovich-Shukin boundary condition is as follows [2]:

$$\vec{E}_{1\tau} = z_{c2} \left[\vec{n}_0, \vec{H}_1\right]. \tag{3.3}$$

From this relationship, it can be seen that the tangential organization of the electric field on the real metal surface of the microwave circular waveguide differs from zero.

Let's use condition (3.3) and determine the constants in formulas (3.1) and (3.2). According to the boundary condition (3.3), it can be assumed that the tangential components of the electric field of the waveguide are zero, i.e.:

$$E_z = 0, x = 0, y = 0, \tag{3.4}$$

$$E_z = 0, x = a, y = b. \tag{3.5}$$

It follows from the condition (3.4) that the coefficients of the cosinusoidal components in Eq. (3.1) become zero, i.e.:

$$E_2 = 0, \tag{3.6}$$

$$E_4 = 0, \tag{3.7}$$

happens. Denote the product of the remaining coefficients of this equation by E_0 and find as follows:

$$\mathrm{E_z} = \mathrm{E_0}(\gamma_{2x}\mathrm{x})\sin\left(\gamma_{2y}\mathrm{y}\right)e^{-\gamma_1 \mathrm{z}}. \tag{3.8}$$

(3.5) is obtained from the boundary condition that

$$sin(\gamma_{2x}a) = 0, \tag{3.9}$$

$$sin(\gamma_{2x}b) = 0. \tag{3.10}$$

Fulfillment of these equalities only

$$\gamma_{2\mathrm{x}} = \frac{\mathrm{n}\pi}{\mathrm{a}}, \tag{3.11}$$

$$\gamma_{2\mathrm{y}} = \frac{\mathrm{n}\pi}{\mathrm{b}}, \tag{3.12}$$

is possible within the conditions. Where, m and n are arbitrary positive integers that define a wave of wave type E_{mn}. Note that none of these numbers can be zero for the E wave. Otherwise $E_z = 0$ and hence there would be no electromagnetic wave, i.e., E_{nm} and E_{mn}. So in conclusion

$$\mathrm{E_z} = \mathrm{E_0}\sin\left(\mathrm{m}\pi\mathrm{x}/\mathrm{a}\right)\sin\left(\mathrm{n}\pi\mathrm{y}/\mathrm{b}\right)\mathrm{e}^{-\gamma_1\mathrm{z}} \tag{3.13}$$

we can write.

The presence of the H wave requires the conditions $H_z \neq 0$, $E_z \neq 0$. Let's use these conditions to find the solution of the wave Eq. 3.2.

According to Maxwell's equations in the Cartesian coordinate system, we can write the boundary conditions as follows:

$$\frac{\partial \mathrm{H_z}}{\partial \mathrm{y}} = 0, \quad \mathrm{y} = 0, \mathrm{y} = \mathrm{b}, \tag{3.14}$$

$$\frac{\partial \mathrm{H_z}}{\partial \mathrm{x}} = 0, \quad \mathrm{x} = 0, \mathrm{x} = \mathrm{a}. \tag{3.15}$$

As before, if we find terms $\frac{\partial \mathrm{H_z}}{\partial \mathrm{y}}$ and $\frac{\partial \mathrm{H_z}}{\partial \mathrm{x}}$ and write for H_z, we get

$$\mathrm{H_z} = \mathrm{H_0}\cos\left(\mathrm{m}\pi\mathrm{x}/\mathrm{a}\right)\cos\left(\mathrm{n}\pi\mathrm{y}/\mathrm{b}\right)\mathrm{e}^{-\gamma_1\mathrm{z}}. \tag{3.16}$$

For E and H waves, the H_x, H_y, E_x and E_y components

$$E_x = \left(-i/\gamma_2^2\right)\left(\omega \mu_\alpha\, \partial H_z/\partial y + \gamma_1\, \partial E_z/\partial x\right), \tag{3.17}$$

$$E_y = \left(i/\gamma_2^2\right)\left(\omega \mu_\alpha\, \partial H_z/\partial x - \gamma_1\, \partial E_z/\partial y\right), \tag{3.18}$$

$$H_x = \left(i/\gamma_2^2\right)\left(\omega\, \varepsilon_\alpha\, \partial H_z/\partial y - \gamma_1\, \partial H_z/\partial x\right), \tag{3.19}$$

$$H_y = \left(-i/\gamma_2^2\right)\left(\omega \varepsilon_\alpha\, \partial H_z/\partial x - \gamma_1\, \partial H_z/\partial y\right). \tag{3.20}$$

should be found from equation. For this, in these equations, it is necessary to write $H_z = 0$ and E_{zn} according to the formula (3.11) for the E wave, and $E_z = 0$ and H_z according to the formula (3.16) for the H wave. Now let's look at the wavelength of the crisis.

The following formulas allow determining the transverse propagation coefficient in a rectangular waveguide:

$$\gamma_{2x}^2 + \gamma_{2y}^2 = \gamma_2^2 \mathrm{v} \cdot \mathrm{ya} \gamma_\mathrm{a} = \sqrt{\gamma_{2x}^2 + \gamma_{2y}^2}\, \mathrm{v} \cdot \gamma_{2x} = m\pi/a, \gamma_{2y} = m\pi/b, \tag{3.21}$$

$$\gamma_2 = \sqrt{(m\pi/a)^2 + (n\pi/b)^2}. \tag{3.22}$$

If we substitute this value of γ_2 in the formula $\gamma_2 = 2\pi/\lambda_{kr}$ or, $\lambda_{kp} = \frac{2\pi}{\gamma}$ we get the critical wavelengths for E and H waves in a rectangular waveguide

$$\lambda_{kr} = 2/\sqrt{(m/a)^2 + (n/b)^2}. \tag{3.23}$$

It can be seen here that λ_{kr} – the critical wavelength – depends on the geometric dimensions of the rectangular waveguide.

The solution of Eqs. 3.17, 3.18, 3.19, and 3.20 is sought as follows:

for E-type wave

$$E_z^{mn} = E_0 \sin\frac{m\pi x}{a} \sin\frac{n\pi y}{b} e^{-i\beta z}, \tag{3.24}$$

$$E_x^{mn} = -iE_0 \frac{\beta \chi_x}{\chi^2} \cos\frac{m\pi x}{a} \sin\frac{n\pi y}{b} e^{-i\beta z}, \tag{3.25}$$

$$E_y^{mn} = -iE_0 \frac{\beta \chi_y}{\chi^2} \sin\frac{m\pi x}{a} \cos\frac{n\pi y}{b} e^{-i\beta z}, \tag{3.26}$$

$$H_x^{mn} = iE_0 \frac{\omega \varepsilon_0 \chi_y}{\chi^2} \sin\frac{m\pi x}{a} \cos\frac{n\pi y}{b} e^{-i\beta z}, \tag{3.27}$$

$$H_y^{mn} = -iE_0 \frac{\omega \varepsilon_0 \chi_x}{\chi^2} \cos\frac{m\pi x}{a} \sin\frac{n\pi y}{b} e^{-i\beta z}, \tag{3.28}$$

$$H_z^{mn} = 0, \tag{3.29}$$

where χ—is the transverse wave number, $\chi_x = \frac{m\pi x}{a}$, $\chi_y = \frac{n\pi x}{b}$, m and $n = 1, 2, 3,\ldots$, E_0 – the amplitude value of the signal determined depending on the excitation condition.

For H-type wave

$$H_z^{mn} = H_0 \cos\frac{m\pi x}{a}\cos\frac{n\pi y}{b}e^{-i\beta z}, \tag{3.30}$$

$$H_x^{mn} = iH_0\frac{\beta\chi_x}{\chi^2}\sin\frac{m\pi x}{a}\cos\frac{n\pi y}{b}e^{-i\beta z}, \tag{3.31}$$

$$H_y^{mn} = iH_0\frac{\beta\chi_y}{\chi^2}\cos\frac{m\pi x}{a}\sin\frac{n\pi y}{b}e^{-i\beta z}, \tag{3.32}$$

$$E_x^{mn} = iH_0\frac{\omega\mu_0\chi_y}{\chi^2}\cos\frac{m\pi x}{a}\sin\frac{n\pi y}{b}e^{-i\beta z}, \tag{3.33}$$

$$E_y^{mn} = -iH_0\frac{\omega\mu_0\chi_x}{\chi^2}\cos\frac{m\pi x}{a}\sin\frac{n\pi y}{b}e^{-i\beta z}, \tag{3.34}$$

$$E_z^{mn} = 0, \tag{3.35}$$

where m and $n = 0, 1, 2, 3,\ldots$, H_0 is the amplitude value of the signal determined depending on the excitation condition.

In real practice, since the medium in which electromagnetic waves propagate behaves as a nonlinear one, the intensities of the electric and magnetic fields inside the waveguide will have different values. Therefore, it is necessary to use the mobile computing system (Fig. 3.2).

The relationship between the stationary and moving coordinate systems shown in Fig. 3.2 is determined by the Lorentz transformation and is as follows in the Cartesian system:

$$x = x', y = y', z = \frac{z' + vt'}{\sqrt{1-(v/c)^2}}, t = \frac{t' + (v/c^2)z'}{\sqrt{1-(v/c)^2}}, \tag{3.36}$$

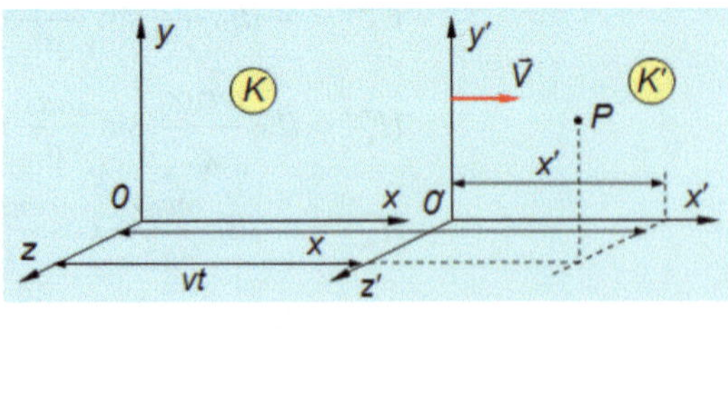

Fig. 3.2 A moving computing system

$$E_x = \frac{E'_x + vB'_y}{\sqrt{1-(v/c)^2}},\ E_y = \frac{E'_y - vB'_x}{\sqrt{1-(v/c)^2}},\ E_z = E'_z, \tag{3.37}$$

$$B_x = \frac{B'_y - (v/c^2)E'_y}{\sqrt{1-(v/c)^2}},\ B_y = \frac{B'_y + (v/c^2)E'_x}{\sqrt{1-(v/c)^2}},\ B_z = B'_z. \tag{3.38}$$

Considering the following equation, we get the following expressions for longitudinal wave number and frequency:

$$\phi = \omega t - \beta z = \omega' t' - \beta' z' = \phi'$$

$$\beta' = \frac{\beta - (v/c^2)\omega}{\sqrt{1-(v/c)^2}},\ \omega' = \frac{\omega - v\beta}{\sqrt{1-(v/c)^2}}. \tag{3.39}$$

Thus, for the moving coordinate we get:
For E-type wave

$$E'^{mn}_z = E_0 \sin\frac{m\pi x'}{a} \sin\frac{n\pi y'}{b} e^{-i\beta z}, \tag{3.40}$$

$$E'^{mn}_x = -iE_0 \frac{\chi_x[\beta - (v/c^2)\omega]}{\chi^2\sqrt{1-(v/c^2)}} \cos\frac{m\pi x'}{a} \sin\frac{n\pi y'}{b} e^{-i\beta z}, \tag{3.41}$$

$$E'^{mn}_y = -iE_0 \frac{\chi_x[\beta - (v/c^2)\omega]}{\chi^2\sqrt{1-(v/c^2)}} \sin\frac{m\pi x'}{a} \cos\frac{n\pi y'}{b} e^{-i\beta z}, \tag{3.42}$$

$$H'^{mn}_x = iE_0 \frac{\varepsilon_0 \chi_y(\omega - v\beta)}{\chi^2\sqrt{1-(v/c^2)}} \sin\frac{m\pi x'}{a} \cos\frac{n\pi y'}{b} e^{-i\beta z}, \tag{3.43}$$

$$H'^{mn}_y = -iE_0 \frac{\varepsilon_0 \chi_x(\omega - v\beta)}{\chi^2\sqrt{1-(v/c^2)}} \cos\frac{m\pi x'}{a} \sin\frac{n\pi y'}{b} e^{-i\beta z}, \tag{3.44}$$

$$H'^{mn}_z = 0, \tag{3.45}$$

for H-type wave

$$E'^{mn}_z = 0, \tag{3.46}$$

$$E'^{mn}_x = iH_0 \frac{\mu_0 \chi_y(\omega - v\beta)}{\chi^2\sqrt{1-(v/c^2)}} \cos\frac{m\pi x'}{a} \sin\frac{n\pi y'}{b} e^{-i\beta z}, \tag{3.47}$$

$$E'^{mn}_y = -iH_0 \frac{\mu_0 \chi_x(\omega - v\beta)}{\chi^2\sqrt{1-(v/c^2)}} \sin\frac{m\pi x'}{a} \cos\frac{n\pi y'}{b} e^{-i\beta z}, \tag{3.48}$$

$$H'^{mn}_x = iH_0 \frac{\chi_x[\beta - (v/c^2)\omega]}{\chi^2\sqrt{1-(v/c^2)}} \sin\frac{m\pi x'}{a} \cos\frac{n\pi y'}{b} e^{-i\beta z}, \tag{3.49}$$

$$H'^{mn}_y = iH_0 \frac{\chi_y[\beta - (v/c^2)\omega]}{\chi^2\sqrt{1-(v/c^2)}} \cos\frac{m\pi x'}{a} \sin\frac{n\pi y'}{b} e^{-i\beta z}, \tag{3.50}$$

$$H'^{mn}_z = H_0 \cos\frac{m\pi x'}{a} \sin\frac{n\pi y'}{b} e^{-i\beta z}. \tag{3.51}$$

Knowing the field components, the Umov-Poynting vector can also be determined for the moving coordinate:

$$S' = \frac{1}{2}\left[E', H'^*\right] = \frac{1}{2}\left\{\begin{array}{c} e_{xr}\left(E'_y H'^*_z - E'_z H'^*_y\right) + \\ +e_y\left(E'_z H'^*_x - E'_x H'^*_z\right) + \\ +e_z\left(E'_x H'^*_y - E'_y H'^*_x\right) \end{array}\right\}. \tag{3.52}$$

For E-type wave

$$\begin{aligned} S'_x = -\frac{1}{2}E'_z H'^*_y = -iE_0^2 \frac{\varepsilon_0 \chi_x(\omega - v\beta)}{4\chi^2\sqrt{1-(v/c)^2}} \times \\ \times \sin 2\frac{m\pi x'}{a} \sin^2\frac{n\pi y'}{b} e^{-i2\beta' z'}, \end{aligned} \tag{3.53}$$

$$\begin{aligned} S'_y = -\frac{1}{2}E'_z H'^*_x = -iE_0^2 \frac{\varepsilon_0 \chi_y(\omega - v\beta)}{4\chi^2\sqrt{1-(v/c)^2}} \times \\ \times \sin^2\frac{m\pi x'}{a} \sin 2\frac{n\pi y'}{b} e^{-i2\beta' z'}, \end{aligned} \tag{3.54}$$

$$\begin{aligned} S'_z = \frac{1}{2}\left(E'_x H'^*_y - E'_y H'^*_x\right) = \frac{\varepsilon_0 E_0^2(\omega - v\beta)\left(\beta - \frac{v}{c^2}\omega\right)}{2\chi^4\left(1-(v/c)^2\right)} \times \\ \times \left\{\begin{array}{l} \left(\frac{m\pi}{a}\right)^2 \cos^2\frac{m\pi x'}{a} \sin^2\frac{n\pi y'}{b} + \left(\frac{n\pi}{b}\right)^2 \times \\ \times \sin^2\frac{m\pi x'}{a} \cos^2\frac{n\pi y'}{b} \end{array}\right\} e^{-i2\beta' z'}. \end{aligned} \tag{3.55}$$

Dependencies of parameters β' and ω' on the speed of the coordinate system in the new coordinate system for the E-type wave are shown in Fig. 3.3*a*, *b*.

For H-type wave

$$\begin{aligned} S'_x = -\frac{1}{2}E'_y H'^*_z = -iH_0^2 \frac{\mu_0 \chi_x(\omega - v\beta)}{4\chi^2\sqrt{1-(v/c)^2}} \times \\ \times \sin 2\frac{m\pi x'}{a} \cos^2\frac{n\pi y'}{b} e^{-i2\beta' z'}, \end{aligned} \tag{3.56}$$

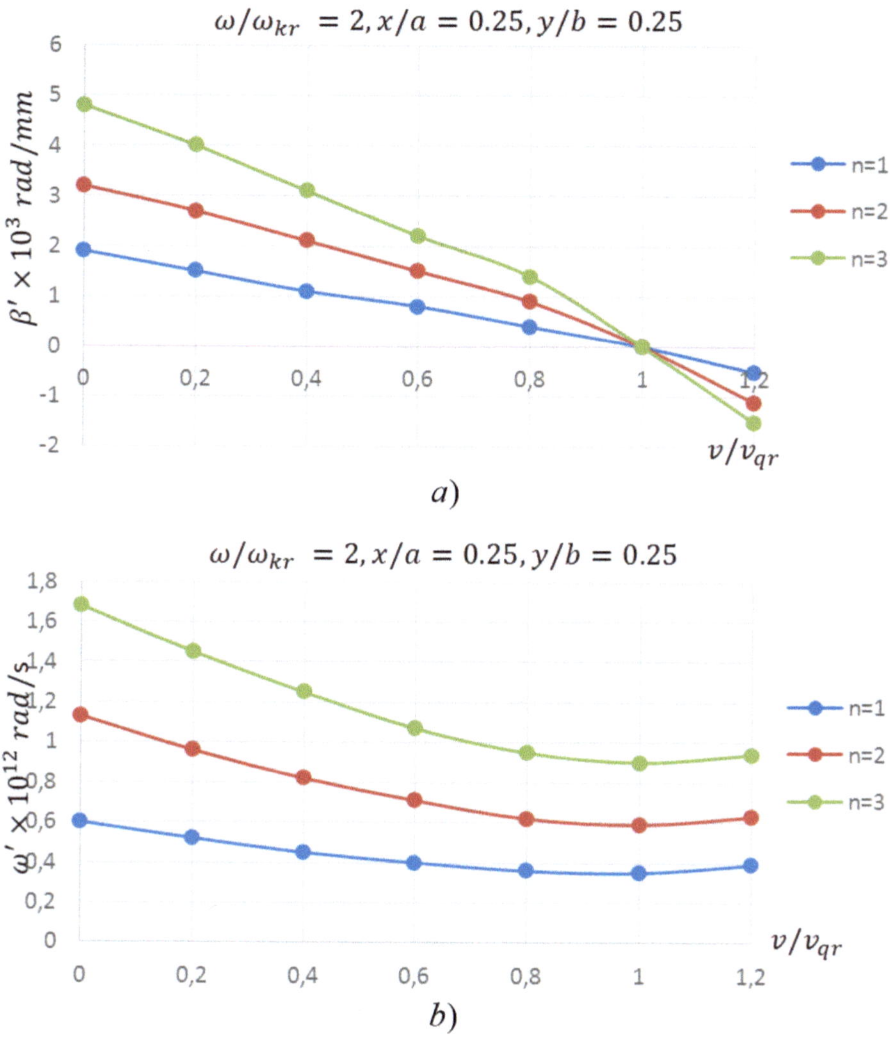

Fig. 3.3 Dependencies of parameters β' (*a*) and ω' (*b*) on the speed of the coordinate system in the new coordinate system for the *E*-type wave

$$S'_y = -\frac{1}{2}E'_x H'^{*}_z = -iH_0^2 \frac{\mu_0 \chi_y (\omega - v\beta)}{4\chi^2 \sqrt{1 - (v/c)^2}} \times \\ \cos^2 \frac{m\pi x'}{a} \sin 2\frac{n\pi y'}{b} e^{-i2\beta' z'}, \tag{3.57}$$

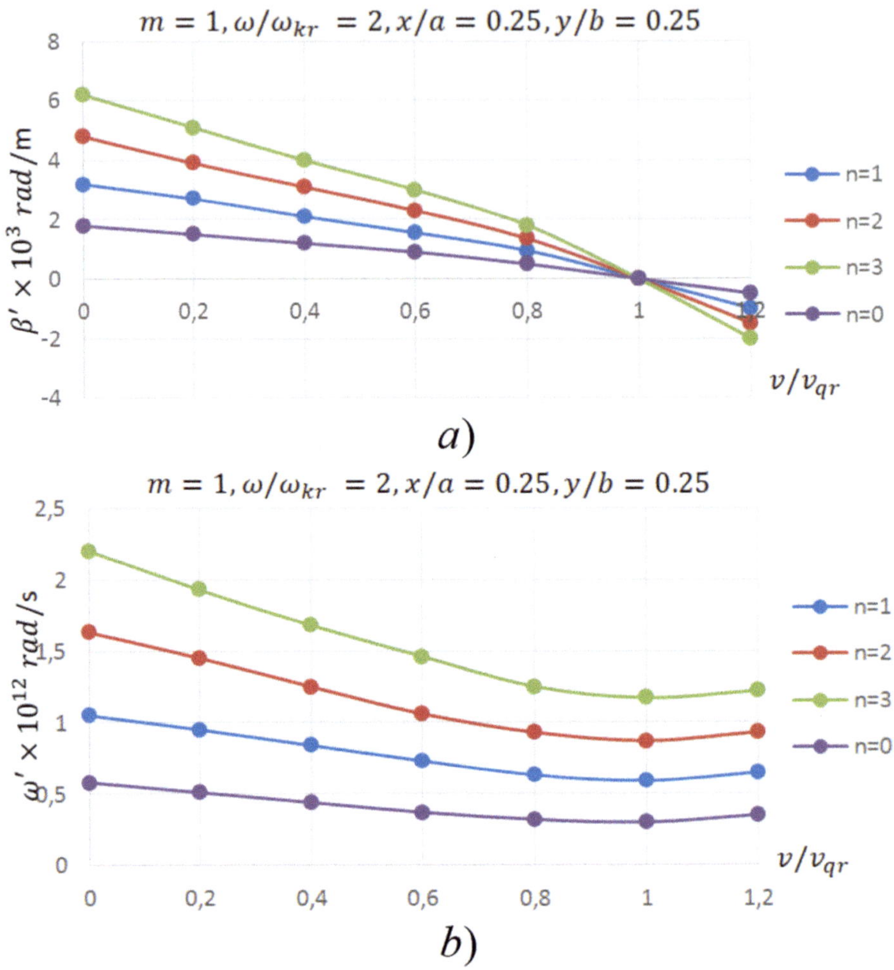

Fig. 3.4 Dependencies of parameters $\beta^{'}$ (*a*) and $\omega^{'}$ (*b*) on the speed of the coordinate system in the new coordinate system for the *H*-type wave

$$S_z^{\prime}=\frac{1}{2}\left(E_x^{\prime}H_y^{\prime *}-E_y^{\prime}H_x^{\prime *}\right)=\frac{\mu_0H_0^2(\omega-v\beta)\left(\beta-\frac{v}{c^2}\omega\right)}{2\chi^4\left(1-(v/c)^2\right)}\times$$
$$\times\left\{\begin{array}{l}\left(\frac{m\pi}{a}\right)^2\sin^2\frac{m\pi x^{\prime}}{a}\cos^2\frac{n\pi y^{\prime}}{b}+\\+\left(\frac{n\pi}{b}\right)^2\cos^2\frac{m\pi x^{\prime}}{a}\sin^2\frac{n\pi y^{\prime}}{b}\end{array}\right\}e^{-i2\beta^{\prime}z^{\prime}}. \tag{3.58}$$

The dependencies of parameters $\beta^{'}$ and $\omega^{'}$ on the speed of the coordinate system in the new coordinate system for the *H*-type wave are shown in Fig. 3.4*a*, *b*.

The directions of the Umov-Poynting vector for different types of waves are shown in Fig. 3.5.

Tables 3.1 and 3.2 show values calculated by formulas (3.40, 3.41, 3.42, 3.43, 3.44, 3.45, 3.46, 3.47, 3.48, 3.49, 3.50, and 3.51) of the intensities of the electric and magnetic fields, respectively, of the rectangular waveguide operating in the frequency range of 4.9–7.05 GHz. Figures 3.6 and 3.7 show the distributions of electric and magnetic field intensities along the Z axis according to elemental regions, corresponding to the microwave rectangular waveguide operating in the frequency range of 4.9–7.05 GHz.

In any specific waveguide (quantities a and b are given), each pair of numbers m and n corresponds to a certain structure of the electromagnetic field and the value of λ_{kr}. These types of electric and magnetic fields are denoted by E_{mn} and H_{mn}. Here, as mentioned above, the numbers m and n are called indices of the given type of oscillation. Again, in line with the above, this means how many stationary half-waves are located when the numbers m and n are traversed across the x and y coordinate axes. Since the indices are arbitrary, any number of E- and H-type waves can be present in a rectangular waveguide. Table 3.3 lists the characteristics of various rectangular waveguides.

In high-type waves, the structure of the E_{11}, H_{10} and H_{11} wave fields is repeated and forms a very complex picture. In this case, it is necessary to repeat the described distribution in one or another coordinate system so that the number is equal to the indices m and n for the given type of wave. Since the coordinate dependence of the components of the magnetic and electric fields is described by harmonic functions, the directions of the force lines must be repeated. The lines of magnetic force of the E wave are located in the cross-sectional plane and are closed. The lines of electric force of this wave are located in the longitudinal cross-sectional plane. Closed consists of an open curve, whose beginning and end are located on the metallic walls of the waveguide.

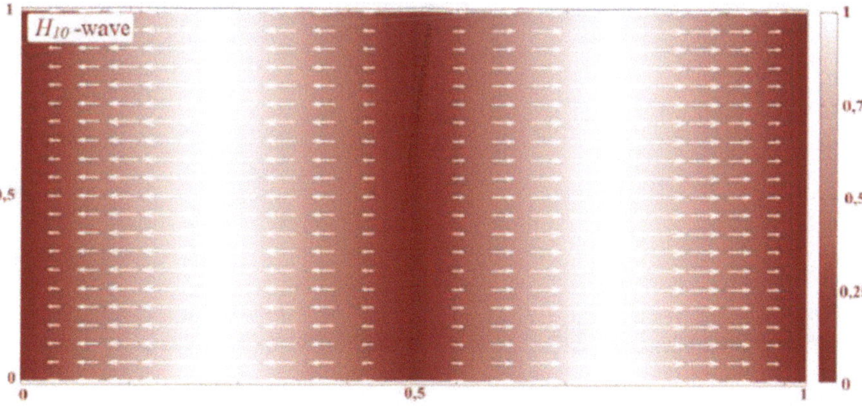

a)

Fig. 3.5 Directions of the Umov-Poynting vector for waves of type H_{10} (*a*), H_{11} (*b*), E_{11} (*v*), E_{21} (*d*)

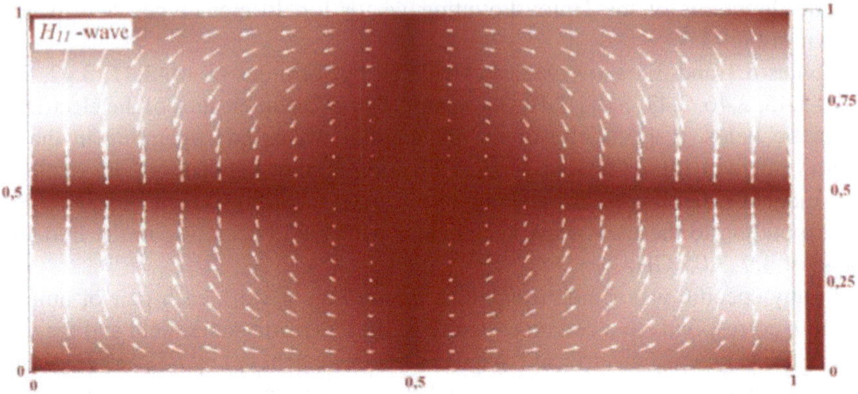

b)

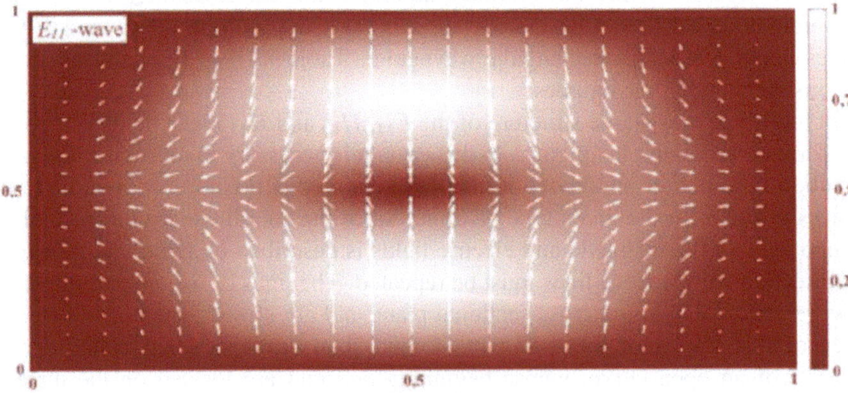

v)

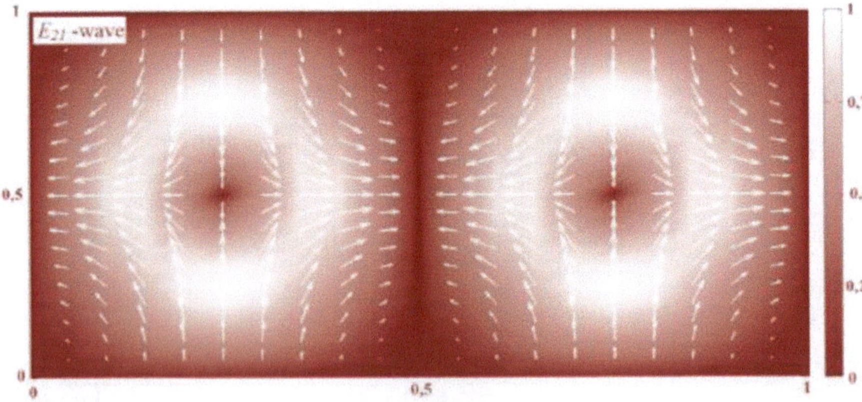

d)

Fig. 3.5 (continued)

Table 3.1 Values of the electric field intensities of the microwave rectangular waveguide operating in the frequency range of 4.9–7.05 GHz

	For *E*-type wave		For *H*-type wave	
No. of elem-entary fields	E_x, (10^5) *V/m*	E_y, (10^5) *V/m*	E_x, (10^5) *V/m*	E_y, (10^5) *V/m*
1	2.18	4.45	4.45	2.45
2	3.48	5.24	1.6	2.59
3	4.07	5.28	1.43	3.39
4	4.47	5.29	5.42	4.13
5	4.44	5.36	5.51	3.44
6	4.43	6.07	4.78	3.37
7	4.43	6.42	2.11	2.42
8	2.83	4.75	1.37	1.40
9	3.85	4.27	4.21	1.19
10	4.15	5.47	5.21	1.23
11	4.19	6.03	6.52	1.60
12	3.16	6.25	4.12	1.54
13	2.67	6.26	4.11	1.79
14	2.55	6.29	7.31	2.19
15	2.43	5.59	7.52	3.61
16	2.27	5.14	1.49	1.57
17	2.91	5.12	1.42	1.19
18	3.16	1.20	2.51	4.1
19	4.37	4.37	3.19	5.49
20	5.24	4.78	4.42	6.49

When looking at the structures of the *H* wave, it should be noted that the lines of electric force are located in the cross-sectional plane or consist of a closed curve or an open curve. The lines of magnetic force are closed curves located in the longitudinal cross-section. Figures 3.8 and 3.9 show the distribution of critical wavelengths in square and rectangular waveguides.

Comparing the graphs, it can be seen that as the height of the waveguide increases, the number of wave types excited in the waveguide increases. This makes it possible to use waveguides with a height of $b \leq 0.5a$ more widely.

3.2 Modeling of a Microwave Band Circular Waveguide Telecommunication Device

The microwave range is now intensively used to transmit information for various purposes. One of the devices operating in this range is the waveguide type telecommunication devices. These devices themselves have different constructions and specifications. They are widely used in telecommunications, television technology,

Table 3.2 Values of magnetic field intensities of a microwave rectangular waveguide operating in the frequency range of 4.9–7.05 GHz

No. of elementary fields	For *E*-type wave		For *H*-type wave	
	H_x, *A/m*	H_y, *A/m*	H_x, *A/m*	H_y, *A/m*
1	51.8	24.3	54	28.3
2	67.8	43.2	25.4	35.6
3	44.5	25.6	35.8	77.1
4	63.7	35.4	46.2	45.7
5	11.6	85.8	66.1	62.7
6	52.6	43.7	34.5	64.7
7	54.7	38.9	76.3	22.6
8	61.2	32.1	24.3	34.7
9	63.4	25.2	12.1	37.7
10	53.2	44.3	32.6	11.7
11	52.9	45.9	36.2	33.7
12	43.5	47.1	74.7	35.4
13	34.7	52.5	36.1	20.6
14	55.7	66.9	66.2	41.9
15	47.5	31.7	34.3	42.5
16	59.2	43.7	34.1	29.5
17	79.3	62.9	64.3	10.2
18	61.7	44.4	75.8	37.7
19	61.8	44.2	41.9	29.5
20	49.5	25.8	41.1	21.6

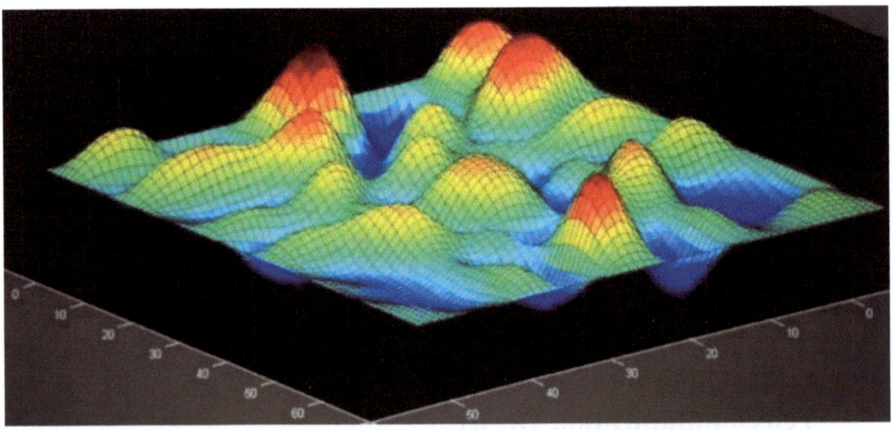

Fig. 3.6 Distribution of the electric field intensities of the microwave rectangular waveguide operating in the frequency range of 4.9–7.05 GHz along the *Z* axis according to elementary areas

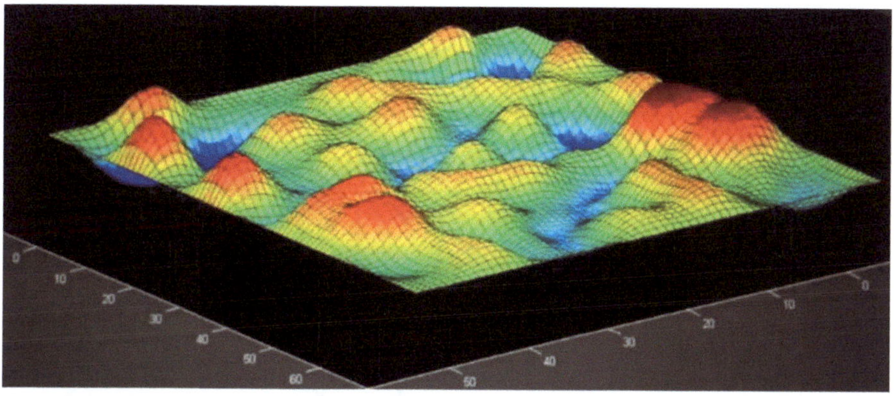

Fig. 3.7 Distribution of the magnetic field intensities of the microwave rectangular waveguide operating in the frequency range of 4.9–7.05 GHz along the *Z* axis according to elementary areas

Table 3.3 Characteristics of various rectangular waveguides

№	Transdu-cer cross section, *axb, m*	Working range, λ_{min}–λ_{max}, *cm*	Extinction coefficient, *α, dB*			P_{max}, *kWt*
			Cop-per	Brass	Alu-mi-num	
1	3.6 ÷ 1.8	0.396 ÷ 0.575	1.61	3.16	2.1	29
2	7.2 ÷ 3.4	0.792 ÷ 1.15	1695	1.36	0.99	108
3	16 ÷ 8	1.76 ÷ 2.56	0.17	0.333	0.22	560
4	23 ÷ 10	2.52 ÷ 3.70	0.11	0.215	0.143	990
5	35 ÷ 15	3.85 ÷ 5.60	0.059	0.115	0.076	2350
6	40 ÷ 20	4.40 ÷ 6.40	0.043	0.0845	0.066	3580
7	72 ÷ 34	7.92 ÷ 11.5	0.022	0.043	0.0286	10,860
8	110 ÷ 55	12.1 ÷ 17.6	0.0096	0.0188	0.0125	27,180
9	130 ÷ 65	14.2 ÷ 20.8	0.0074	0.0145	0.0069	36,600

Fig. 3.8 Critical values of wavelengths in a microwave rectangular waveguide

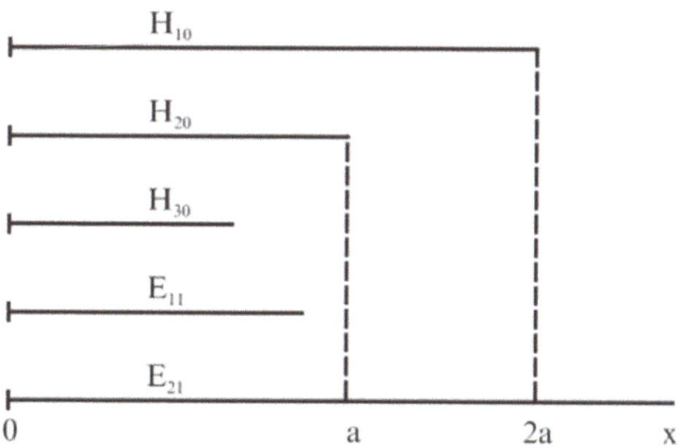

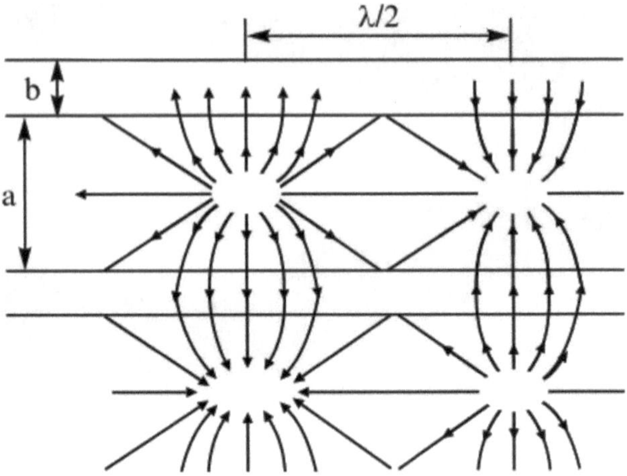

Fig. 3.9 Distribution picture of surface currents in a microwave rectangular waveguide for H_{10} type wave

radiolocation, radionavigation, medical technology, radio control, communication technology, antenna technology, and other fields of science and technology. Among the waveguides used for information transmission, the most widespread are the microwave circular waveguides. Microwave circular waveguides have several advantages. Thus, they have high reliability, stability of parameters and long service life in the process of operation. At the same time, microwave circular waveguides are completely isolated from the environment, that is, when transmitting information with such waveguides, radiation of electromagnetic waves (radio waves) to the external environment almost does not occur. Despite the very wide application of microwave circular waveguides, there are not enough scientific works that fully reflect the electromagnetic processes propagated in such transmission lines and are of experimental importance. In known works in this field, the medium in which electromagnetic waves (radio waves) propagate is considered linear. Until now, when designing such waveguides, the medium in which electromagnetic waves (radio waves) propagate was considered linear. But in real practice, the medium behaves nonlinearly. At the same time, when modeling such waveguides, complex classical mathematical apparatuses were used, which increase the error of calculations and make them difficult to perform. Therefore, taking into account the nonlinearity of the environment, in order to improve the electrical, magnetic, technical, structural and operating parameters and characteristics of microwave circular waveguides, a great deal of attention has recently been paid to the development of modeling methods that allow the development and algorithmization of new mathematical models that reflect the specific characteristics of these devices. is needed and this issue is of great practical importance.

On the other hand, since microwave band circular waveguides are widely used in production, improving the electrical, magnetic, technical, structural and operating parameters and characteristics of these devices and designing new devices with more optimal structural dimensions are very important from both a scientific point of view and a production point of view is relevant.

E-type (*TH*) and *H*-type (*TE*) waves can exist separately in a microwave circular waveguide. It should be noted that these waveguides are transmission lines of a special design that ensure the transmission of electromagnetic waves in one direction without radiating energy to the environment. Transmission lines are divided into two parts: regular transmission lines; irregular lines.

Regular transmitting lines are those lines in which the property of the electromagnetic wave in the direction of their propagation axes remains unchanged or changes according to a periodic law. In the first case, transmission lines are called longitudinal-homogeneous, and in the second case, periodic transmission lines. If the energy of the electromagnetic wave varies along the axis of the transmission line, the transmission lines are called longitudinally inhomogeneous or irregular transmission lines. Transmission lines that channel electromagnetic energy are open and closed. If the electromagnetic energy in the transmission line is unlimited in the transverse direction, it is called an open transmission line. In a closed transmission line, the electromagnetic energy exists only within the closed metal cover. A circular waveguide is a closed type transmission line.

Depending on the material that forms the transmission line, they are metal, dielectric or metal-dielectric. The properties of the transmission lines are strongly dependent on the relationship of their cross section. In this respect, transmission lines are divided into two parts, single-connection or multi-connection. A circular waveguide is called a closed metal duplex transmission line. Research shows that the solution of Maxwell's equations in the desired transmission line consists of the solution of the Helmholtz equation for the longitudinal intensity vectors of the electric and magnetic fields.

The membrane equation or the two-dimensional Helmholz equation is written in the following way [3]:

$$\nabla_{\perp}^{2} + \mathrm{k}_{\mathrm{c}}^{2}\psi = 0, \tag{3.59}$$

where $\mathrm{k}_{\mathrm{c}}^{2} + \mathrm{k}_{\mathrm{z}}^{2} = \mathrm{k}^{2}$ is called the characteristic equation of the transmission line.

The quantities k_c and k_z included in this equation are called transverse and longitudinal wave numbers. The number k_z is also called the propagation constant, and k_c is called the crisis wave number. The function ψ is called the membrane function. Because it satisfies the membrane equation, that is, the two-dimensional Helmholtz equation, and $\nabla_{\perp}$ is the Hamiltonian operator.

As noted, a circular wave is a metallic pipe with the circle circle (Fig. 3.10*a*, *b*).

In order to determine the parameter and characteristics of the electromagnetic area of round wave, the Leontovich-Shukin boundary condition should be considered. This condition is one of the important issues to determine the area when there are metallic bodies with finite sizes.

This issue is included in the approximate conditions called the Leontovich-Shukin boundary. Unlike the usual border terms, Leontovich-Shukin shook the connection between the organization of the organization in an environment of the border terms and vectors. The statement of the Leontovich-Shukin boundary condition is as follows [4]:

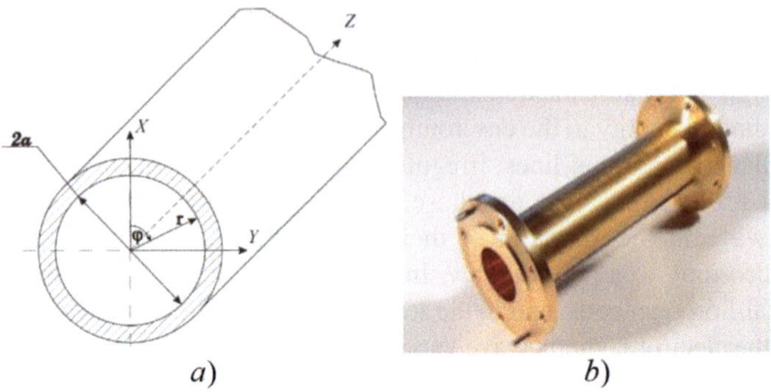

Fig. 3.10 Microwave circular waveguide: (*a*) area of the most cut; (*b*) the spatial model

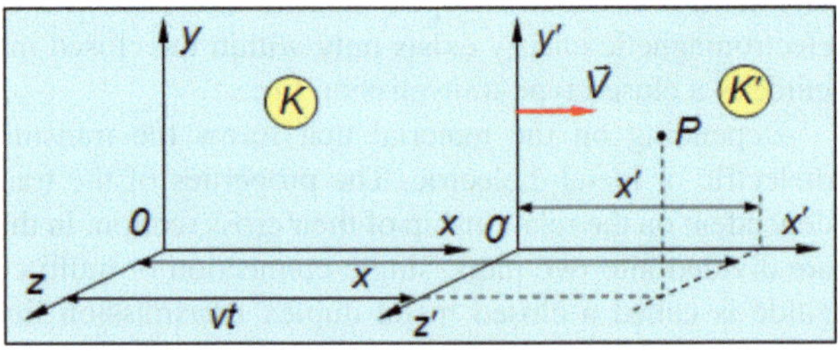

Fig. 3.11 The moving coordinate system

$$\vec{E}_{1\tau} = z_{c2}\left\lfloor \vec{n}_0, \vec{H}_1 \right\rfloor. \tag{3.60}$$

It seems that it seems that the area of the electricity touches the electric field on the real metal surface of the circular wave differs from scratch.

In real practice, the environment spread by electromagnetic waves will receive various prices due to electricity and magnetic fields inside the wave. Therefore, it is necessary to use the moving coordinate system (Fig. 3.11).

Let's look at the circular wave via the circular wave shown in Fig. 3.10. This wave is the area of the field to be solved in (r, θ, Z) cylindrical coordinate systems ($x_1 = r$, $x=\theta$, $h_1 = 1$, $h_2 = r$). $\nabla_\perp^2 + k_c^2\psi$ equations in this coordinate system [5]:

$$\frac{1}{r}\frac{\partial}{\partial r}\left(r\frac{\partial\psi}{\partial r}\right) + \frac{1}{r^2}\frac{\partial^2\psi}{\partial\theta^2} + k_c^2\psi = 0, \tag{3.61}$$

written in the form. When the circular wave, the *E*-tip and *H*-type electromagnetic waves are spread out as follows [6]:

For *E*-type wave

$$E_r^{nm} = -i\frac{\beta}{\chi_{nm}}J_n'(\chi_{nm}r)\psi_n(\theta)\exp(-i\beta z), \tag{3.62}$$

$$E_\theta^{nm} = -i\frac{n\beta}{\chi_{nm}^2 r} J_n(\chi_{nm} r)\psi_n'(\theta)\exp(-i\beta z), \quad (3.63)$$

$$E_z^{nm} = J_n(\chi_{nm} r)\psi_n(\theta)\exp(-i\beta z), \quad (3.64)$$

$$H_r^{nm} = i\frac{n\omega\varepsilon_0}{\chi_{nm}^2 r} J_n(\chi_{nm} r)\psi_n'(\theta)\exp(-i\beta z), \quad (3.65)$$

$$H_\theta^{nm} = -i\frac{\omega\varepsilon_0}{\chi_{nm}} J_n'(\chi_{nm} r)\psi_n(\theta)\exp(-i\beta z), \quad (3.66)$$

$$H_z = 0, \quad (3.67)$$

$$\beta = \beta_{nm}^E = \sqrt{\omega^2\varepsilon_0\mu_0 - \left(\frac{\nu_{nm}}{a}\right)^2}, \quad (3.68)$$

where

$\psi_n(\theta) = C_1 \cos n\theta + C_2 \sin n\theta$, $\psi_n(\theta) = -C_1 \sin n\theta + C_2 \cos n\theta$, $J_n'(\chi_{nm} r) - \chi_{nm} r$.

derivative of the function according to the argument,

$\chi_{nm} = \nu_{nm}/a-$ the width of the wave, ν_{nm} – n the root of the first Bessel function of compiled, a – is the radius of waveguide.

For H-type wave

$$H_r^{nm} = -i\frac{\beta}{\chi_{nm}} J_n'(\chi_{nm} r)\psi_n(\theta)\exp(-i\beta z), \quad (3.69)$$

$$H_\theta^{nm} = -i\frac{n\beta}{\chi_{nm}^2 r} J_n(\chi_{nm} r)\psi_n'(\theta)\exp(-i\beta z), \quad (3.70)$$

$$H_z = J_n(\chi_{nm} r)\psi_n(\theta)\exp(-i\beta z), \quad (3.71)$$

$$E_r^{nm} = i\frac{n\omega\mu_0}{\chi_{nm}^2 r} J_n(\chi_{nm} r)\psi_n'(\theta)\exp(-i\beta z), \quad (3.72)$$

$$E_\theta^{nm} = -i\frac{\omega\mu_0}{\chi_{nm}} J_n'(\chi_{nm} r)\psi_n(\theta)\exp(-i\beta z), \quad (3.73)$$

$$H_z^{nm} = 0, \quad (3.74)$$

$$\beta = \beta_{nm}^H = \sqrt{\omega^2\varepsilon_0\mu_0 - \left(\frac{\nu_{nm}'}{a}\right)^2}, \quad (3.75)$$

where $\chi_{nm} = \nu_{nm}'/a$ – the width of the wave, ν_{nm}' – n is the root root of the first Bessel function of compiled.

The connection between stationary and moving coordinate systems is determined by the transformation of Lorens and the cylindrical system is as follows (Fig. 3.11*b*):

$$r=r',\theta=\theta',z=\frac{z'+vt'}{\sqrt{1-(v/c)^2}},t=\frac{t'+(v/c^2)z'}{\sqrt{1-(v/c)^2}},\tag{3.76}$$

$$E_r=\frac{E'_r+vB'_\theta}{\sqrt{1-(v/c)^2}},E_\theta=\frac{E'_\theta-vB'_r}{\sqrt{1-(v/c)^2}},E_z=E'_z,\tag{3.77}$$

$$B_r=\frac{B'_r-(v/c^2)E'_\theta}{\sqrt{1-(v/c)^2}},B_\theta=\frac{B'_\theta+(v/c^2)E'_r}{\sqrt{1-(v/c)^2}},B_z=B'_z.\tag{3.78}$$

Given that there are $\phi=\omega t-\beta z=\omega' t'-\beta' z'=\phi'$, we will receive the following statements for a long-term wave number and frequency:

$$\beta'=\frac{\beta-(v/c^2)\omega}{\sqrt{1-(v/c)^2}},\omega'=\frac{\omega-v\beta}{\sqrt{1-(v/c)^2}}.\tag{3.79}$$

Thus, we will receive the following for moving coordinates:
for E-type wave

$$E'^{nm}_z=J_n(\chi_{nm}r')\psi_n(\theta')\exp(-i\beta z'),\tag{3.80}$$

$$\begin{aligned}E'^{nm}_r=&-i\frac{\beta-(v/c^2)\omega}{\chi_{nm}\sqrt{1-(v/c^2)}}\times\\&\times J'_n(\chi_{nm}r')\psi_n(\theta')\exp(-i\beta z'),\end{aligned}\tag{3.81}$$

$$\begin{aligned}E'^{nm}_\theta=&-in\frac{\beta-(v/c^2)\omega}{\chi^2_{nm}r'\sqrt{1-(v/c^2)}}\times\\&\times J_n(\chi_{nm}r')\psi'_n(\theta')\exp(-i\beta z'),\end{aligned}\tag{3.82}$$

$$H'_z=0,\tag{3.83}$$

$$\begin{aligned}H'^{nm}_r=&i\frac{n\varepsilon_0(\omega-v\beta)}{\chi^2_{nm}r'\sqrt{1-(v/c^2)}}\times\\&\times J_n(\chi_{nm}r')\psi'_n(\theta')\exp(-i\beta z'),\end{aligned}\tag{3.84}$$

$$\begin{aligned}H'^{nm}_\theta=&-i\frac{\varepsilon_0(\omega-v\beta)}{\chi_{nm}\sqrt{1-(v/c^2)}}\times\\&\times J'_n(\chi_{nm}r')\psi_n(\theta')\exp(-i\beta z'),\end{aligned}\tag{3.85}$$

for H-type wave

$$E'^{nm}_{z} = 0, \tag{3.86}$$

$$\begin{aligned} E'^{nm}_{r} = & -i\frac{n\mu_0(\omega - v\beta)}{\chi^2_{nm} r' \sqrt{1-(v/c^2)}} \times \\ & \times J_n(\chi_{nm} r')\psi'_n(\theta') \exp(-i\beta z'), \end{aligned} \tag{3.87}$$

$$\begin{aligned} E'^{nm}_{\theta} = & \, i\frac{\mu_0(\omega - v\beta)}{\chi_{nm} \sqrt{1-(v/c^2)}} \times \\ & \times J'_n(\chi_{nm} r')\psi_n(\theta') \exp(-i\beta z'), \end{aligned} \tag{3.88}$$

$$H'^{nm}_{z} = J_n(\chi_{nm} r')\psi_n(\theta') \exp(-i\beta z'), \tag{3.89}$$

$$\begin{aligned} H'^{nm}_{r} = & -i\frac{\beta - (v/c^2)\omega}{\chi^2_{nm} \sqrt{1-(v/c^2)}} \times \\ & \times J'_n(\chi_{nm} r')\psi_n(\theta') \exp(-i\beta z'), \end{aligned} \tag{3.90}$$

$$\begin{aligned} H'^{nm}_{\theta} = & -in\frac{\beta - (v/c^2)\omega}{\chi^2_{nm} r' \sqrt{1-(v/c^2)}} \times \\ & \times J_n(\chi_{nm} r')\psi'_n(\theta') \exp(-i\beta z'). \end{aligned} \tag{3.91}$$

The Umov-Poynting vector can also be determined for the coordinates who acts as follows:

$$S' = \frac{1}{2}\left[E', H'^*\right] = \frac{1}{2}\left\{ \begin{array}{c} e_r\left(E'_\theta H'^*_z - E'_z H'^*_\theta\right) + \\ +e_\theta\left(E'_z H'^*_r - E'_r H'^*_z\right) + \\ +e_z\left(E'_r H'^*_\theta - E'_\theta H'^*_r\right) \end{array} \right\}. \tag{3.92}$$

For E-type wave

$$\begin{aligned} S'_r = & -\frac{1}{2}E'_z H'^*_\theta = -\frac{1}{2}i\frac{\varepsilon_0(\omega - v\beta)}{\chi_{nm}\sqrt{1-(v/c)^2}} \times \\ & \times J_n(\chi_{nm} r') J'_n(\chi_{nm} r')\psi^2_n(\theta'), \end{aligned} \tag{3.93}$$

$$\begin{aligned} S'_\theta = & -\frac{1}{2}E'_z H'^*_r = -\frac{1}{2}i\frac{n\varepsilon_0(\omega - v\beta)}{\chi^2_{nm} r'\sqrt{1-(v/c)^2}} \times \\ & \times J^2_n(\chi_{nm} r')\psi_n(\theta')\psi'_n(\theta'), \end{aligned} \tag{3.94}$$

$$\begin{aligned} S'_z = & \frac{1}{2}\left(E'_r H'^*_\theta - E'_\theta H'^*_r\right) = \frac{1}{2}\frac{\varepsilon_0(\omega - v\beta)\left(\beta - \frac{v}{c^2}\omega\right)}{\chi^2_{nm}\left(1-(v/c)^2\right)} \times \\ & \times \left\{ -\frac{n^2}{\chi^2_{nm} r'^2} J^2_n(\chi_{nm} r')\psi'^2_n(\theta') - J'^2_n(\chi_{nm} r')\psi^2_n(\theta') \right\}. \end{aligned} \tag{3.95}$$

For H-type wave

$$S'_r = -\frac{1}{2}E'_\theta H'^*_z = \frac{1}{2}i\frac{\mu_0(\omega - v\beta)}{\chi_{nm}\sqrt{1-(v/c)^2}} \times \\ \times J_n(\chi_{nm}r')J'_n(\chi_{nm}r')\psi_n^2(\theta'), \tag{3.96}$$

$$S'_\theta = -\frac{1}{2}E'_r H'^*_z = \frac{1}{2}i\frac{n\mu_0(\omega - v\beta)}{\chi^2_{nm}r'\sqrt{1-(v/c)^2}} \times \\ \times J_n^2(\chi_{nm}r')\psi_n(\theta')\psi'_n(\theta'), \tag{3.97}$$

$$S'_z = \frac{1}{2}\left(E'_r H'^*_\theta - E'_\theta H'^*_r\right) = \frac{1}{2}\frac{\mu_0(\omega - v\beta)\left(\beta - \frac{v}{c^2}\omega\right)}{\chi^2_{nm}\left(1-(v/c)^2\right)} \times \\ \times \left\{\frac{n^2}{\chi^2_{nm}r'^2}J_n^2(\chi_{nm}r')\psi'^2_n(\theta') - J'^2_n(\chi_{nm}r')\psi_n^2(\theta')\right\}. \tag{3.98}$$

Figure 3.12 shows the density of power floods of E_{01} (a), E_{11} (b), H_{01} (c), H_{11} (d) types of waves in the coordinates moving with v_{qr} speeds. Table 3.4 has been given v_{mn} and χ_{mn} prices for various m and n indices. Tables 3.5 and 3.6 provide calculated prices by the formulas of electrical and magnetic fields (3.80, 3.81, 3.82, 3.83, 3.84, 3.85, 3.86, 3.87, 3.88, 3.89, 3.90 and 3.91) formulas. Figures 3.13 and 3.14, the microwave circular wave transplant, in accordance with the microwave circular wave transplantation of the electricity and magnetic fields in Fig. 3.14, was shown distributions along Z reads for elementary oblasts.

On the above phrases, the status of the extreme electromagnetic waves (stucture of electromagnetic waves, epurents of the power lines of the electromagnetic field) are described (Fig. 3.15).

In Fig. 3.15, the stuke of the H_{11}-type wave area appears like a rectangular wave of the H_{10}-type wave. Therefore, the H_{11}-type wave is a key type of wave.

The price of ψ functions for H_{11}-type waves in the wallflow of waves, $\left(\frac{\partial\psi}{\partial\theta} = \frac{\partial\psi}{\partial l} = 0\right)$ of them are declining as frequency increases. This creates the H_{11}-type wave, the microwave signals, i.e., to provide distant communication. The price of a fleet coefficient of 37.5 GHz for H_{11} wave is 0.2 dB/km (for copper wave of copper with 50 mm). This means that the capacity of the wave transmitted in a distance of 10 km is reduced 1ç6 times. It should be noted that in this situation, along with the radio transmission, along with the waves of workers, otherwise waves are spread. But their weakening is very strong.

It is known that when it is designed for radio, they are usually, technological defects, including non-smoothness, and other shapes and sizes. As a result, the energy of the H_{11}-type wave becomes the energy of another type of waves. To overcome this event, the non-working waves should not be spread within the radio transmission. For this purpose, the radio transmission is prepared in the form of separate circles made of each other, and is transmitted to its internal special absorber. Figure 3.16 specified the specifics of the specific dances of the circular waveguide.

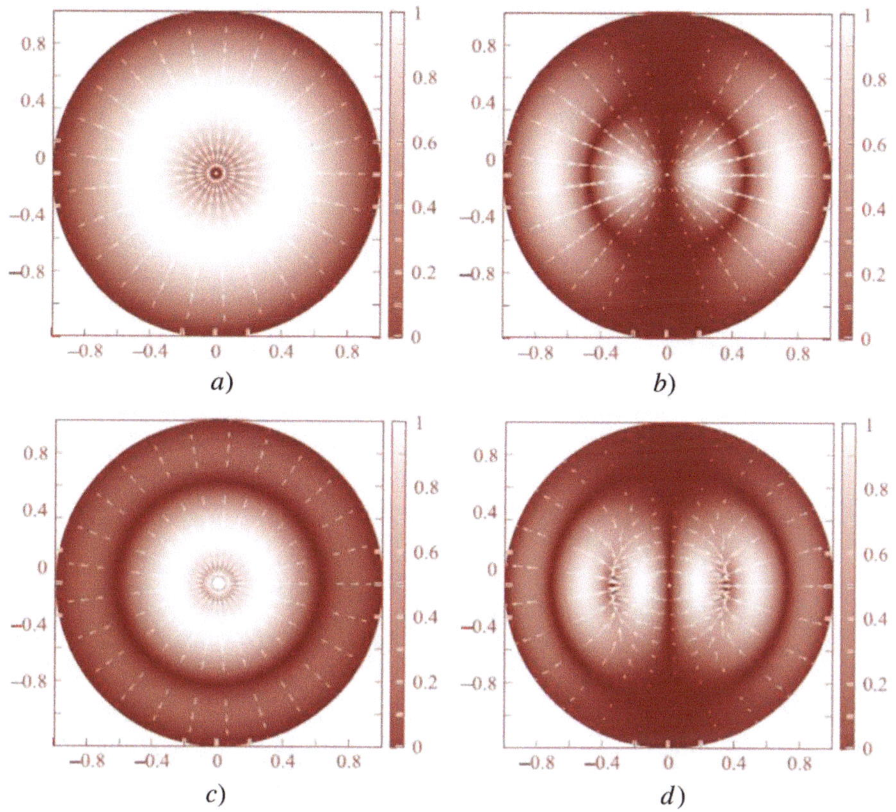

Fig. 3.12 Density of power floods of power floods in the coordinates moving with group speed E_{01} (*a*), E_{11} (*b*), H_{01} (*c*), H_{11} (*d*) type waves

Table 3.4 ν_{mn} and χ_{mn} prices for various *m* and *n* indices

	ν_{mn} in different prices of *n*			χ_{mn} in different prices of *n*		
M	$n = 1$	$n = 2$	$n = 3$	$n = 1$	$n = 2$	$n = 3$
0	2405	5520	8654	3382	7016	10,174
1	3832	7016	10,174	1840	5335	6705
2	5135	8417	11,620	3054	6705	9969
3	6380	9761	13,015	4201	8015	11,346

The following can be noted from the analysis of the stagger of the waves:

1. The low-type wave (H_{11}) in the microwave circular waveguide is similar to the H_{10}-type wave of the microwave rectangular waveguide. The H_{11} type wave can be represented as the sum of two mutually perpendicular waves.
2. Both waves have the same distribution coefficient and are distorted. If a microwave circular waveguide has some non-uniform cross-sectional shape, including a small elliptical or other distortion, then the amplitudes and phase velocities of both waves will be equal. This causes the polarization plane of the final wave to

Table 3.5 Prices for the intensity of electric fields of microwave wave transplant operating in 9 GHz

Number of elementary areas	*E*-type wave		*H*-type wave	
	$E_z^{\prime nm}$, (10^5), *V/m*	$E_r^{\prime nm}$, (10^5), *V/m*	$E_r^{\prime nm}$, (10^5), *V/m*	$E_\theta^{\prime nm}$, (10^5), *V/m*
1	10.21	12.42	44.80	80.64
2	12.58	19.59	42.92	85.72
3	13.47	25.81	37.67	90.84
4	75.59	72.92	49.89	27.68
5	34.61	10.68	58.77	19.53
6	21.90	10.50	93.68	20.30
7	14.79	18.72	27.68	25.79
8	12.71	17.55	33.59	28.28
9	14.79	25.60	42.60	34.64
10	16.51	14.68	45.79	10.92
11	18.47	9.31	58.27	26.69
12	26.91	8.59	65.15	40.68
13	33.21	16.17	90.96	68.79
14	70.32	18.85	33.68	76.16
15	11.29	24.37	28.35	80.05
16	10.95	32.89	26.71	71.29
17	19.88	45.43	16.56	39.57
18	22.59	57.71	45.42	35.06
19	39.49	23.49	49.87	18.72
20	87.21	12.95	68.23	13.95

rotate. It is difficult to control the plane of polarization, so the H_{11} wave is used to transmit energy in a circular waveguide.

3. The H_{11} wave, which has an azimuthal homogeneous field due to its structure, is also interesting because there are no H_φ components on the walls of the waveguide and only the longitudinal component of the H vector. It approaches zero in $\frac{\lambda_b}{R} \to 0$ cases. So, in the limit, there is no field on the walls of the waveguide and no current flows through the walls. This shows that the H_{11} wave in the circular waveguide has weak attenuation.
4. In a microwave circular waveguide, the H_{11} type wave has circular symmetry. Due to this, the H_{11} type wave is used in rotating joints of antenna devices. In addition, due to the existence of the longitudinal arrangement of the electric field concentrated along the axis of the waveguide, it allows to apply the waveguide in linear electron accelerators.
5. It is advantageous to use a microwave circular waveguide from the H_{11} type wave in power transmission. H_{11} type wave has weak attenuation. The main cause of extinction is the spread of H_{01}, E_{01}, E_{11}, H_{21} type waves. Therefore arbitrary inhomogeneities lead to conversion of different types. As a result, signals are distorted, energy loss increases, etc. Therefore, transducers and devices based on them must be designed with great precision.

Table 3.6 Prices for the intensity of magnetic fields of microwave wave transplant operating in 9 GHz

	E-type wave		*H*-type wave	
Number of elementary areas	H'^{nm}_{r}, *A/m*	H'^{nm}_{θ}, *A/m*	H'^{nm}_{r}, *A/m*	H'^{nm}_{θ}, *A/m*
1	42.36	44.49	35.29	23.39
2	25.45	52.23	23.66	34.15
3	27.27	65.35	36.36	85.56
4	64.13	75.36	41.04	57.51
5	62.76	65.40	23.50	17.25
6	33.44	96.23	28.56	25.56
7	47.27	63.27	32.54	46.24
8	39.34	24.35	22.46	19.26
9	42.53	41.24	96.28	49.28
10	50.24	12.33	17.12	55.16
11	94.57	25.04	12.28	31.12
12	66.29	38.19	41.36	30.07
13	82.39	34.47	54.87	63.42
14	73.35	57.31	24.13	12.35
15	48.32	78.06	85.23	17.53
16	36.57	57.38	56.72	45.95
17	64.59	19.25	23.13	79.30
18	92.25	19.20	23.34	56.97
19	83.94	76.46	82.67	96.18
20	75.53	25.33	13.72	15.69

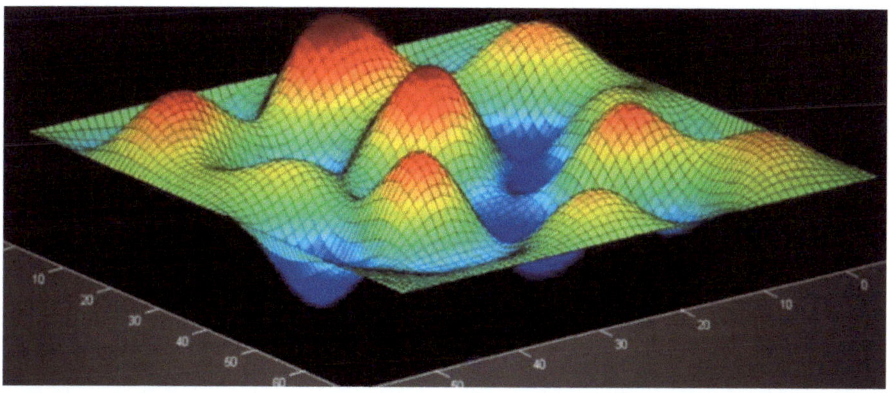

Fig. 3.13 9 Distribution of microwave circular waves operated in 9 GHz frequency of the electric field of electric fields along *Z*

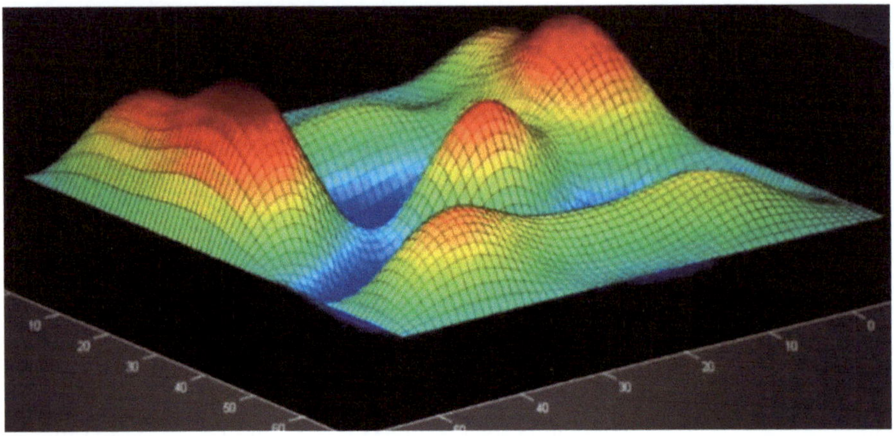

Fig. 3.14 9 Distribution of microwave circular waves operated in 9 GHz frequency of the magnetic field of electric fields along Z

Fig. 3.15 The power lines of the electromagnetic field in the circular wave: (*a*) H_{11}; (*b*) E_{01}; (*v*) H_{01}; (*d*) the visual distribution of the area of the H_{11} wave: electric field – the force lines of the magnetic field

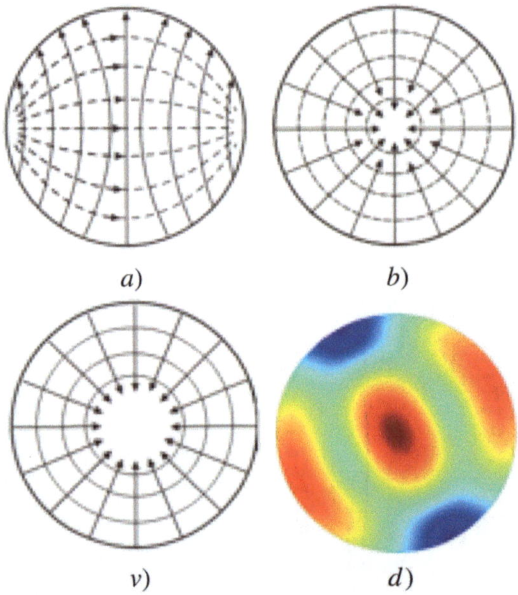

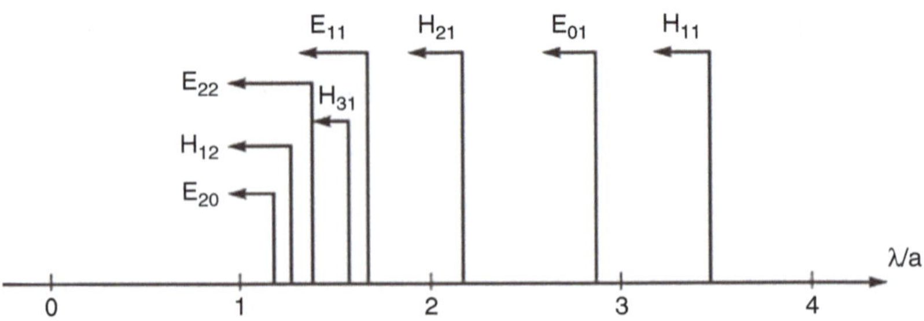

Fig. 3.16 Spectrum of specific waves of microüave circular waveguide

3.3 Mathematical Models of Impedance Characteristics of Microwave Rectangular and Circular Waveguides

Currently, the frequency range up to 60 GHz has been studied for information transmission in telecommunications. However, with the emergence of other priority areas, it became necessary to study other ranges. The main elements of any microwave system are the guiding systems, examples of which are waveguides with rectangular and circular cross-sections. During the transmission of microwave electromagnetic waves through waveguides, the high-frequency current settles mainly on the inner surface of the waveguide (Fig. 3.17) [7].

Since the inner surface of waveguides is not ideal, the energy transmitted here is subject to attenuation. Therefore, the calculation of the impedance characteristics of microwave and circular waveguides is of great importance. The electromagnetic wave propagates inside the waveguides by reflecting from its inner walls (Fig. 3.18).

As can be seen from Fig. 3.18, the electromagnetic wave propagates inside the microwave rectangular waveguide by reflecting from its walls *A* and *B*. At this time, electromagnetic waves undergo two types of polarization: *s*-polarization and *p*-polarization (Fig. 3.19).

Fig. 3.17 When microwave electromagnetic waves are transmitted through waveguides, the high frequency current settles on the inner surface of the waveguide

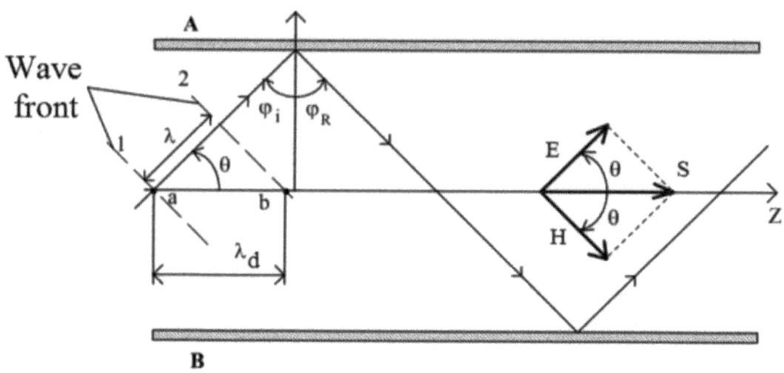

Fig. 3.18 Propagation of the electromagnetic wave inside the waveguide by reflection from its inner walls

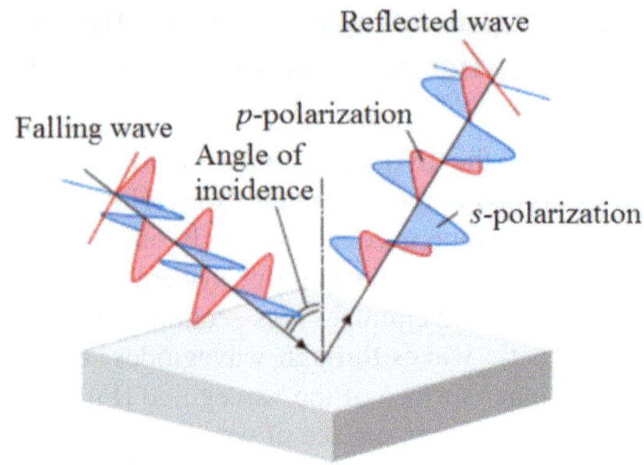

Fig. 3.19 Exposure of electromagnetic wave to *s*-polarization and *p*-polarization inside the waveguide

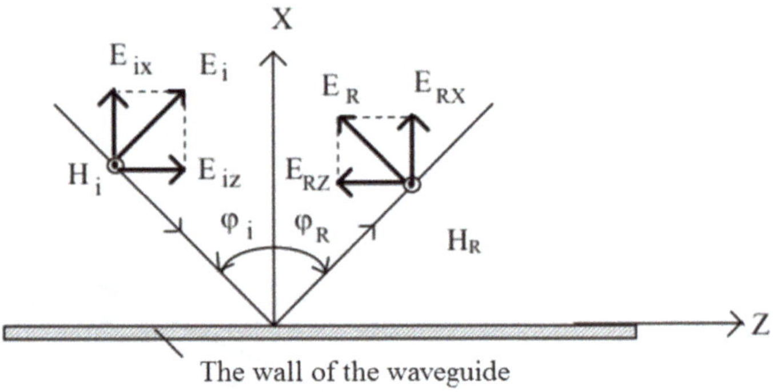

Fig. 3.20 Generation of an *E*-type wave in a waveguide

As can be seen from Fig. 3.19, in the case of *s*-polarization, the intensity vector of the electric field of the electromagnetic wave is perpendicular to the plane of incidence, and in the case of *p*-polarization, the intensity vector of the electric field of the electromagnetic wave is located in the plane of incidence.

On the other hand, the electromagnetic wave propagates at the separation boundary of the two media inside the microwave rectangular and circular waveguides.

These environments are metal and vacuum. Therefore, it is appropriate to use the Leontevich-Shukin boundary condition in this case. The generation of *E*-type and *H*-type waves in the waveguide is shown in Figs. 3.20 and 3.21, respectively.

Let's look at two cases, that is, when the electromagnetic wave propagates in stationary and moving computing systems. A normally incident electromagnetic wave in a stationary computing system is shown in Fig. 3.22*a*, and a normally incident electromagnetic wave in a moving computing system is shown in Fig. 3.22*b*.

The expressions for a normally incident electromagnetic wave in a stationary computing system would be:

- for *s*-polarization $\vec{E}_{1d} = -E_0 \vec{x}_0, \vec{H}_{1d} = H_0 \vec{z}_0;$,
- for *p*-polarization $\vec{E}_{21d} = E_0 \vec{z}_0, \vec{H}_{2d} = H_0 \vec{x}_0.$.

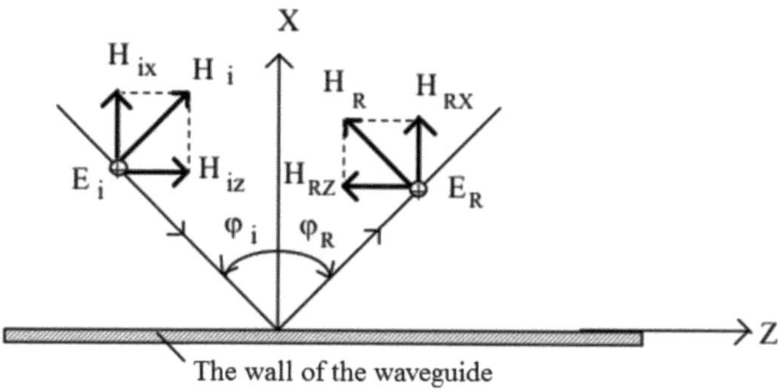

Fig. 3.21 Generation of an *H*-type wave in a waveguide

Fig. 3.22 Normally incident electromagnetic wave in a stationary computing system (*a*), normally incident electromagnetic wave in a moving computing system (*b*)

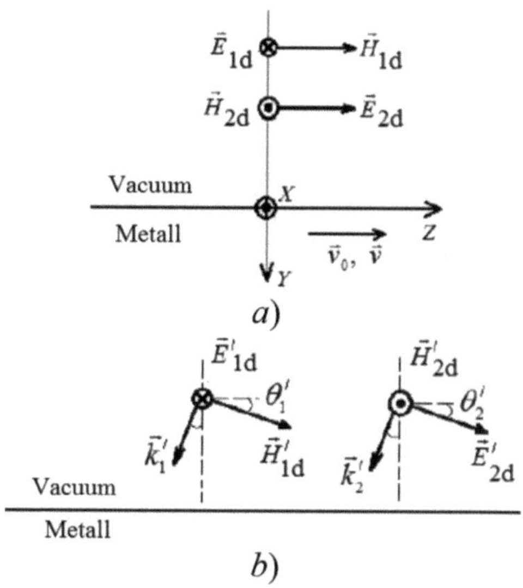

It follows that the following expressions are satisfied for the components of the incident wave in the initial calculation system:

$$\begin{gathered} E_{2dx}=0, E_{2dy}=0, E_{2dz}=E_0, H_{2dx}=H_0, \\ H_{2dy}=0, H_{2dz}=0. \\ E_{1dx}=-E_0, E_{1dy}=0, E_{1dz}=0, H_{1dx}=0, \\ H_{1dy}=0, H_{1dz}=H_0, \end{gathered} \tag{3.99}$$

Using the Lorentz transformation, we obtain the following expressions for the components of the electromagnetic field in a moving computing system:

– for s-polarization

$$E'_{1dx} = \frac{E_{1dx} - v_0 B_{1dy}}{\sqrt{1-(v_0/c)^2}} = -\frac{E_0}{\sqrt{1-(v_0/c)^2}}, \tag{3.100}$$

$$\begin{aligned} E'_{1dy} &= \frac{E_{1dy} + v_0 B_{1dx}}{\sqrt{1-(v_0/c)^2}} = 0, \quad E'_{1dz} = E_{1nz} = 0, \\ H'_{1dx} &= \frac{H_{1dx} + v_0 D_{1dy}}{\sqrt{1-(v_0/c)^2}} = 0, \end{aligned} \tag{3.101}$$

$$H'_{1dy} = \frac{H_{1dy} - v_0 D_{1dx}}{\sqrt{1-(v_0/c)^2}} = \frac{v_0 \varepsilon_0 E_0}{\sqrt{1-(v_0/c)^2}}, \quad H'_{1dz} = H_{1nz} = H_0,$$

– for p-polarization

$$E'_{2dn} = 0, \quad E'_{2dy} = \frac{v_0 \mu_0 H_0}{\sqrt{1-(v_0/c)^2}}, \quad E'_{2dz} = E_0, H'_{2dx} = \frac{H_0}{\sqrt{1-(v_0/c)^2}}, \quad H'_{2dy} = 0, H'_{2nz} = 0. \tag{3.102}$$

Then, if we denote the reflection coefficient from the surface of the waveguide walls for the s-polarized wave by δ_1, and the reflection coefficient from the surface of the waveguide walls for the p-polarized wave by δ_2 then we get for the final field on the conducting wall of the waveguide [45]:

$$\begin{aligned} E'_{1\sum x} &= E'_{1x}(1+\delta_1), \quad H'_{1\sum y} = H'_{1y}(1-\delta_1), \\ H'_{1\sum z} &= H'_{1z}(1-\delta_1), \end{aligned} \tag{3.103}$$

$$\begin{aligned} E'_{2\sum y} &= E'_{1y}(1+\delta_1), \quad E'_{2\sum z} = E'_{1z}(1+\delta_1), \\ H'_{2\sum x} &= H'_{2x}(1-\delta_1). \end{aligned} \tag{3.104}$$

It should be noted that in a moving computing system, electric and magnetic waves fall on the surface of non-moving metal at angles θ'_1 and θ'_2 (Fig. 3.22*b*). These angles are defined as follows:

$$tg\theta'_1 = \frac{H'_{1dy}}{H'_{1dz}} = \frac{v_0 \varepsilon_0 E_0}{H_0\sqrt{1-(v_0/c)^2}} = \frac{Z_0 v_0 \varepsilon_0}{\sqrt{1-(v_0/c)^2}}, \tag{3.105}$$

$$tg\theta_2' = \frac{E_{2dy}'}{E_{2dz}'} = \frac{v_0\mu_0 E_0}{E_0\sqrt{1-(v_0/c)^2}} = \frac{v_0\mu_0}{Z_0\sqrt{1-(v_0/c)^2}}, \tag{3.106}$$

where $Z_0 = E_0/H_0$ is the wave resistance of the wave propagating medium. If we use the expression $Z_0\varepsilon_0 = Z_0^{-1}\mu_0 = c^{-1}$, we get

$$tg\theta_1' = tg\theta_2' = \frac{v_0/c}{\sqrt{1-(v_0/c)^2}}. \tag{3.107}$$

Accordingly, other angles of incidence are defined as follows:

$$\cos\theta' = \sqrt{1-(v_0/c)^2},\ \sin\theta' = v_0/c. \tag{3.108}$$

As a result of refraction of the electromagnetic wave from the inner metal surface of the waveguide, it turns into a non-homogeneous plane wave. Thus, the phases and amplitudes of the equal surfaces of this wave do not coincide. At this time, the actual refraction angle of the electromagnetic wave from the inner metal surface of the waveguide is determined as follows [8]:

$$tg\theta_m = \frac{k\sin\theta_0}{\operatorname{Re}\sqrt{k_m - k^2\sin^2\theta_0}}, \tag{3.109}$$

where are the wave numbers of the environment and metal, which fell in accordance with k and k_m.

(3.109) in the statement, the hole in the phrase can be converted as follows:

$$\operatorname{Re}\sqrt{X} = \sqrt{(|X| + \operatorname{Re}X)/2}. \tag{3.110}$$

Given the statement (3.110) in the statement (3.109), we will receive

$$tg\theta_m = \frac{k\sin\theta_0}{\operatorname{Re}\sqrt{\frac{1}{2}\left(\left|k_m^2 - k^2\sin^2\theta_0\right|\right) + \operatorname{Re}\left(k_m^2 - k^2\sin^2\theta_0\right)}} = \frac{k\sin\theta_0}{\sqrt{\frac{1}{2}\left[\sqrt{\left(\operatorname{Re}k_m^2 - k^2\sin^2\theta_0\right)^2 + \left(\operatorname{Im}k_m^2\right)^2} + \operatorname{Re}k_m^2\right]}}, \tag{3.111}$$

where $k_m = k_0\sqrt{\varepsilon_k} = k_0\sqrt{\varepsilon_m - j\frac{\sigma}{\omega\varepsilon_0}}$, k metal conductivity, ω – a rounded frequency of the electromagnetic field.

On the other hand, it is known that

$$\operatorname{Re} k_m^2 = k_0^2 \varepsilon_m, \operatorname{Im} k_m^2 = -k_0^2 \frac{\sigma}{\omega \varepsilon_0}. \tag{3.112}$$

Then the angle of the breaking for the computing system considered is determined as follows:

$$\theta'_m = arctg \frac{v_0/c}{\sqrt{\frac{1}{2}\left[\sqrt{\left(\varepsilon_m - (v_0/c)^2\right)^2 + \frac{\sigma^2}{\omega^2 \varepsilon_0^2}} + \left(\varepsilon_m - (v_0/c)^2\right)\right]}}. \tag{3.113}$$

According to the Frenel formula, the reflection and breaking angles are determined as follows:

$$\delta_1 = \frac{Z_m \cos\theta' - Z_0 \cos\theta'_m}{Z_m \cos\theta' + Z_0 \cos\theta'_m}, \tag{3.114}$$

$$\delta_2 = \frac{Z_0 \cos\theta' - Z_m \cos\theta'_m}{Z_0 \cos\theta' + Z_0 \cos\theta'_m}. \tag{3.115}$$

The components of the final electromagnetic wave on the internal metal surface of the wave in the moving calculation system are determined as follows:

$$\begin{gathered} E'_{1x} = \frac{E_0}{\sqrt{1-(v_0/c)^2}}(1+\delta_1), \\ H'_{1y} = H_0 \frac{v_0/c}{\sqrt{1-(v_0/c)^2}}(1+\delta_1), \;\; H'_{1z} = H_0(1-\delta_1), \end{gathered} \tag{3.116}$$

$$\begin{gathered} E'_{2y} = \frac{E_0(v_0/c)}{\sqrt{1-(v_0/c)^2}}(1-\delta_2), \;\; E'_{2z} = \frac{E_0}{\sqrt{1-(v_0/c)^2}}(1-\delta_2), \\ H'_{2x} = H_0 \frac{1}{\sqrt{1-(v_0/c)^2}}(1+\delta_2). \end{gathered} \tag{3.117}$$

Tangential components of the electromagnetic field in the initial calculation system for the transformation of Lorens are determined as follows:

$$\begin{gathered} E_{1x} = \frac{E_0\left(1+(v_0/c)^2\right)}{\sqrt{1-(v_0/c)^2}}(1+\delta_1), \;\; H_{1z} = H_0(1-\delta_1), \\ E_{2z} = \frac{E_0}{\sqrt{1-(v_0/c)^2}}(1-\delta_2), H_{2x} = H_0(1+\delta_2), \end{gathered} \tag{3.118}$$

Thus, the impredans on the internal metal surface of the wave transmission in the calculation system moving for various polyarizations are determined as follows:

$$
\begin{aligned}
Z_{s1} &= \frac{E_{1x}}{H_{1z}} = \frac{1+\delta_1}{1-\delta_1}\left(1+(v_0/c)^2\right)Z_1 = \\
&= \frac{Z_m}{\cos\theta'_m}\sqrt{1-(v_0/c)^2}\left(1+(v_0/c)^2\right),
\end{aligned}
\tag{3.119}
$$

$$
\begin{aligned}
Z_{s2} &= \frac{E_{2z}}{H_{2x}} = \frac{1-\delta_1}{1+\delta_1}\frac{1}{\sqrt{1-(v_0/c)^2}}Z_1 = \\
&= Z_m \cos\theta'_m \frac{1}{1-(v_0/c)^2}.
\end{aligned}
\tag{3.120}
$$

Figure 3.23 shows the dependencies of the surface impedances on the speed of the computing system for *s*-polarized and *p*-polarized electromagnetic waves in the waveguide.

Let's assume that the thickness of the conductive wall of the microwave circular waveguide is much greater than the depth of penetration of the electromagnetic wave into it. The components of the electromagnetic field in a microwave circular waveguide are defined as:

$$
\begin{aligned}
E_\delta = &-j\left(\frac{A\beta}{\chi}J'_n(\chi\delta) + \frac{Bn\omega\mu_0}{\chi^2\delta}J_n(\chi\delta)\right)\times \\
&\times \cos n\theta \exp[j(\omega t - \beta z)],
\end{aligned}
\tag{3.121}
$$

$$
\begin{aligned}
E_\theta = &j\left(\frac{An\beta}{\chi^2\delta}J_n(\chi\delta) + \frac{B\omega\mu_0}{\chi}J'_n(\chi\delta)\right)\times \\
&\times \sin n\theta \exp[j(\omega t - \beta z)],
\end{aligned}
\tag{3.122}
$$

$$
E_z = AJ_n(\chi\delta)\cos n\theta \exp[j(\omega t - \beta z)],
\tag{3.123}
$$

$$
\begin{aligned}
H_\delta = &-j\left(\frac{An\omega\varepsilon_0}{\chi^2\delta}J_n(\chi\delta) + \frac{B\beta}{\chi}J'_n(\chi\delta)\right)\times \\
&\times \sin n\theta \exp[j(\omega t - \beta z)],
\end{aligned}
\tag{3.124}
$$

$$
\begin{aligned}
H_\theta = &-j\left(\frac{A\omega\varepsilon_0}{\chi}J'_n(\chi\delta) + \frac{Bn\beta}{\chi^2\delta}J_n(\chi\delta)\right)\times \\
&\times \cos n\theta \exp[j(\omega t - \beta z)],
\end{aligned}
\tag{3.125}
$$

$$
H_z = BJ_n(\chi\delta)\sin n\theta \exp[j(\omega t - \beta z)].
\tag{3.126}
$$

It should be noted that the received expressions can be applied to any conductive surface. A microwave waveguide operating at 9 GHz is shown in Fig. 3.24.

The dispersion equation for the electromagnetic wave of a microwave circular waveguide is defined as:

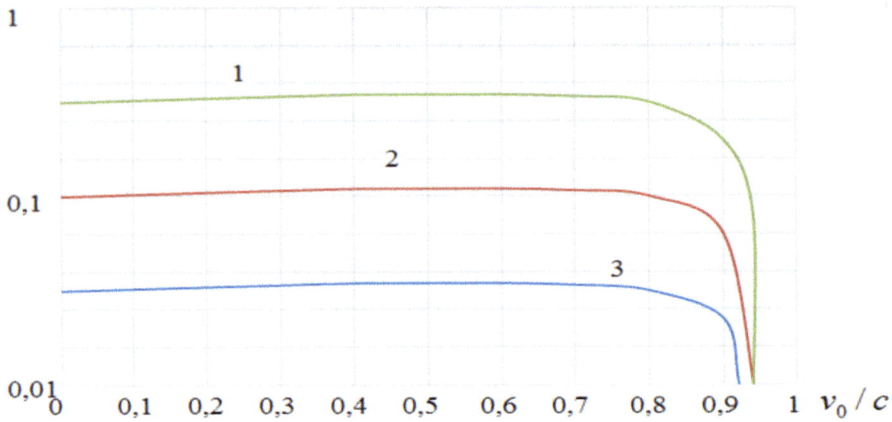

a)

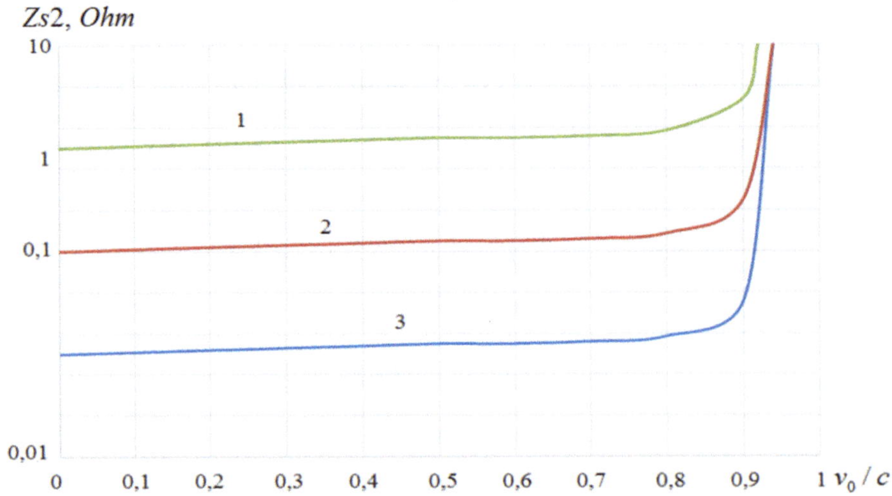

b)

Fig. 3.23 Dependencies on the speed of the calculation system of surface impedances for *s*-polarized (*a*) and *p*-polarized (*b*) electromagnetic waves in the waveguide: when $f = 7$ GHz: $1 - \sigma = 10^6 Sim/m$; $2 - \sigma = 10^7 Sim/m$; $3 - \sigma = 10^8 Sim/m$

$$\left[J_n(\chi a) - jZ_s \frac{\omega\varepsilon_0}{\chi} J'_n(\chi a)\right]\left[j\frac{\omega\mu_0}{\chi} J'_n(\chi a) - Z_s J_n(\chi a)\right] = \\ = Z_s \left[\frac{n\beta}{\chi^2 a} J_n(\chi a)\right]^2, \tag{3.127}$$

where β is the propagation coefficient (in this case, it is a complex number: $\beta = \operatorname{Re}\beta + j\operatorname{Im}\beta$, $\operatorname{Re}\beta$ is the phase coefficient, $\operatorname{Im}\beta$ is the attenuation coefficient, the impedance of the conducting wall of the waveguide).

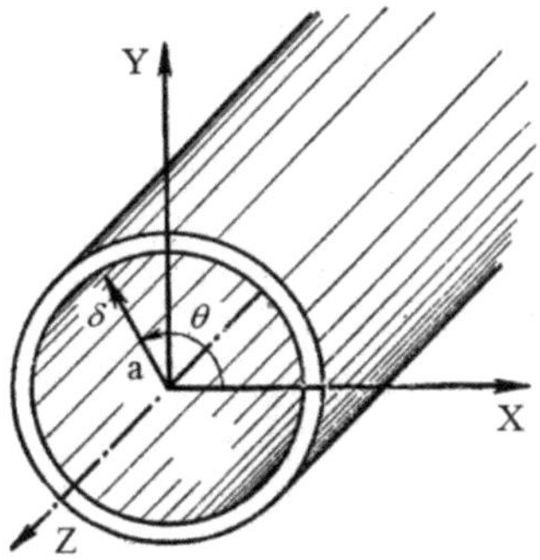

Fig. 3.24 Microwave circular operating at 9 GHz cross section of the waveguide

The solution of Eq. 3.127 [9] was reviewed. However, this dispersion equation was obtained by considering the impedance boundary conditions. Thus, the components of the electromagnetic field in the cylindrical coordinate system in the moving calculation system, taking into account the Leontovich-Shukin boundary conditions, are determined as follows:

$$E'_\delta = -j\left(\frac{A\beta'}{\chi'}J'_n(\chi'\delta') + \frac{Bn\omega'\mu_0}{\chi'^2\delta'}J_n(\chi'\delta')\right)\times \\ \times\cos n\theta' \exp[\mathrm{Im}(\beta' z')]\exp[j\omega' t'], \tag{3.128}$$

$$E'_\theta = j\left(\frac{An\beta'}{\chi'^2\delta'}J_n(\chi'\delta') + \frac{B\omega'\mu_0}{\chi'}J'_n(\chi'\delta')\right)\times \\ \times\sin n\theta' \exp[\mathrm{Im}(\beta' z')]\exp[j\omega' t'], \tag{3.129}$$

$$E'_z = jAJ_n(\chi'\delta')\cos n\theta' \exp[\mathrm{Im}(\beta' z')]\exp[j\omega' t'], \tag{3.130}$$

$$H'_\delta = -j\left(\frac{An\omega'\varepsilon_0}{\chi'^2\delta'}J_n(\chi'\delta) + \frac{B\beta'}{\chi'}J'_n(\chi'\delta)\right)\times \\ \times\sin n\theta \exp[\mathrm{Im}(\beta' z')]\exp[j\omega' t'], \tag{3.131}$$

$$H'_\theta = -j\left(\frac{A\omega'\varepsilon_0}{\chi'}J'_n(\chi'\delta) + \frac{Bn\beta'}{\chi'^2\delta'}J_n(\chi'\delta)\right)\times \\ \times\cos n\theta \exp[\mathrm{Im}(\beta' z')]\exp[j\omega' t'], \tag{3.132}$$

$$H'_z = BJ_n(\chi'\delta')\sin n\theta' \exp[\mathrm{Im}(\beta' z')]\exp[j\omega' t']. \tag{3.133}$$

In this case, the boundary conditions should be of the following form:

$$\frac{E'_z}{H'_\theta} = -Z_s, \ \frac{E'_\theta}{H'_z} = Z_s. \tag{3.134}$$

Thus, Z_{1s} is taken for H-type wave and Z_{2s} for E-type wave. As a result, we get the dispersion equation in the following form

$$\left[J_n(\chi' a) - jZ_s \frac{\omega' \varepsilon_0}{\chi'} J'_n(\chi' a)\right]\left[j\frac{\omega' \mu_0}{\chi'} J'_n(\chi' a) - Z_s J_n(\chi' a)\right] = \\ = Z_s \left[\frac{n \mathrm{Im}(\beta'')}{\chi'^2 a} J_n(\chi' a)\right]^2, \tag{3.135}$$

where $\chi'^2 - \mathrm{Im}^2\beta' = (\omega'/c)^2$ allows us to define the $\beta'(\omega')$ dependence of equality.

It is important to use the Lorentz transformation when converting from a moving to a stationary computing system. At this time, the wave vector and frequency change as follows:

$$\chi = \chi', \omega = \frac{\omega' + v\beta'}{\sqrt{1 - (v/c)^2}}. \tag{3.136}$$

The longitudinal wave number is determined using the following expression:

$$\beta = \sqrt{\left(\frac{\mathrm{Re}(\omega)}{c}\right) - \chi^2}. \tag{3.137}$$

The attenuation per unit length of the waveguide is determined using the following expression:

$$L(dB/m) = 20\lg(\exp(\mathrm{Im}(\beta))). \tag{3.138}$$

Figure 3.25 shows the frequency dependence of the extinction coefficient for H_{11} and H_{01} modes in a microwave circular waveguide.

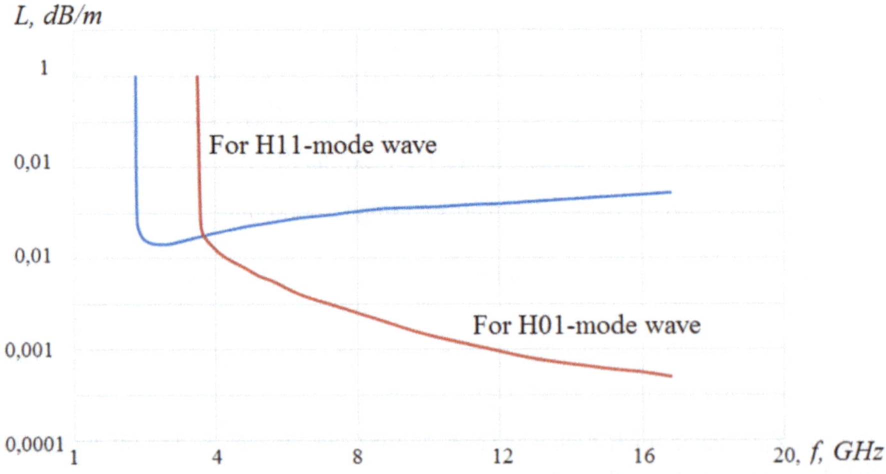

Fig. 3.25 Frequency dependence of extinction coefficient for modes H_{11} and H_{01} in microwave circular waveguide ($a = 6$ cm)

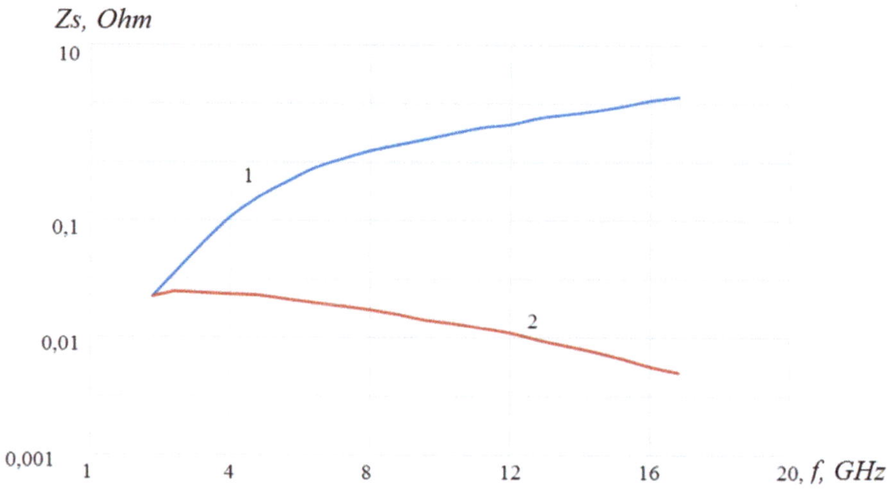

Fig. 3.26 Frequency dependence of the impedance of the conductive surface in a microwave rectangular waveguide: for a 1-*p*-polarized wave; for a 2-*s*-polarized wave

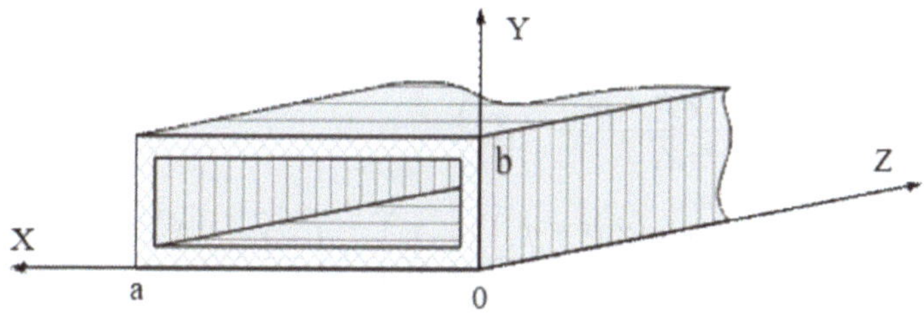

Fig. 3.27 Cross section of a microwave rectangular waveguide operating in the 4.9–7.05 GHz frequency range

As can be seen from the graphs of Fig. 3.25, the attenuation of both wave modes decreases as we move away from the critical frequency. However, the nature of later addictions differs sharply. So, the H_{01} fashion starts to decrease, and the H_{11} fashion starts to increase after taking its minimum price. The frequency dependence of the impedance of the conductive surface of the waveguide is shown in Fig. 3.26.

Now let's perform similar operations for a microwave rectangular waveguide. A microwave rectangular waveguide operating in the frequency range of 4.9–7.05 GHz is shown in Fig. 3.27.

As we know, *E*-type and *H*-type waves are propagated in microwave rectangular waveguide. The components of the electromagnetic field of these waves are determined as follows [10]:

$$E_x = -\frac{j}{\chi^2}\left(E_0\gamma\chi_x - H_0\omega\mu_0\chi_y\right) \times \times \cos(\chi_x x + \phi_x)\sin\left(\chi_y y + \phi_y\right)e^{-j\gamma z}, \tag{3.139}$$

$$\begin{aligned} E_y = & -\frac{j}{\chi^2}\left(E_0\gamma\chi_y + H_0\omega\mu_0\chi_x\right) \times \\ & \times \sin(\chi_x x + \phi_x)\cos\left(\chi_y y + \phi_y\right)e^{-j\gamma z}, \end{aligned} \tag{3.140}$$

$$E_z(x,y) = E_0 \sin(\chi_x x + \phi_x)\sin\left(\chi_y y + \phi_y\right)e^{-j\gamma z}, \tag{3.141}$$

$$\begin{aligned} H_x = & \frac{j}{\chi^2}\left(H_0\gamma\chi_x + E_0\omega\varepsilon_0\chi_y\right) \times \\ & \times \sin(\chi_x x + \phi_x)\cos\left(\chi_y y + \phi_y\right)e^{-j\gamma z}, \end{aligned} \tag{3.142}$$

$$\begin{aligned} H_y = & \frac{j}{\chi^2}\left(H_0\gamma\chi_y + E_0\omega\varepsilon_0\chi_x\right) \times \\ & \times \cos(\chi_x x + \phi_x)\sin\left(\chi_y y + \phi_y\right)e^{-j\gamma z}, \end{aligned} \tag{3.143}$$

$$H_z(x,y) = H_0 \cos(\chi_x x + \phi_x)\cos\left(\chi_y y + \phi_y\right)e^{-j\gamma z}, \tag{3.144}$$

where $\phi_x = \frac{1}{2}(m\pi - \chi_x a), \phi_y = \frac{1}{2}\left(n\pi - \chi_y b\right)$.

The dispersion equation of the electromagnetic waves of the microwave rectangular waveguide with the inner surface as an ideal conductor is determined as follows [11]:

$$\begin{aligned} & \left\{\left(\frac{j\omega\mu_0\chi_y tg\left(\chi_y b + \phi_y\right)}{\chi_x^2 + \chi_y^2} + Z_s\right) \times \right. \\ & \times \left(\frac{j\omega\varepsilon_0\chi_y ctg\left(\chi_y b + \phi_y\right)}{\chi_x^2 + \chi_y^2} - \frac{1}{Z_s}\right) = \left(\frac{\gamma\chi_x}{\chi_x^2 + \chi_y^2}\right)^2, \end{aligned} \tag{3.145}$$

$$\begin{aligned} & \left\{\left(\frac{j\omega\mu_0\chi_x tg(\chi_x a + \phi_x)}{\chi_x^2 + \chi_y^2} + Z_s\right)\left(\frac{j\omega\varepsilon_0\chi_x ctg\left(\chi_y a + \phi_x\right)}{\chi_x^2 + \chi_y^2} - \frac{1}{Z_s}\right) = \right. \\ & = \left(\frac{\gamma\chi_y}{\chi_x^2 + \chi_y^2}\right)^2. \end{aligned} \tag{3.146}$$

The dispersion equations shown in expressions (3.145), (3.146) are approximate equations. These equations are satisfied when electromagnetic waves are normal to the surface. Impedance boundary conditions were used when deriving Eqs. (3.145 and 3.146). Let's use the expressions we got earlier for the surface impedance to get more accurate dispersion equations. Let's write the expressions that define the electromagnetic field for a computing system moving along the axis of the transducer. In this case, the real part of the wave propagation constant will be zero. Then the $H10$ mode will fall normal to the conducting surface of the waveguide and the Leontovich-Shukin boundary conditions will be satisfied. For the considered case, we get for the components of the electromagnetic field of the rectangular waveguide in the moving calculation system:

$$E'_x = -\frac{j}{\chi^2}\left(E_0\gamma'\chi'_x - H_0\omega'\mu_0\chi'_y\right)\cos\left(\chi'_x x + \phi'_x\right)\times \times \sin\left(\chi'_y y + \phi'_y\right)\exp[\mathrm{Im}(\gamma' z')]\exp[j\omega' t'], \tag{3.147}$$

$$E'_y = -\frac{j}{\chi'^2}\left(E_0\gamma'\chi'_y - H_0\omega'\mu_0\chi'_x\right)\sin\left(\chi'_x x + \phi'_x\right)\times \times \cos\left(\chi'_y y + \phi'_y\right)\exp[\mathrm{Im}(\gamma' z')]\exp[j\omega' t'], \tag{3.148}$$

$$E'_z = E_0\sin\left(\chi'_x x + \phi'_x\right)\cos\left(\chi'_y y + \phi'_y\right)\times \times \exp[\mathrm{Im}(\gamma' z')]\exp[j\omega' t'], \tag{3.149}$$

$$H'_x = \frac{j}{\chi'^2}\left(H_0\gamma'\chi'_x - E_0\omega'\varepsilon_0\chi'_y\right)\sin\left(\chi'_x x + \phi'_x\right)\times \times \cos\left(\chi'_y y + \phi'_y\right)\exp[\mathrm{Im}(\gamma' z')]\exp[j\omega' t'], \tag{3.150}$$

$$H'_y = \frac{j}{\chi'^2}\left(H_0\gamma'\chi'_y - E_0\omega'\varepsilon_0\chi'_x\right)\cos\left(\chi'_x x + \phi'_x\right)\times \times \sin\left(\chi'_y y + \phi'_y\right)\exp[\mathrm{Im}(\gamma' z')]\exp[j\omega' t'], \tag{3.151}$$

$$H'_z = H_0\cos\left(\chi'_x x + \phi'_x\right)\cos\left(\chi'_y y + \phi'_y\right)\times \times \exp[\mathrm{Im}(\gamma' z')]\exp[j\omega' t']. \tag{3.152}$$

In this case, the boundary conditions should be of the following form:

$$y' = 0, b : \frac{E'_z}{H'_ч} = -\frac{E'_x}{} = Z_s, x' = 0, a : \frac{E'_y}{H'_z} = -\frac{E'_z}{H'_y} = Z_s. \tag{3.153}$$

Thus, Z_{1s} is taken for H-type wave and Z_{2s} for E-type wave.

Thus, for a rectangular waveguide, we get the dispersion equations in the moving calculation system as follows:

$$\left\{\left(\frac{j\omega''\mu_0\chi'_y tg\left(\chi'_y b + \phi'_y\right)}{\chi'^2_x + \chi'^2_y} + Z_s\right)\left(\frac{j\omega'\varepsilon_0\chi'_y ctg\left(\chi'_y b + \phi'_y\right)}{\chi'^2_x + \chi'^2_y} - \frac{1}{Z_s}\right) = = \left(\frac{\mathrm{Im}(\gamma')\chi'_x}{\chi'^2_x + \chi'^2_y}\right)^2, \right. \tag{3.154}$$

$$\left\{\left(\frac{j\omega'\mu_0\chi'_x tg\left(\chi'_x a + \phi'_x\right)}{\chi'^2_x + \chi'^2_y} + Z_s\right)\left(\frac{j\omega'\varepsilon_0\chi'_x ctg\left(\chi'_x a + \phi'_x\right)}{\chi'^2_x + \chi'^2_y} - \frac{1}{Z_s}\right) = = \left(\frac{\mathrm{Im}(\gamma')\chi'_y}{\chi'^2_x + \chi'^2_y}\right)^2. \right. \tag{3.155}$$

where $\chi'^2 - \mathrm{Im}^2\gamma' = (\omega'/c)^2$ allows us to define the $\gamma'(\omega')$ dependence of equality.

It is important to use the Lorentz transformation when converting from a moving to a stationary computing system. At this time, the wave vector and frequency change as follows:

$$\chi = \chi', \omega = \frac{\omega' + v\gamma'}{\sqrt{1 - (v/c)^2}}. \tag{3.156}$$

The longitudinal wave number is determined using the following expression:

$$\gamma = \sqrt{\left(\frac{\mathrm{Re}(\omega)}{c}\right) - \chi^2}. \tag{3.157}$$

Figure 3.28 shows the frequency dependence of the extinction coefficient for H_{10} and E_{11} modes in a microwave rectangular waveguide operating in the frequency range of 4.9–7.05 GHz. Conductivity of the inner conducting surface of the waveguide $\sigma = 5.8 \cdot 10^8$Sim/m accepted as:

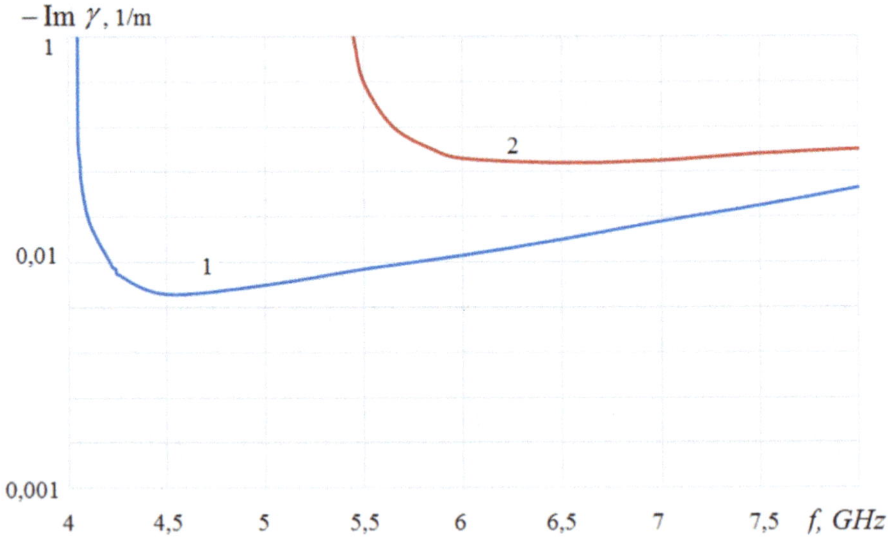

Fig. 3.28 Frequency dependence of the extinction coefficient for modes H_{10} (1) and E_{11}(2) in a microwave rectangular waveguide operating in the frequency range of 4.9–7.05 GHz:$\sigma = 5.8 \cdot 10^8$Sim/m

3.4 Conclusions

1. Taking into account the nonlinearity of the environment and the types of waves, new mathematical models of the electromagnetic field moving calculation system of the rectangular waveguide operating in the frequency range of 4.9–7.05 GHz and the circular waveguide operating at the frequency of 9 GHz from the microwave telecommunication devices were developed.
2. Proposing effective algorithms for solving new mathematical models of the electromagnetic field in the moving computing system of a rectangular waveguide operating in the frequency range of 4,9–7.05 GHz and a circular waveguide operating at a frequency of 9 GHz from microwave telecommunication devices, taking into account the nonlinearity of the environment, the types of waves has been done. This allowed to improve the electrical, magnetic, constructive and operating parameters and characteristics of microwave rectangular and circular waveguides.
3. For E-type and H-type waves, $3D$ models of distribution of frequencies of the electromagnetic field in elementary regions were obtained in the moving computer system of the rectangular waveguide operating in the frequency range of 4.9–7.05 GHz and the circular waveguide operating at the frequency of 9 GHz.
4. It is determined that: (*a*) the smaller the wavelength of the excited oscillations, the more types of waves can be excited in the microwave rectangular waveguide; (*b*) in a microwave rectangular waveguide, the H_{10} wave has the largest critical wavelength. This wave is called low type or main type wave. All other waves are waves with shorter crisis length. These waves are called high type waves; (*v*) in a microwave rectangular waveguide, the field is distributed differently, but waves with the same critical length may exist (E_{11}-H_{11}, E_{21}-H_{21}, etc.). These waves are called torn; (*q*) only one type of H_{10} wave can be excited in the region whose borders are limited by H_{10} and H_{20} critical length waves.
5. New mathematical models were developed in the moving coordinate system of the electromagnetic field of the rectangular waveguide telecommunication device operating in the frequency range of 4.9–7.05 GHz. The distribution of the electromagnetic field intensities of the microwave rectangular waveguide along the Z axis according to the elemental areas is shown.
6. The problem of normal incidence on the inner conductive surface of the waveguide in the moving calculation system of electromagnetic waves with mutual orthogonal polarization was posed and solved.
7. As a result of the solution of this problem, it was shown that the dependence of the surface impedances on the speed of the moving computing system has different characteristics for waves with orthogonal polarization.
8. The attenuation of electromagnetic waves propagating in microwave rectangular and circular waveguides with conductive inner layers was determined using the Lorentz transformation.

9. New mathematical models of the electromagnetic field of microwave rectangular and circular waveguides have been developed in the polar coordinate system.
10. The Ritz and Galerkin methods were applied to solve the developed mathematical models, and the algorithms of these methods were developed to solve the problem under consideration.
11. As a result of the solution of the developed mathematical models, the curves of the equipotential lines of the electromagnetic field of the microwave rectangular and circular waveguide were established.
12. As a result of the established curves, it is possible to control the distribution of the electromagnetic field of the microwave rectangular and circular waveguide.
13. The developed methodology is distinguished by its universality. So, this methodology can be applied to other telecommunications devices.

References

1. Rahman, B. M. A. (1984). Penalty function improvement of waveguide solution by finite elements. *IEEE Transactions on Microwave Theory and Techniques, 32*(8), 922–928.
2. Rahmatillah, R. (2014). *Numerical analysis for wave propagation in circular waveguide using cylindrical coordinate system-based FDTD method.* In International conference on advanced informatics: Concept, theory and application (ICAICTA) (pp. 209–213).
3. Railton, C. J., & Paul, D. L. (2010). Analysis of circular waveguide filter using enhanced FDTD. *International Journal of Numerical Modelling-Electronic Networks Devices and Fields, 15*(5–6), 535–547.
4. Rakhshani, M. R., & Rashki, M. (2022). Numerical simulations of metamaterial absorbers employing vanadium dioxide. *Plasmonics, 17*, 1107–1117.
5. Reimann, F., et al. (2018). Rayleigh-Ritz based expansion method for wakefields in dielectrically lined rectangular waveguides. *Journal of Computational Physics, 372*, 299–315.
6. Rong, Y., & Zaki, K. A. (2010). Characteristics of generalized rectangular and circular ridge waveguides. *IEEE Transactions on Microwave Theory and Techniques, 48*(2), 258–265.
7. San-Blas, A. A., et al. (2015). Flexible and efficient computer-aided design tool for advanced comb-line rectangular waveguide filters. *International Journal of RF and Microwave Computer-Aided Engineering, 25*(8), 696–708.
8. Scott, M. M., et al. (2016). Permittivity and permeability determination for high index specimens using partially filled shorted rectangular waveguides. *Microwave and Optical Technology Letters, 58*(6), 1298–1301.
9. Shafi, K. T. M., et al. (2017). *Applied microwave imaging of composite structures using open-ended circular waveguide.* In: IEEE international conference on imaging systems and techniques, pp 479–483
10. Shah, I. A., et al. (2019). Design and analysis of a compact-sized multiband spiral-shaped implantable antenna for scalp implantable and leadless pacemaker systems. *IEEE Transactions on Antennas and Propagation, 67*(6), 4230–4234.
11. Shao, X., et al. (2021). Frequency diverse array beampattern synthesis with modified sinusoidal frequency offset. *IEEE Antennas and Wireless Propagation Letters, 20*(9), 1780–1788.

Chapter 4
Modeling of Microwave Waveguide Systems of Complex Structure

4.1 Mathematical Model of the Surface Impedance of a Microwave Rectangular Waveguide with Irregular Walls

The interconnection of complex waveguide tracts by opening slits of different sizes is currently widely used in the microwave technique [1]. Since such devices have different waveguides and communication gaps, calculating their electrical characteristics is a rather complicated matter. Therefore, in addition to the relevance of constructing a mathematical model of the surface impedance parameters of microwave rectangular waveguides with irregular narrow walls, it is also of great practical importance.

For this, the impedance method, which is widely used in the problems of diffraction and radio wave propagation, as a mathematical apparatus [2] is more convenient to use.

An microwave rectangular waveguide with a slit on its narrow wall for communication is the device under study (Fig. 4.1). Circular waveguides whose diameter is equal to the diameter of this slit are placed in the opening slit. In an microwave rectangular waveguide, the main type of wave is the H_{10} wave.

Let's accept the following notations*:* a – the size of the wide wall of the waveguide; b – the size of the narrow wall of the waveguide; d – the distance along the y axis from the coordinate origin to the center of the crack; the distance along the *l-z* axis from the coordinate origin to the center of the slit; h is the period of the structure.

As can be seen from Fig. 4.1, the value of the surface impedance in the narrow wall of the waveguide is equal to zero in other places except the surface slits. Assume that the surface impedance coincides with the wave resistances of circular waveguides connected by slits.

I. Islamov, *Radio Engineering and Telecommunications Waveguide Systems in the Microwave Range*, https://doi.org/10.1007/978-3-031-37916-1_4

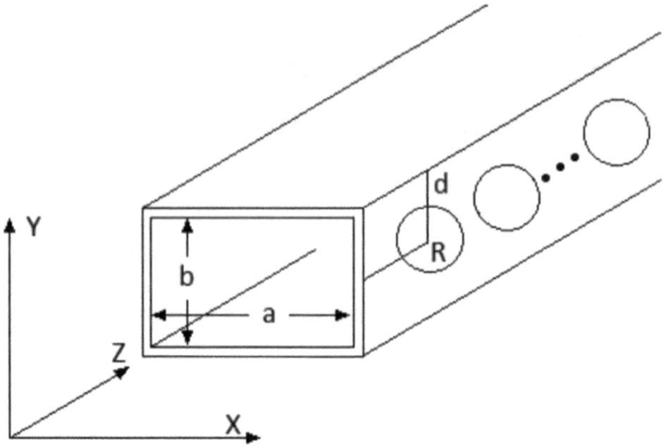

Fig. 4.1 Narrow wall irregular microwave rectangular waveguide

Let's divide the surface impedance in the *YOZ* plane into a double Fourier series [3]:

$$Z_s(y,z)=\sum_{m=0}^{\infty}\sum_{n=0}^{\infty}\begin{matrix}\omega_{mn}a_{mn}\cos\alpha_y y\cdot\cos\alpha_z z\\ +b_{mn}\sin\alpha_y y\cdot\cos\alpha_z z\\ +c_{mn}\cos\alpha_y y\cdot\sin\alpha_z z+d_{mn}\sin\alpha_y y\cdot\sin\alpha_z z,\end{matrix} \tag{4.1}$$

where ω_{mn} is determined by the following equations:

$$\omega_{mn}\begin{cases}\frac{1}{4},m=0,n=0\\ \frac{1}{2},m\geq 0,n\leq 0\\ 1,m\geq 0,n\geq 0\end{cases} \tag{4.2}$$

$\alpha_y=\frac{m\pi}{b},\alpha_z=\frac{n\pi}{h}$ are the transverse wave numbers along the *Y* and *Z* axes, respectively.

Given that the fundamental wave type in an microwave rectangular waveguide is an H_{10} wave, then the surface impedance will be a double function along the *Y* axis. In this case, the coefficients b_m and d_m in (4.1) are equal to zero. The unknown coefficients a_{mn}, c_{mn} in (4.1) are determined as follows.

$$\begin{aligned}a_{mn}&=\frac{4}{bh}\iint_s Z_s\cos(\alpha_y y)\cos(\alpha_z z)dydz,\\ c_{mn}&=\frac{4}{bh}\iint_s Z_s\cos(\alpha_y y)\sin(\alpha_z z)dydz.\end{aligned} \tag{4.3}$$

Between the rectangular coordinate system and the cylindrical coordinate system

$$z = l + \rho\cos\phi, \\ y = d + \rho\sin\phi, \tag{4.4}$$

Let's determine the mathematical expressions for the coefficients a_{mn}, c_{mn} in (4.3) using the relations First, let's look at the definition of the mathematical expression of the a_{mn} coefficient. Taking into account (4.4), we get for a_{mn}.

$$a_{mn} = \frac{4}{bh}\int_0^R\int_0^{2\pi} Z_s\cos\left[\alpha_y(d+\rho\sin\phi)\right]\cdot\times\cos[\alpha_z(l+\rho\cos\phi)]\rho d\rho d\phi. \tag{4.5}$$

By making certain transformations in (4.5), we get the following four integrals.

$$\begin{aligned} I_1 &= \frac{4Z_s}{bh}\int_0^R\int_0^{2\pi}\cos(\alpha_y d)\cos(\alpha_y\rho\sin\phi)\cdot\times\cos(\alpha_z l)\cos(\alpha_z\rho\cos\phi)\rho d\rho d\phi,\\ I_2 &= \frac{4Z_s}{bh}\int_0^R\int_0^{2\pi}\cos(\alpha_y d)\cos(\alpha_y\rho\sin\phi)\cdot\times\sin(\alpha_z l)\sin(\alpha_z\rho\cos\phi)\rho d\rho d\phi,\\ I_3 &= \frac{4Z_s}{bh}\int_0^R\int_0^{2\pi}\sin(\alpha_y d)\sin(\alpha_y\rho\sin\phi)\cdot\times\cos(\alpha_z l)\cos(\alpha_z\rho\cos\phi)\rho d\rho d\phi,\\ I_4 &= \frac{4Z_s}{bh}\int_0^R\int_0^{2\pi}\sin(\alpha_y d)\sin(\alpha_y\rho\sin\phi)\cdot\times\sin(\alpha_z l)\sin(\alpha_z\rho\cos\phi)\rho d\rho d\phi. \end{aligned} \tag{4.6}$$

Yokobi-Anger [4] let's use the formula

$$e^{iz\sin\phi} = \sum e^{in\phi}J_n(z), \\ e^{iz\cos\phi} = \sum i^n e^{in\phi}J_n(z). \tag{4.7}$$

Using (4.7), the sub-integral expression of the integral I_1 in (4.6) can be shown as follows:

$$I_1 = \frac{4Z_s}{bh}\cos(\alpha_y d)\cos(\alpha_z l)\int_0^R\int_0^{2\pi}\left[J_0(\alpha_y\rho)+2\sum_{p=1}J_{2p}(\alpha_y\rho)\cos 2\rho\phi\right] \\ \cdot\left[J_0(\alpha_z\rho)+\sum(-1)^p J_{2p}(\alpha_z\rho)\cos 2\rho\phi\right]\rho d\rho d\phi. \tag{4.8}$$

The following functions in (4.8) are orthogonal on the interval $0 < \phi < 2\pi$.

$\cos 2\rho\phi$, $\cos(2\rho - 1)\phi$, $\sin 2\rho\phi$, $\sin(2\rho - 1)\phi$.

Considering their orthogonality, let's express the integral I_1 as follows:

$$I_1 = \frac{4Z_s}{bh}\cos(\alpha_y d)\cos(\alpha_z l)\int_0^R \left[\frac{1}{2}J_0(\alpha_y\rho)\cdot J_0(\alpha_z\rho) - J_2(\alpha_y\rho)J_2(\alpha_z\rho) + J_4(\alpha_y\rho)J_4(\alpha_z\rho) - \ldots\right]\rho d\phi. \tag{4.9}$$

Each limit included in (4.9) can be expressed as follows

$$\int \rho J_p(i\rho)J_p(j\rho)d\rho = \frac{j\rho J_p(i\rho)J_{p-1}(j\rho) - i\rho J_{p-1}(i\rho)J_p(j\rho)}{i^2 - j^2}, \tag{4.10}$$

where p is the index corresponding to the degree of the Bessel function in the subintegral limits of the expression (4.9).

If we carry out the above analogous transformations with the integrals I_2, I_3, I_4, it can be shown that they are equal to zero.

Thus, the coefficient amn in (4.3) coincides with the expression of the integral I_1 in (4.9), that is, $a_{mn} = I_1$.

For the c_{mn} coefficient, we get

$$c_{mn} = \frac{4Z_s}{bh}\int_0^R\int_0^{2\pi} \cos\left[\alpha_y(d + \rho\sin\phi)\right] \times \sin[\alpha_z(l + \rho\cos\phi)]\rho d\rho d\phi. \tag{4.11}$$

If we carry out the analogous conversion performed in determining the a_{mn} coefficient, we will get the following mathematical expression for c_{mn}:

$$c_{mn} = \frac{4Z_s}{bh}\cos(\alpha_y d)\sin(\alpha_z l)\int_0^R \left[\frac{1}{2}J_0(\alpha_y\rho)J_0(\alpha_z\rho) - J_2(\alpha_y\rho)J_2(\alpha_z\rho) + J_4(\alpha_y\rho)J_4(\alpha_z\rho) - \ldots\right]\rho d\rho d\phi. \tag{4.12}$$

Let the expression for the surface impedance be as follows:

$$Z_s(y,z) = \sum_{p,r}\alpha_{pr}Z_{pr}\cos\left(\frac{p\pi y}{b}\right)\cos\left[\frac{r\pi z}{h} + \phi_r\right], \tag{4.13}$$

where is $Z_{pr} = \sqrt{a_{pr}^2 + c_{pr}^2}$; $\phi_r = \frac{r\pi l}{h}$; p, r are integer indices corresponding to the interval of change of m and n indices.

If we substitute the values of coefficients a_{pr} and C_{pr} in the expression obtained for Z_{pr}, we get

$$Z_{pr} = \frac{4Z_s}{bh} \cos\left(\frac{p\pi}{b} d\right) \sum_{p=0}^{\infty} S_p, \tag{4.14}$$

where S_p are the values of the integrals in (4.10).

4.2 Modeling of the Electromagnetic Field in a Microwave Rectangular Waveguide to Circular Waveguide Converter

Currently, devices with an microwave range are widely used in production. Therefore, improving the electrical, magnetic, structural, technical and operational parameters and characteristics of these devices and designing new devices with more optimal constructive dimensions is very important and relevant both from the scientific point of view and from the point of view of production. For this reason, let's look at the mathematical modeling of the electromagnetic field of microwave devices. Figure 4.2 shows the transition (*a*) and spatial view (*b*) from an microwave rectangular waveguide to a circular waveguide.

This section has two main objectives:

1. Analyzing and comparing existing analytical and numerical methods for mathematical modeling of microwave devices, developing a more effective algorithm.
2. To develop operationally more sophisticated microwave range devices by applying the developed algorithms.

If we denote by the area where the electromagnetic field spreads in the microwave waveguide, then the equation expressing the field can be shown in integral form as follows [5]:

$$J(E + \alpha\eta) \geq J(E).$$

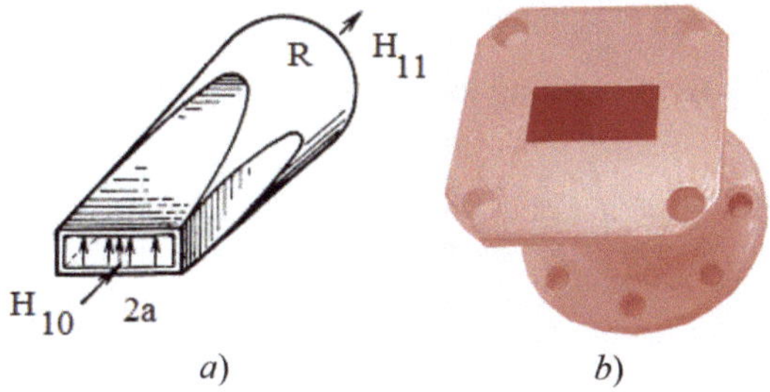

Fig. 4.2 Microwave rectangular to circular waveguide transition (**a**) and spatial view (**b**)

$$J=\int_S\left[\left(\frac{\partial E}{\partial x}\right)^2+\left(\frac{\partial E}{\partial y}\right)^2+2F(x,y)E\right]ds, \tag{4.15}$$

where $F(x,y)$ is the given function.

Since the value of the integral (4.15) depends on the function, it can be called functional. To clarify the physical nature of the problem, let's assume that the function E is the minimum of the functional (4.15). So let's look at the following function:

$$E(x,y)+\alpha\eta(x,y),$$

where α is a very small limit, $\eta(x,y)$ is a continuous function, and its first two differentials approach zero in the domain of S.

Then within the condition of $\alpha \neq 0$.

$$J(E+\alpha\eta)\geq J(E).$$

It can be seen from here that the functional (4.15) takes a minimum value when it is $\alpha = 0$, i.e.

$$\left.\frac{dJ(E+\alpha\eta)}{d\alpha}\right|_{\alpha=0}=0.$$

Let's calculate this differential:

$$\frac{aJ(E+\alpha\eta)}{d\alpha}=\frac{d}{d\alpha}\int_S\left[\left(\frac{\partial E}{\partial x}+\alpha\frac{\partial\eta}{\partial x}\right)^2+\left(\frac{\partial E}{\partial y}+\alpha\frac{\partial\eta}{\partial y}\right)^2+2F(x,y)(E+\alpha\eta)\right]ds.$$

Now assuming that it is $\alpha = 0$, we define:

$$\int_S\left[\frac{\partial E}{\partial x}\frac{\partial\eta}{\partial x}+\frac{\partial E}{\partial y}\frac{\partial\eta}{\partial y}+F\eta\right]ds=0. \tag{4.16}$$

By converting the obtained integral (4.16), we can write it as follows:

$$\int_S\left[\frac{\partial}{\partial x}\left(\eta\frac{\partial E}{\partial x}\right)+\frac{\partial}{\partial y}\left(\eta\frac{\partial E}{\partial y}\right)\right]ds-\int_s\eta\left[\frac{\partial^2 E}{\partial x^2}+\frac{\partial^2 E}{\partial y^2}-F\right]ds=0,$$

və ya

$$\int_S div(\eta \nabla E)ds - \int_S \eta(\Delta E - F)ds = 0. \tag{4.17}$$

According to Ostrogradsky's theorem, we can write:

$$\int_S div(\eta \nabla E)ds = \int_S \eta \frac{\partial E}{\partial n} dl,$$

where C is the contour bounded by the S surface, normal to the contour $\vec{n} - C$.

It should be noted that since the parameter η approaches zero along the C contour, we get

$$\int_S div(\eta \nabla E)ds = 0,$$

$$\int_S \eta(\nabla E - F)ds = 0,$$

This derived expression is true regardless of the function η under the following condition:

$$\Delta E = \frac{\partial^2 E}{\partial x^2} + \frac{\partial^2 E}{\partial y^2} = F(x, y). \tag{4.18}$$

Thus, from the above, it can be concluded that the minimum of the integral (4.15) is equivalent to the solution of the Poisson Eq. (4.18). It should be noted that the minimum of the integral (4.15) is a necessary condition that it is equivalent to the solution of Eq. 4.18.

In a special case, that is, when $F(x, y) = 0$, the integral (4.15) takes the form of the Dirichlet integral. Therefore, the solution of the Laplace equation is equivalent to the determination of the minimum function of the Dirichlet integral. In the general case, the determination of the minimum function of the integral of the type (4.15) is performed by applying the Ritz and Galerkin methods. Therefore, let's apply these methods to model the electromagnetic field of an microwave rectangular waveguide to circular waveguide converter.

Let's assume that $E(x, y)$ is the exact solution of the given problem and the minimum value of functional (4.15) is $J(U) = m$.If we can construct a function $\vec{E}(x, y)$ that satisfies the boundary conditions for the function E, then the constructed function E will be close to the true solution of E. On the other hand, if we have defined the sequence of the E_n function, that is, $J\left(\bar{E_n}\right)_{n\to\infty} \to m$, then this sequence will accumulate until the solution of E.

Thus, according to the Ritz method, the function $\_E_n$ in the considered problem will depend on the n parameters as follows:

$$\overline{E}_n = F(x, y, a_1, a_2, \ldots, a_n).$$

If we put the function $\bar{E_n}$ instead of the function E in the functional (4.15) and perform the integration operation, then the obtained result will be a function of the parameters $a_k(k = 1, 2, \ldots, n)$.So that,

$$J\left(\bar{E_n}\right) = J(a_1, a_2, \ldots, a_n).$$

Since the problem is to determine the minimum of the functional J, then the parameters a_k must satisfy the following system of equations:

$$\frac{\partial J}{\partial a_k} = 0 \quad (k = 1, 2, \ldots, n). \tag{4.19}$$

It should be noted that the system of $\_E_n$ functions is complete under the following condition:

$$\lim_{n \to \infty} J\left(\bar{E_n}\right) = J(E) = m.$$

According to the Ritz method, the problem of minimization should be sought in the form of linear functions depending on the parameters a_k:

$$\overline{E}_n = \sum_{k=1}^{n} \alpha_l \gamma_k(x, y). \tag{4.20}$$

It should be noted that the E parameter fulfills the homogeneous boundary conditions on the C contour. Otherwise, it is necessary to find any function $E_1(x, y)$ which also satisfies the following condition:

$$E_1(x, y) = E(x, y).$$

Therefore, it is necessary to include the new function $11V(x, y)$.This function, in turn, must fulfill the following condition:

$$E_1(x, y) = V(x, y) + E_1(x, y).$$

So that,

$$\Delta V = F - \Delta E_1.$$

If C is $\omega(x, y) = 0$ along the contour, then ω_k can be written as a system of functions:

$$\gamma_1 = \omega, \gamma_2 = \omega x, \gamma_3 = \omega y, \gamma_4 = \omega x^2, \omega_5 = \omega xy, \gamma_6 = \omega y^2.$$

If C is $F(x, y) = 0$ along the contour, then we can choose the function $\omega(x, y)$ as follows:

$$\omega(x, y) = \pm F(x, y).$$

In this case (4.18), the solution of the Poisson equation will be as follows:

$$\begin{aligned} J(\overline{E}_n) &= \int\limits_S \left[\left(\frac{\partial \overline{E}_n}{\partial x} \right)^2 + \left(\frac{\partial \overline{E}_n}{\partial y} \right)^2 + 2F\overline{E}_n \right] ds \\ &= \int\limits_S \left[\left(\sum_{k=1}^n a_k \frac{\partial \phi_k}{\partial x} \right)^2 + \left(\sum_{k=1}^n a_k \frac{\partial \phi_k}{\partial \gamma} \right)^2 + 2F \sum_{k=1}^n a_k \phi_k \right] ds \qquad (4.21) \\ &= \sum_{S=1}^n \sum_{k=1}^n \alpha_{ks} a_k a_s + 2 \sum_{k=1}^n \beta_k a_k, \end{aligned}$$

where

$$\alpha_{ks} = \alpha_{sk} = \int\limits_s \left(\frac{\partial \varphi_k}{\partial x} \frac{\partial \varphi_s}{\partial x} + \frac{\partial \varphi_k}{\partial y} \frac{\partial \varphi_s}{\partial y} \right) ds,$$

$$\beta_k = \int\limits_s F \varphi_k ds. \qquad (4.22)$$

The system of equations defining the parameters a_k can be written as follows:

$$\frac{1}{2} \frac{\partial J\left(\bar{E}_n\right)}{\partial a_k} = \sum_{S=1}^n \alpha_{ks} a_s + \beta_k = 0 \quad (k = 1, 2, \ldots .., n), \qquad (4.23)$$

or

$$\frac{1}{2} \frac{\partial J\left(\bar{E}_n\right)}{\partial a_k} = \int\limits_S \left(\frac{\partial\_E_n}{\partial x} \frac{\partial \varphi_k}{\partial x} + \frac{\partial\_E_n}{\partial y} \frac{\partial \varphi_k}{\partial y} + F \varphi_k \right) ds = 0 \quad (k = 1, 2, \ldots \ldots n). \qquad (4.24)$$

Let's denote the solution of the system (4.23) or (4.24) by a_k^n. Then, according to the Ritz method:

$$\underset{n}{\bar{E}}(x,y) = \sum_{k=1}^{n} a_k^n \, \varphi_k(x,y). \tag{4.25}$$

According to Galerkin's method, the solution of the given problem is sought in the form of linear functions depending on a_k. However, in this case, the equations for their determination have a different form. In this case, if we make transformations in the Eq. 4.24, we get:

$$\begin{aligned}
&\int_S \left[\left(\frac{\partial \overline{E}_n}{\partial x}\frac{\partial \phi_n}{\partial x}\right) + \left(\frac{\partial \overline{E}_n}{\partial y}\frac{\partial \phi_n}{\partial y}\right) + F\phi_k\right] ds - \\
&= \int_S \left[\frac{\partial}{\partial x}\left(\phi_k \frac{\partial \overline{E}_n}{\partial x}\right) + \frac{\partial}{\partial y}\left(\phi_k \frac{\partial \overline{E}_n}{\partial y}\right)\right] ds - \\
&- \int_S \left[\left(\frac{\partial^2 \overline{E}_n}{\partial x^2}\right) + \left(\frac{\partial^2 \overline{E}_n}{\partial y^2}\right) - F\right] \phi_k ds = \\
&= \int_C \varphi_k \frac{\partial _E_n}{\partial \upsilon} dl - \int_S \left(\frac{\partial^2 _E_n}{\partial x^2} + \frac{\partial^2 _E_n}{\partial y^2} - F\right) \varphi_k ds = 0,
\end{aligned} \tag{4.26}$$

where v is the external normal to the contour C.

If the sought solution satisfies the homogeneous boundary conditions, then the function γ_k along the contour C will be equal to zero. In this case, we get:

$$\int_S \left(\frac{\partial^2 \bar{E}_n}{\partial x^2} + \frac{\partial^2 \bar{E}_n}{\partial y^2} - F\right) \varphi_k ds = 0 \ \ (k = 1, 2, \ldots, n). \tag{4.27}$$

If the considered equation is of the form $L(E) = 0$, then according to the Galerkin method, the coefficients a_k in the n-th approximation

$$\bar{E}_n\,(x,y) = \sum_{k=1}^{n} a_k^n \varphi_k(x,y),$$

$$\int_S L\left(\bar{E}_n\right) \varphi_k ds = 0 \ \ (k = 1, 2, \ldots, n). \tag{4.28}$$

Now let's solve the developed mathematical models for a specific device. As shown in Fig. 4.2, the device is a circular and H_{11} type waveguide on one side and an microwave rectangular waveguide on the other side. The radius of a circular waveguide is R, and the side of a rectangular waveguide is $2a$ (Fig. 4.2). In an microwave circular waveguide, the electric field intensity is E_0. The problem should be solved by integrating equation $\Delta E = 0$. First, let's make the boundary conditions homogeneous. Because of this

$$E = V + E_1$$

Let's accept. At the same time, let's choose the function E_1 so that it approaches E_0 in the circular part of the waveguide and zero in the rectangular part. Let's look at the issue in the (ρ, γ) polar system. In this case, we can write the function E_1 as:

$$E_1 = E_0 \frac{\rho - F(\gamma)}{R - F(\gamma)},$$

where

$$F(\phi) = \begin{cases} \dfrac{a}{\cos\phi}, & -\dfrac{\pi}{4} \leq \phi \dfrac{\pi}{4}, \\ -\dfrac{a}{\cos\phi}, & \dfrac{3\pi}{4} \leq \phi \leq \dfrac{5\pi}{4}, \\ \dfrac{a}{\sin\phi}, & \dfrac{\pi}{4} \leq \phi \leq \dfrac{3\pi}{4}, \\ -\dfrac{a}{\sin\phi}, & \dfrac{5\pi}{4} \leq \phi \leq \dfrac{7\pi}{4}. \end{cases}$$

At this time, let's solve the following equation for the function V under homogeneous boundary conditions:

$$\Delta V = -\Delta E_1.$$

Considering that

$$V_n = \sum_{k=1}^{n} a_k \varphi_k(\rho, \varphi),$$

then the solution of the equation considered in the n-th approximation is in the following form:

$$E = E_1 + V_n.$$

In the expression (4.27), we can basically write for the determination of coefficients a_k:

$$\int_S \sum_{k=1}^{n} a_k \varphi_m \Delta\varphi_k ds + \int_S \varphi_m \Delta E_1 ds = 0 \quad (m = 1, 2, \ldots, n),$$

or

$$\sum_{k=1}^{n} a_k \int_0^{\pi/4} d\phi \int_R^{\frac{a}{\cos\phi}} \phi_m \Delta\phi_k \rho d\rho + \int_0^{\pi/4} d\phi \int_R^{\frac{a}{\cos\phi}} \phi_m \Delta E_1 \rho d\rho = 0.$$

We can write the following functions within the symmetry condition of the field as ϕ_k functions that approach zero in the considered computational domain:

$$\begin{cases} \varphi_1(\rho,\varphi) = (\rho - R)\ [\rho - F(\varphi)], \\ \varphi_2(\rho,\varphi) = \varphi_1(\rho,\varphi)\rho^2 \cos^2\varphi, \\ \varphi_3(\rho,\varphi) = \varphi_1(\rho,\varphi)\rho^2 \sin^2\varphi, \\ \varphi_4(\rho,\varphi) = \varphi_1(\rho,\varphi)\rho^4 \cos^2\varphi \sin^2\varphi. \\ \ldots\ldots\ldots\ldots\ldots\ldots\ldots\ldots \end{cases}$$

Let's look at the solution of function ϕ_1 for the first approximation in concrete case. In this case, for the calculation of the coefficient a_1, we get:

$$a_1 \int_0^{\pi/4} d\phi \int_R^{a/\cos\phi} \phi_1 \Delta\phi_1 \rho d\rho + \int_0^{\pi/4} d\phi \int_R^{a/\cos\phi} \phi_1 \Delta E_1 \rho d\rho = 0.$$

Considering that the Laplacian in the polar coordinate system has the following form:

$$\Delta\psi = \frac{\partial}{\partial\rho}\left(\rho \frac{\partial\psi}{\partial\rho}\right) + \frac{1}{\rho}\frac{\partial^2\psi}{\partial\varphi^2},$$

Then we get for the coefficient a_1:

$$a_1 = -\frac{\alpha}{a^2},$$

where

$$\alpha = \frac{p(1,11 - 1,42p + 0,524p^2)}{0,518 - 3,73p - 6,36p^2 - 3,11p^3 - 0,514p^4 + 0,734p^5},$$

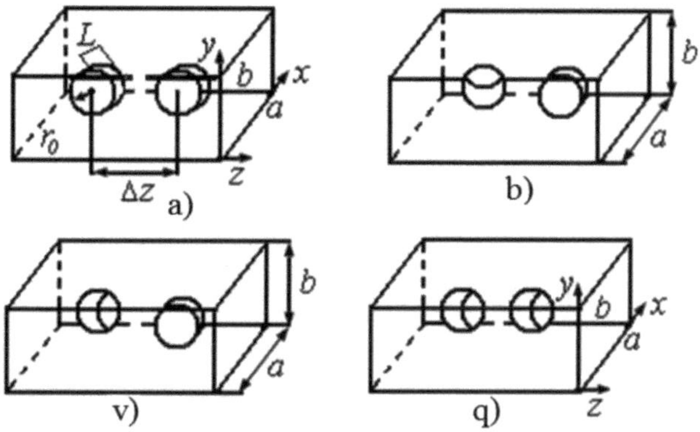

Fig. 4.3 Field distribution in an microwave square waveguide to circular waveguide converter

$$p = \frac{R}{a}.$$

In the first approximation, we get for the intensity:

$$\frac{E_1}{E_0} = \frac{\frac{\rho}{a} - \frac{1}{\cos\phi}}{p - \frac{1}{\cos\phi}} + \alpha\left(\frac{\rho}{a} - \frac{1}{\cos\phi}\right)\left(\frac{\rho}{a} - p\right), -\left(\frac{\pi}{4} \le \phi \le \frac{\pi}{4}\right)$$

The dashed curve in square$0 \le \phi \le \pi/2$ in Fig. 4.3 corresponds to the part of the equipotential line. The calculation was made when it was $p = \frac{R}{a} = 0,5$. It seems clear that the first approximation is unsatisfactory for use. Therefore, the following are taken for further approximation.

$$E = E_1 + a_1\phi_1(\rho, \phi) + a_2\phi_2(\rho, \phi), \tag{4.29}$$

where ϕ_1 remains as before, it becomes $\phi_2 = \phi_1 \cos 2\phi$.

Such a solution of the problem was chosen to simplify the calculations, and at this time the functions ϕ_2 and $\phi_3$were combined in the Eq. 4.29. At the same time,ρ^2 shots were fired. So, if we had kept the multiplication ρ^2, then it would be necessary to calculate 12 intervals. This would make it difficult to solve the problem. To determine the coefficients a_1 and a_2, we use the following system of equations:

$$\begin{cases} a_1 \int\limits_S \varphi_1 \Delta\varphi_1 ds + a_2 \int\limits_S \varphi_1 \Delta\varphi_2 ds = -\int\limits_S \varphi_1 \Delta E_1 ds, \\ a_1 \int\limits_S \varphi_2 \Delta\varphi_1 ds + a_2 \int\limits_S \varphi_2 \Delta\varphi_2 ds = -\int\limits_S \varphi_2 \Delta E_1 ds. \end{cases}$$

If we calculate these integrals analogously to the above, we get:

$$a_1\psi_{11}(p) + a_2\psi_{12}(p) = \psi_{13}(p),$$
$$a_1\psi_{21}(p) + a_2\psi_{22}(p) = \psi_{23}(p),$$

where $p = \frac{R}{a}$ and

$$\psi_{11}(p) = a^2\left(0,52 - 3,73p + 6,36p^2 - 3,11p^3 - 0,51p^4 + 0,73p^5\right),$$
$$\psi_{12}(p) = -a^2\left(2,93 - 9,52p + 12,08p^2 - 9,39p^3 + 5,08p^4 - p^5\right),$$
$$\psi_{13}(p) = -p\left(1,11 - 1,43p + 0,52p^2\right),$$
$$\psi_{21}(p) = a^2\left(0,08 - 1,17p + 1,43p^2 - 1,27p^3 - 1,32p^4 + 0,47p^5\right),$$
$$\psi_{22}(p) = -a^2\left(1,09 - 3,94p + 5,07p^2 - 5,26p^3 + 1,89p^4 - 0,78p^5\right),$$
$$\psi_{23}(p) = -p\left(0,51 - 0,51p + 0,43p^2\right).$$

Assuming it is $p = 0,5$,we get:

$$\psi_{11} = -0,15a^2, \psi_{12} = -0,31a^2, \psi_{13} = -0,26,$$
$$\psi_{21} = -0,38a^2, \psi_{22} = 0,17a^2, \psi_{23} = -0,18,$$
$$a_1 = \frac{0,708}{a^2}, a_2 = \frac{0,508}{a^2}.$$

Beləliklə, axtarılan həll aşağıdakı formada olar:

$$\frac{E}{E_0} = \frac{\rho - F}{R - F} + (0,708 + 0,508\cos 2\phi)\frac{(\rho - F)(\rho - R)}{a^2}.$$

The equipotential lines of the field defined by this formula are shown by curves given by solid lines in Fig. 4.3.

4.3 Determination of the Interaction Coefficient of Circular Waveguides Placed in an Air-Filled Microwave Rectangular Waveguide

Air-filled microwave rectangular and circular waveguides, including their interconnections, are widely used in microwave telecommunications. When designing such systems, it is necessary to know the coefficient of interaction in the transition device from a rectangular waveguide to a circular waveguide. Assume that an air-filled microwave waveguide has a circular waveguide as shown in Fig. 4.4. In such

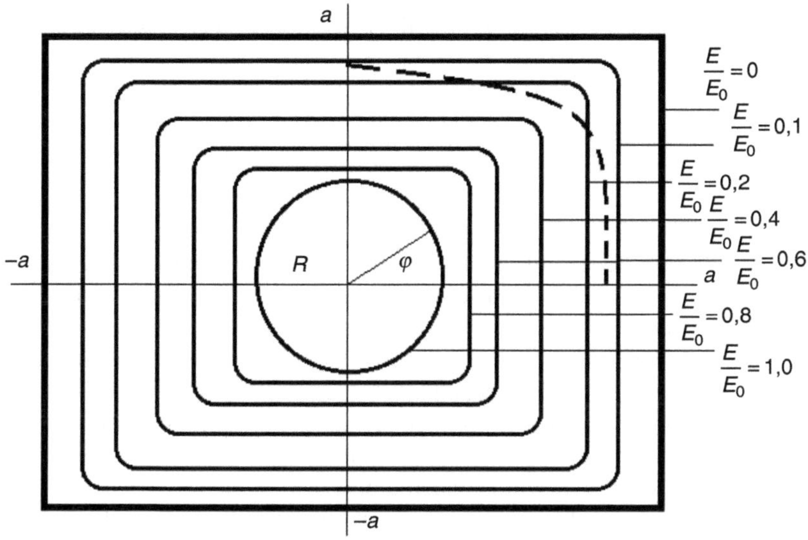

Fig. 4.4 Micowave circular waveguides placed on a rectangular waveguide and having different axis directions

systems, the propagation of electromagnetic waves and the determination of the interaction coefficient have not been considered. So let's look at this issue.

Let's assume that the circular waveguides shown in Fig. 4.4 propagate independent waves of ω_0 frequencies. Let's define the correlation coefficient as follows:

$$\rho = \frac{\omega^+ - \omega^-}{\omega_0}, \tag{4.30}$$

where ω^+, ω^- are the mutual wave frequencies of two circular waveguides.

The expression (4.30) can be written as follows:

$$\rho = \sqrt{\rho_{1,2}\rho_{2,1}}, \tag{4.31}$$

$$\rho_{1,2} = \frac{-1}{\omega_0 \chi_2}\left[\sum_{i=1}^{\infty} p_i^{1+}\left(p_i^{2-}\right)^* e^{-k\Delta z} - \infty \sum q_i^{1+}\left(q_i^{2-}\right)^* e^{-k\Delta z}\right], \tag{4.32}$$

$$\rho_{2,1} = \frac{-1}{\omega_0 \chi_1}\left[\sum_{i=1}^{\infty} p_i^{2-}\left(p_i^{1+}\right)^* e^{-k\Delta z} - \infty \sum q_i^{2-}\left(q_i^{1+}\right)^* e^{-k\Delta z}\right], \tag{4.33}$$

where Z_1, Z_2 – longitudinal coordinates of circular waveguides in a rectangular waveguide $\Delta z = |z_2 - z_1|$; k – longitudinal wave number; $p_i^{j\pm}$; $q_i^{j\pm}$ – external electric and magnetic fields ($\vec{E}_m^{i\pm}$, $\vec{H}_m^{i\pm}$) of the *j*-th ($j = 1, 2$) circular waveguide [6] are the coefficients.

$$\begin{pmatrix} p_i^{j\pm} \\ q_i^{j\pm} \end{pmatrix} = (\pm)i/2 \int_{j_c} \left\{ \left[\vec{e}_{m}^{\,j}, \vec{n} \right] \left(\vec{H}_m^{i\mp} \right)^* + \left[\vec{n}, \vec{h}_m^{\,i} \right] \left(\vec{E}_m^{i\mp} \right)^* \right\} ds, \tag{4.34}$$

where ($\vec{e}_{m}^{\,i}$, $\vec{h}_{m}^{\,i}$) is the area of specific oscillation of the *j*-th circular waveguides in the rectangular waveguide; is the normal directed to the surface of the $n - j$-th circular waveguides.

According to expressions (4.31) to (4.34), let's calculate the interaction coefficient of circular waveguides in a rectangular waveguide. At this time, let's look at the excitation of uniform magnetic waves in circular waveguides. Assume that two circular waveguides have a radius of r_0 and a height of L. Let the relative dielectric constant of the circular waveguide material be ε_{1r}.

If the axis of the circular waveguide is orthogonal to the axis of the rectangular waveguide, parallel to its larger side (Fig. 4.4, *a*), then the position of the circular waveguides located in the rectangular waveguide is as consider the position of the circular waveguides as *b*, if the axis of the circular waveguide is parallel to the axis of the rectangular waveguide, then consider the position of the circular waveguides located on the rectangular waveguide as *v* (Fig. 4.4, *q*). Thus, the state of the system of circular waveguides located on a rectangular waveguide can be characterized as four independent geometric states: *aa* (Fig. 4.4, *a*); *ab* (Fig. 4.4, *b*); *av* (Fig. 4.4, *v*); and *vv* (Fig. 4.4, *q*).

Assume that coherent waves are excited in circular waveguides. At the same time, let's assume that the relative dielectric constant of the material of circular waveguides is very large ($\varepsilon_{1r} \geq 1$). Then, using expressions (4.31) to (4.34), we get the following expressions for wave $H_{n0l}^{\pm}$:.

In the case of *aa* (Fig. 4.4, *a*):

$$\rho = \frac{128\pi k_0 \beta_z \left(\rho_1^2 - \rho_0^2\right)^2}{\beta^2 abs_{n0l}^H (p_x, p_z)} \sum_{s,u=(0)}^{\infty} \left\{ \begin{array}{l} \left[\dfrac{|\gamma| \chi_{sx}^2}{1 + \delta_{s0} + \delta_{u0}} - \dfrac{\chi_{uy}^2}{|\gamma|} (1 - \delta_{s0} - \delta_{u0}) \right] \times \\ \times \dfrac{e^{-|\gamma| k_0 \Delta z}}{\left[\chi_{u_{sx}} \left(\beta^2 + u_{sx}^2\right) \left(\beta_z^2 - \chi_{sx}^2\right) \right]^2} \left[\sin \chi_{sx} x_1 \sin \chi_{sx} x_2 \cos \chi_{sx} x_1 \cos \chi_{sx} x_2 \right] \end{array} \right\}$$
$$\times \left\{ \begin{array}{l} \left\{ \cos \chi_{uy} y_1 \cos \chi_{uy} y_2 \right\} \left\{ \beta J_0(p_x) I_1(u_{sx} r_0) - u_{sx} J_0(p_x) I_0(u_{sx} r_0) \right\}^2 \times \\ \times \begin{bmatrix} \beta_z \sin p_z \cos \chi_{sx} \dfrac{L}{2} - \chi_{sx} \cos p_z \sin \chi_{sx} \dfrac{L}{2} \\ \beta_z \cos p_z \sin \chi_{sx} \dfrac{L}{2} - \chi_{sx} \sin p_z \cos \chi_{sx} \dfrac{L}{2} \end{bmatrix}^2 \end{array} \right\}, \tag{4.35}$$

In the case of *ab* (Fig. 4.4, *b*):

$$\rho=\frac{128\pi k_0\beta_z\left(\rho_1^2-\rho_0^2\right)^2}{\beta^2 abs_{n0l}^H(p_x,p_z)}\sum_{s,u=1}^{\infty}\left\{\begin{array}{l}\frac{\chi_{sx}\chi_{uy}}{\chi^2}\frac{|\gamma|\left(1-1/|\gamma|^2\right)e^{-|\gamma|k_0\Delta z}}{u_{sx}u_{uy}\left(\beta^2+u_{sx}^2\right)\left(\beta^2+u_{uy}^2\right)\left(\beta_z^2-\chi_{sx}^2\right)\left(\beta_z^2-\chi_{uy}^2\right)}\\ \times\{\beta J_0(p_x)I_1(u_{sx}r_0)-u_{sx}J_1(p_x)I_0(u_{sx}r_0)\}\end{array}\right\}$$

$$\left\{\begin{array}{l}\left\{\beta J_0(p_x)I_1\left(u_{uy}r_0\right)-u_{uy}J_1(p_x)I_0\left(u_{uy}r_0\right)\right\}\times\left[\sin\chi_{sx}x_1\left(\beta_z\sin p_z\cos\chi_{sx}\frac{L}{2}-\chi_{sx}\cos p_z\sin\chi_{sx}\frac{L}{2}\right)\right.\\ \left.\times\cos\chi_{sx}x_1\left(\beta_z\cos p_z\sin\chi_{sx}\frac{L}{2}-\chi_{sx}\sin p_z\cos\chi_{sx}\frac{L}{2}\right)\right][\cos\chi_{sx}x_2 0][\cos\chi_{uy}y_1 0]\end{array}\right\}$$

$$\left\{\begin{array}{l}\left\{\beta J_0(p_x)I_1\left(u_{uy}r_0\right)-u_{uy}J_1(p_x)I_0\left(u_{uy}r_0\right)\right\}\times[\sin\chi_{sx}x_1\left(\beta_z\sin p_z\cos\chi_{sx}\frac{L}{2}-\chi_{sx}\cos p_z\sin\chi_{sx}\frac{L}{2}\right)\\ \left.\times\cos\chi_{sx}x_1\left(\beta_z\cos p_z\sin\chi_{sx}\frac{L}{2}-\chi_{sx}\sin p_z\cos\chi_{sx}\frac{L}{2}\right)\right][\cos\chi_{sx}x_2 0][\cos\chi_{uy}y_1 0]\end{array}\right\}$$

$$\times\left\{\left[\sin\chi_{uy}y_2\left(\beta_z\sin p_z\cos\chi_{sx}\frac{L}{2}-\chi_{sx}\cos p_z\sin\chi_{sx}\frac{L}{2}\right)\cos\chi_{uy}y_2\left(\beta_z\cos p_z\sin\chi_{sx}\frac{L}{2}-\chi_{sx}\sin p_z\cos\chi_{sx}\frac{L}{2}\right)\right]\right\} \tag{4.36}$$

In the case of *av* (Fig. 4.4, *v*):

$$\rho=\frac{128\pi\beta_z\left(\rho_1^2-\rho_0^2\right)^2}{\beta p_x abs_{n0l}^H(p_x,p_z)}\sum_{s,u=(0)}^{\infty}\frac{1}{(1+\delta_{s0}+\delta_{u0})\chi_{u_{sx}}}\frac{\chi_{sx}e^{-|\gamma|k_0\Delta z}}{\left(\beta^2+u_{sx}^2\right)\left(\beta^2-\chi^2\right)\left(\beta_z^2-\chi_{sx}^2\right)\left(\beta_z^2+Q^2\right)}$$
$$\times\begin{bmatrix}\beta_z\sin p_z chQ\frac{L}{2}+Q\cos p_z shQ\frac{L}{2}\\ Q\sin p_z chQ\frac{L}{2}-\beta_z\cos p_z shQ\frac{L}{2}\end{bmatrix}$$
$$\times\{\beta J_0(p_x)I_1(u_{sx}r_0)-u_{sx}J_1(p_x)I_0(u_{sx}r_0)\}\{p_xJ_0(p_x)J_1(\chi r_0)-\chi r_0J_1(p_x)J_0(\chi r_0)\}$$
$$\times\left[\sin\chi_{sx}x_1\left(\beta_z\sin p_z\cos\chi_{sx}\frac{L}{2}-\chi_{sx}\cos p_z\sin\chi_{sx}\frac{L}{2}\right)\cos\chi_{sx}x_1\left(\beta_z\cos p_z\sin\chi_{sx}\frac{L}{2}-\chi_{sx}\sin p_z\cos\chi_{sx}\frac{L}{2}\right)\right] \tag{4.37}$$

In the case of *vv* (Fig. 4.4, *q*):

$$\rho=\frac{128\pi\beta_z\left(\rho_1^2-\rho_0^2\right)^2}{k_0p_x^2abs_{n0l}^H(p_x,p_z)}\times\sum_{s,u=(0)}^{\infty}\frac{e^{-|\gamma|k_0\Delta z}}{(1+\delta_{s0}+\delta_{u0})|\gamma|\left[\left(\beta-\chi^2\right)\left(\beta_z^2+Q^2\right)\right]^2}$$
$$\times\cos\chi_{sx}x_1\cos\chi_{sx}x_2\cos\chi_{uy}y_1\cos\chi_{uy}y_2 \tag{4.38}$$

where

$$s_{n0l}^H(p_x,p_z)=\left\{\left(\frac{\rho_1}{\beta}\right)^2\left[J_1^2(p_x)-J_0(p_x)J_2(p_x)\right]+J_0^2(p_x)+J_1^2(p_x)\right\}$$
$$\times\begin{bmatrix}2p_z+\sin 2p_z\\ 2p_z-\sin 2p_z\end{bmatrix}+\left(\frac{\beta_z}{\beta}\right)^2\left\{J_1^2(p_x)-J_0(p_x)J_2(p_x)\right\}$$
$$\times\begin{bmatrix}2p_z-\sin 2p_z\\ 2p_z+\sin 2p_z\end{bmatrix}, \tag{4.39}$$

x_1, y_1 – coordinates of the first circular transducer (state a); x_2, y_2 – coordinates of the second circular transducer (state b); $\gamma = Q/\rho_0$ – relative diffusion coefficient;$\chi_{sx} = S\pi/a$; $\chi_{uy} = u\pi/b$ – transverse wave number of rectangular waveguide;$u_{sx} = \left[\chi_{sx}^2 - \rho_0^2\right]^{1/2}$; $u_{uy} = \left[\chi_{uy}^2 - \rho_0^2\right]^{1/2}$; β, β_z – area propagation coefficient of the system of circular waveguides located on a rectangular waveguide; $p_x = \beta r_0$; $p_z = \beta_z L/2$; δ_{su} – Kronecker symbol; $J_n(x)[I_n(x)]$ – Bessel function; $\rho_1 = \sqrt{\varepsilon_{1r}}\rho_0$; $\chi = \left[\chi_{sx}^2 + \chi_{uy}^2\right]^{1/2}$.

Figure 4.5 shows the dependencies based on the numerical values of the interaction coefficient of the circular waveguides located in the microwave rectangular waveguide calculated by the expressions (4.35) to (4.39). Where, the circular

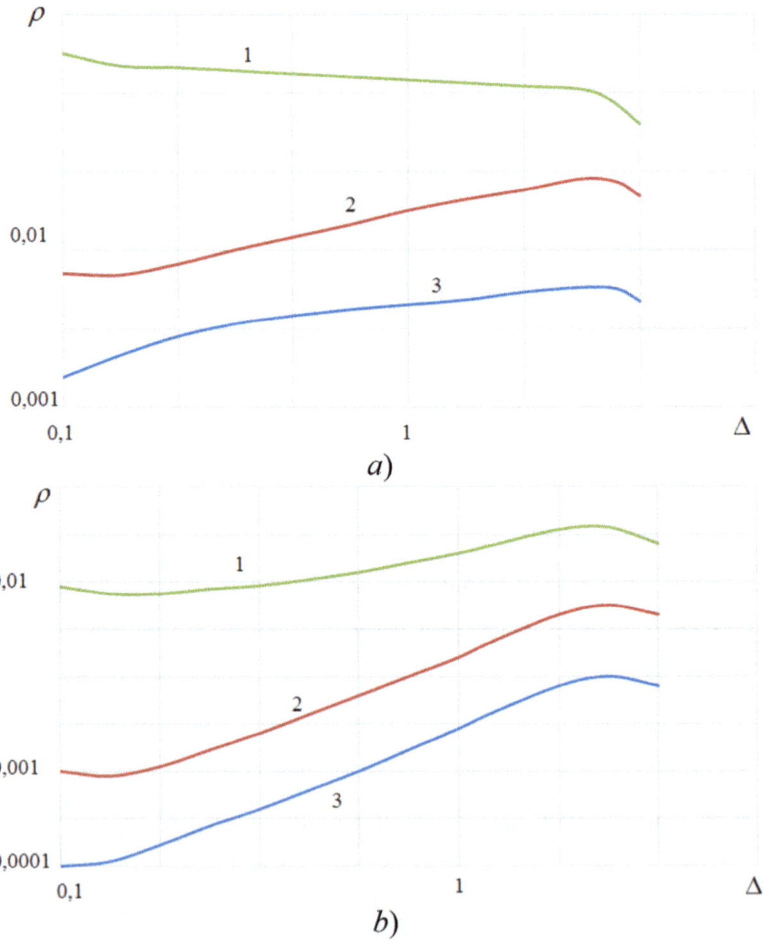

Fig. 4.5 Dependencies of the interaction coefficient of circular waveguides located on an microwave rectangular waveguide on $\Delta = L/2r_0$ parameters

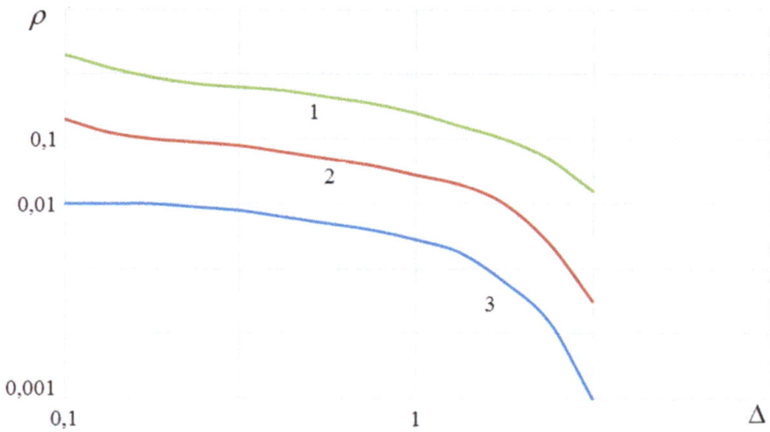

v)

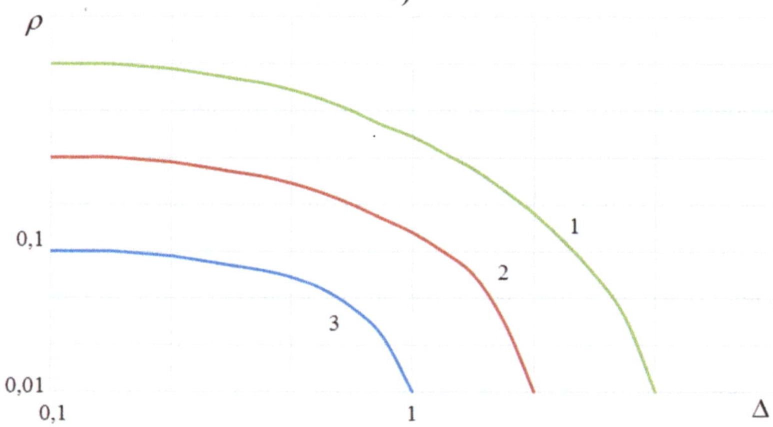

q)

Fig. 4.5 (continued)

waveguides operate at $f_0 = 9$ *GHz* frequency, and the rectangular waveguide operates in the frequency range of 4.9–7.05 *GHz* (cross-sectional area is 40 × 20 *mm*). The working wave is wave H^+_{101}.Dependencies are built depending on $\Delta = L/2r_0$ parameters. At this time, it was taken into account: $\Delta z/2r_0 = \text{const}$, in the case of *ab* (Fig. 4.4, *b*), $\Delta z/L = \text{const}$ in the case of *vv* (Fig. 4.4, *v*).

From the dependencies shown in Fig. 4.5, it can be seen that the interaction coefficient depends not only on the dimensions of the circular waveguides, but also on the different arrangement of the axes of the circular waveguides located on the rectangular waveguide. Thus, increasing the value of the interaction coefficient when placing one or two circular waveguides coaxially (Fig. 4.4, *v*, *q*), reducing the value of the interaction coefficient by placing the axes of one or two circular waveguides orthogonally (Fig. 4.4, *b*) can.

It can be seen from expression (4.36) that the interaction coefficient in the case of ab approaches zero (when $x_1 = x_2 = a/2$, or $y_1 = y_2 = b/2$), expression (4.37) shows that the interaction coefficient in the case of ab coefficient approaches zero in the case of the plane of symmetry ($x_1 = x_2 = a/2$). Calculations were made in $\varepsilon_r = 81$, $\Delta = 0, 5$ cases.

For a circular waveguide $y_1 = y_2 = b/2; \Delta z/2r_0 = 1, 5$; in the case of *ab* (Fig. 4.4, *b*) $y_1 = b/4$; $y_2 = 3b/4; \Delta z/2r_0 = 1, 5$; in the case *av* (Fig. 4.4, *v*) $y_1 = y_2 = b/2; \Delta z/(r_0 + L/2) = 2$, in the case *vv* (Fig. 4.4, *q*) $y_1 = y_2 = b/2; \Delta z/L = 2$.

It can be seen from the dependencies that the correlation coefficient is in the case of *aa*.

4.4 Electrodynamic Modeling of the Electromagnetic Field in a Microwave Circular Waveguide

The study of physical and electrodynamic processes in microwave devices is one of the important issues. As a result of the study of these processes, the creation of new devices, increasing their power, reducing the wavelengths of generators and amplifiers is considered one of the priority areas of telecommunications.

Among the microwave devices, one of the most widely used in practice is the microwave circular waveguide. These transducers have high technical and economic characteristics. This is due to the fact that the field of application of such devices is constantly expanding. Thus, microwave circular waveguides are effectively used in the creation of new types of transmitting tracts, for the transmission of information over long distances in millimeter wave radar systems, in the accurate determination of the coordinates of targets, as well as in the expansion of space exploration. Microwave circular waveguides are also effectively used in other fields of science and technology.

Currently, the expansion of the field of application of microwave circular waveguides makes the issue of propagation of the electromagnetic field in such waveguides relevant. Transducers form the electromagnetic field in modern microwave devices. Therefore, the calculation of the electromagnetic field in waveguides with a complex structure is one of the most important issues.

In some cases, *H*- and *T*-shaped waveguides are also used. Mathematical modeling of such waveguides is the most powerful mathematical tool for creating a deeper understanding of electromagnetic field propagation in these waveguides. Thus, the electrodynamic modeling of the electromagnetic field of circular waveguides provides complete and accurate information about the electromagnetic field and its structure. That is why the electrodynamic modeling of the electromagnetic field in the microwave circular waveguide is not only relevant but also of serious practical importance.

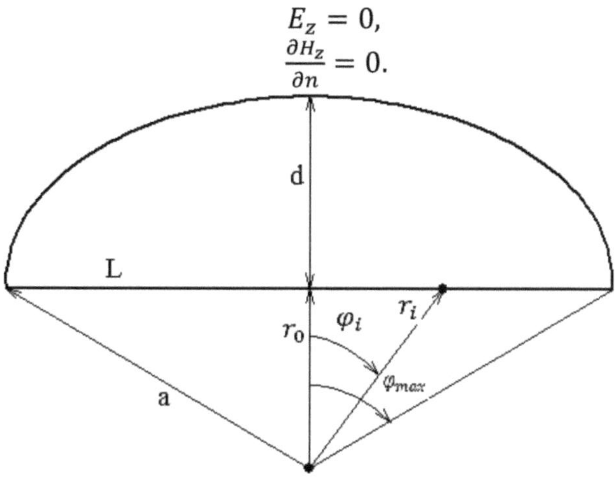

Fig. 4.6 Microwave circular waveguide in polar coordinate system

The determination of the propagation constant and the structure of the electromagnetic field in the microwave circular waveguide shown in Fig. 4.6 is brought to the following homogeneous Helmholtz equation:

$$\nabla^2 E_z + g^2 E_z = 0, \tag{4.40}$$
$$\nabla^2 H_z + g^2 H_z = 0.$$

At this time, the following homogeneous boundary conditions should be included:

$$E_z = 0, \tag{4.41}$$

$$\frac{\partial H_z}{\partial n} = 0. \tag{4.42}$$

To solve this problem, it is necessary to choose orthogonal coordinates so that their coordinate surface coincides with the surface of the microwave circular waveguide.

(4.40) and (4.41) let's use collocation and finite difference methods to solve the boundary problem [7].

In the cylindrical coordinate system, Eq. 4.40 can be written as follows:

$$\frac{\partial^2 E_z}{\partial r^2} + \frac{1}{r} + \frac{\partial E_z}{\partial r} + \frac{1}{r^2}\frac{\partial^2 E_z}{\partial \varphi^2} + g^2 E_2 = 0, \tag{4.43}$$

$$\frac{\partial^2 H_z}{\partial r^2} + \frac{1}{r} + \frac{\partial H_z}{\partial r} + \frac{1}{r^2}\frac{\partial^2 H_z}{\partial \varphi^2} + g^2 H_2 = 0. \tag{4.44}$$

Where, E_z and H_z mean $E_z = E_z(r, \varphi)$ and $H_z = H_z(r, \varphi)$, respectively. The solution of these equations by the method of differentiation leads to the following expressions:

$$E_z(r,\varphi) = \sum_{m=0}^{\infty} [C_{Em} J_m(gr) + D_{Em} N_m(gr)] \times [A_{Em} \cos(m\varphi) + B_{Em} \sin(m\varphi)], \quad (4.45)$$

$$H_z(r,\varphi) = \sum_{m=0}^{\infty} [C_{Em} J_m(gr) + D_{Em} N_m(gr)] \times [A_{Em} \cos(m\varphi) + B_{Em} \sin(m\varphi)], \quad (4.46)$$

where $J_m(gr)$ is the Bessel function or m-order cylindrical function, $N_m(gr)$ is the Neumann function or second-order cylindrical function, g is the transverse wave number.

As can be seen from Fig. 4.6, the cross-sectional area of the microwave circular waveguide and its outline consist of two parts: the radius $r = a$ and the line L.

If we consider that the boundary condition (4.41) on the boundary $r = a$ is satisfied at arbitrary values of φ, we get

$$D_{Em} = -C_{Em} \frac{J_m(ga)}{N_m(ga)}, \quad (4.47)$$

$$D_{Hm} = -C_{Hm} \frac{J'_m(ga)}{N'_m(ga)}. \quad (4.48)$$

Let's assume the following substitutions:

$$Z_{Em}(gr) = J_m(gr) N_m(ga) - J_m(ga) N_m(gr), \quad (4.49)$$

$$Z_{Hm}(gr) = J_m(gr) N_m(ga) - J'_m(ga) N_m(gr), \quad (4.50)$$

Then we will buy

$$E_z(r,\varphi) = \sum_{m-0}^{\infty} Z_{Em}(gr) \, [A_{Em} \cos(m\varphi) + B_{Em} \sin(m\varphi)], \quad (4.51)$$

$$H_z(r,\varphi) = \sum_{m-0}^{\infty} Z_{Hm}(gr) \, [A_{Hm} \cos(m\varphi) + B_{Hm} \sin(m\varphi)]. \quad (4.52)$$

Where, the constants $C_{Em}/N_m(ga)$ and $C_{Em}/J'_m(ga)$ are included in the coefficients A_{Em} and B_{Em}, A_{Hm} and B_{Hm}, respectively. The boundary condition (4.41) on the boundary L cannot be satisfied analytically. For this, let's use the collocation method [8]. The essence of the collocation method is as follows: the parameters A_{Em} and B_{Em} for the E-wave or the parameters A_{Hm} and B_{Hm} for the H-wave are selected in such a way that the functions (4.51) and (4.52) correspond to (4.41) and (4.42) pay the conditions.

Using the collocation method, the following dispersion equations for E-type and H-type waves were obtained:

For E-type wave

$$\left|\left[Z_k(gr_i)\sin k\varphi\right]\right| = 0, \tag{4.53}$$

where $k = 1, 2, . . , n-$ row indices, $i = 1, 2, . . , n-$column indices, n is the number of points located on the border of L.

$$\left|\left[Z_k(gr_i)\cos k\varphi_i\right]\right| = 0, \tag{4.54}$$

where $k = 1, 2, . . . , n; i = 1, 2, . . , n$.

By solving Eqs. 4.53 and 4.54, the transverse wavenumber and critical wavelength can be determined.

The dispersion equations for an H-type wave are:

$$\left\|\left[\cos\,(k+1)_{\varphi i}\left(gZ'_{Hk}(gr_i) - \frac{k}{r_i}Z_{Hk}(gr_i)\right) \quad + \cos\,(k - -1)_{\varphi i}\left(gZ'_{Hk}(gr_i) + \frac{k}{r_i}Z_{Hk}(gr_i)\right)\right]\right\| = 0, \tag{4.55}$$

$$\left\|\left[\sin\,(k+1)_{\varphi i}\left(gZ'_{Hk}(gr_i) - \frac{k}{r_i}Z_{Hk}(gr_i)\right) \quad + \sin\,(k - -1)_{\varphi i}\left(gZ'_{Hk}(gr_i) + + \frac{k}{r_i}Z_{Hk}(gr_i)\right)\right]\right\| = 0. \tag{4.56}$$

We can determine the transverse wave number g by solving Eqs. 4.55 and 4.56. At the same time, after determining the transverse wave number, we can determine the critical wavelength. The solution of Eqs. 4.53 and 4.54, including 4.55 and 4.56 shows that increasing the root of the solution of the equations leads to accumulation at any value.

Figure 4.7a, b shows the dependence of the transverse wave number g obtained from the solution of Eqs. 4.53 to 4.56 on the cross-sectional dimensions r_0/a of the microwave circular waveguide.

If we extrapolate the obtained graphs to the region $r_0/a \to 0$, we can achieve the accumulation of the obtained solutions: for E-type wave $\overline{g}_{c1}$ will asymptotically approach the root of E_{11}-type wave (3832), $\overline{g}_{s1}$-root to E_{21}-type wave (5,52), $\overline{g}_{c2}$-root to E_{31}-type wave (6,38), $\overline{g}_{s2}$-root to E_{41}-type wave (7588); for H-type wave, $\overline{g}_{c1}$ root asymptotically corresponds to H_{21}-type wave (3054), $\overline{g}_{c2}$-root to H_{01}-type wave (3832), $\overline{g}_{s1}$-root to H_{11}-type wave (1841), $\overline{g}_{s2}$-root to H_{31}-type wave (4201).

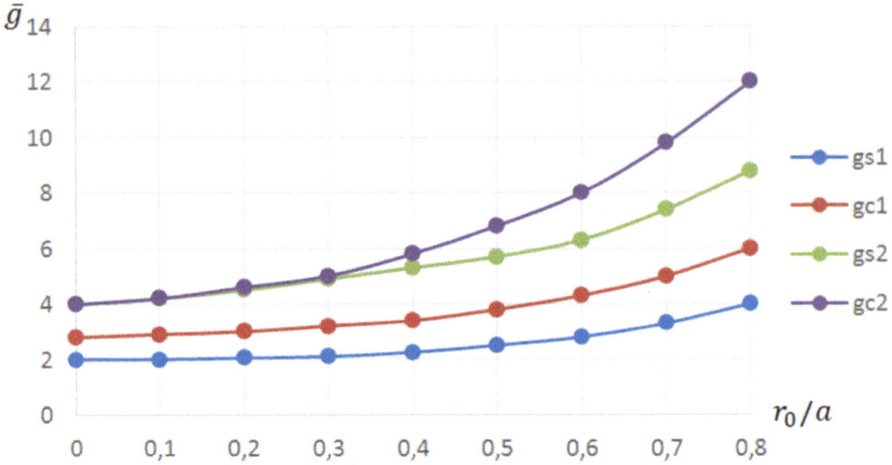

a)

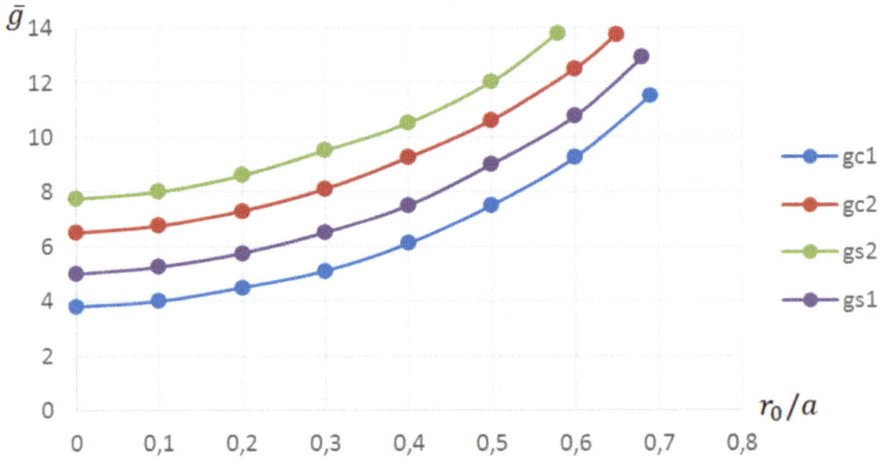

b)

Fig. 4.7 Dependencies of transverse wave number g on the ratio r_0/a of the cross section of microwave circular waveguide: (**a**) for *E*-type wave; (**b**) for *H*-type wave

4.5 Conclusions

1. New mathematical models of the electromagnetic field of microwave rectangular and circular waveguides have been developed in the polar coordinate system.
2. The Ritz and Galerkin methods were applied to solve the developed mathematical models, and the algorithms of these methods were developed to solve the problem under consideration.

3. As a result of the solution of the developed mathematical models, the curves of the equipotential lines of the electromagnetic field of the microwave rectangular and circular waveguide were established.
4. Constructed curves made it possible to control the electromagnetic field distribution of the microwave rectangular and circular waveguide.
5. The developed methodology is distinguished by its universality. So, this methodology can be applied to other microwave devices.
6. A new mathematical expression for the surface impedance of an microwave rectangular waveguide with irregular walls is obtained. This expression allows determining the surface impedance of an microwave rectangular waveguide with any accuracy.
7. The method of obtaining a new mathematical expression for the surface impedance of an microwave rectangular waveguide with irregular walls can be used for the study of more complex waveguide systems (for example, when the communication gap is located on the wide wall of the rectangular waveguide).
8. Dispersion equations for *E*-type and *H*-type waves in the microwave circular waveguide were obtained and these equations were solved by applying the collocation method, and the dependence of the cross-sectional dimensions of the microwave circular waveguide on the r_0/a ratio was established.

References

1. Sharififar, M. (2015). Simulation and optimization of hot extrusion process to produce rectangular waveguides. *International Journal of Advanced Manufacturing Technology, 79*(9–12), 1961–1973.
2. Shcherbinin, V. I., et al. (2019). Cutoff frequencies of a dielectric-loaded rectangular waveguide with arbitrary anisotropic surface impedance. *IEEE Transactions on Microwave Theory and Techniques, 67*(2), 577–583.
3. Sheina, E.A. et al. (2017). FDTD solution of reconstructing permittivity of a dielectric inclusion in a waveguide taking into account measurement inaccuracy. *Progress in Electromagnetics Research Symposium – Spring (PIERS)* (pp. 3188–3195).
4. Shestopalov, Y. V., & Kuzmina, E. A. (2018). On a rigorous proof of the existence of complex waves in a dielectric waveguide of circular cross section. *Progress In Electromagnetics Research, 82*, 137–164.
5. Shestopalov, Y. V., & Kuzmina, E. (2018). Symmetric surface waves along a metamaterial dielectric waveguide and a perfectly conducting cylinder covered by a metamaterial layer. *Advanced Electromagnetics, 7*(2), 91–98.
6. Shestopalov, Y. V. et al. (2016). Numerical-analytical method for reconstructing tensor permittivity and permeability of a diaphragm in a rectangular waveguide. In *Progress in Electromagnetic Research Symposium (PIERS)* (pp. 217–221).
7. Sheu, S.-T., et al. (2007). DDAS: Distance and direction awareness system for intelligent vehicles. *Journal of Information Science and Engineering, 23*, 709–722.
8. Shi, D. Y., et al. (2018). A nonconforming finite element method for the stationary Smagorinsky model. *Applied Mathematics and Computation, 353*, 308–319.

Chapter 5
Experimental Study of Microwave Waveguide Systems of Complex Structure in a Nonlinear Medium

5.1 Experimental Research of the Characteristics of Rectangular Waveguide

To increase the density of information transmitted by radio communication channels, it is necessary to improve the characteristics of the antenna system, including the device for switching from the rectangular waveguide to the horn. The main characteristics of a microwave rectangular transducer-to-loupe converter are reflection coefficient Q, transmission coefficient P and directivity diagram. Thus, the transmission coefficient P determines the shape and width of the orientation diagram and its side lobes. The reflection coefficient Q determines the nature of the mode of propagation of the electromagnetic wave transmitted from the transmitting device to the antenna device.

In real practice, in order for the power transmitted from the transmitting device to reach the antenna to the maximum extent, it is necessary to provide the escaping wave mode in the antenna-feeder tract. In connection with this, there is a need to reduce the reflection coefficient Q along the entire tract. In addition, the number of side lobes of the directional diagram should be minimal for uniform radiation of the parabolic antenna in the transmitter-receiver system to create microwave communication. This, in turn, is achieved by reducing the transmission coefficient P and smoothing the phase front of the electromagnetic wave.

As you can see, fulfilling these conditions at the same time is a complex issue. Therefore, the minimization of reflection coefficient Q and transmission coefficient P can be performed by applying Galerkin and Ritz methods. In order to minimize the reflection coefficient Q and the transmission coefficient P, a device for switching from a rectangular waveguide to a horn radiating microwave electromagnetic waves has been studied here.

Figure 5.1 shows the space model of the microwave rectangular waveguide working in its wave, and Fig. 5.2 shows its connection to the loudspeaker. This rectangular waveguide operates in the frequency range of 4.9–7.05 GHz.

I. Islamov, *Radio Engineering and Telecommunications Waveguide Systems in the Microwave Range*, https://doi.org/10.1007/978-3-031-37916-1_5

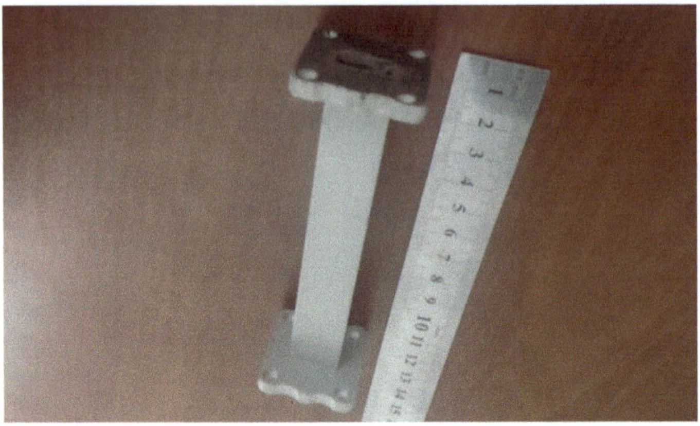

Fig. 5.1 Spatial model of a microwave rectangular waveguide operating at H_{10} wave

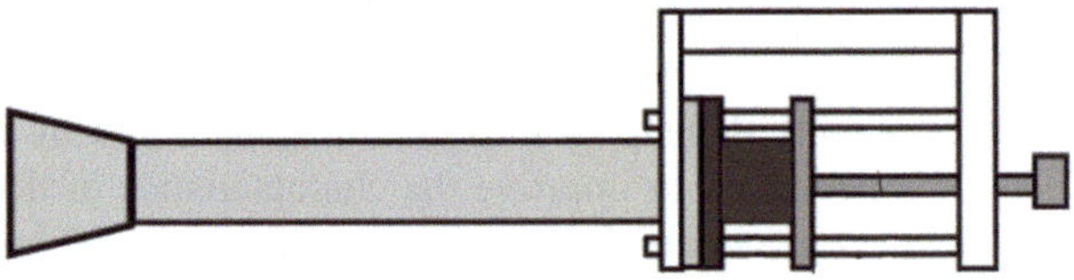

Fig. 5.2 Connection diagram of a microwave rectangular waveguide operating in the H_{10} wave to a loudspeaker

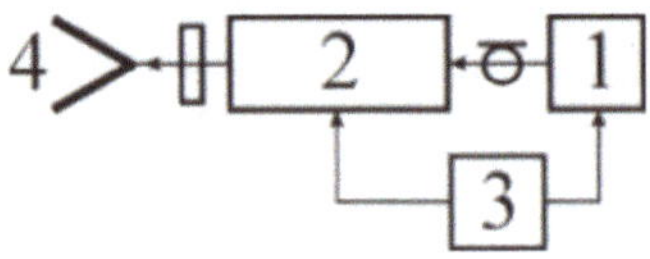

Fig. 5.3 Block diagram of the device for experimental determination of the standing wave coefficient Q_{dd}: 1 – SG type microwave signal generator; 2 – *P*1-28 type waveguide measurement line; 3 – food block; 4 – mouthpiece radiator

Its cross-sectional area measures 40 × 20 mm and is made of brass material. The geometric length of the microwave rectangular waveguide is $L = 105$ mm.

An experimental investigation of the characteristics of a microwave rectangular waveguide-to-loupe converter is related to the determination of the standing wave factor Q_{dd} and the directivity diagram. The research was carried out at $\lambda = 3$cm values of the wavelength and the measurements were performed with a waveguide measuring line. The measurement of the orientation diagram was performed with the help of the developed measurement complex. The block diagram of the device for the experimental determination of the standing wave coefficient Q_{dd} is shown in Fig. 5.3.

Measurements of the following parameters were performed.

Standing wave coefficient directivity diagram in a microwave rectangular waveguide operating at H_{11} wave.

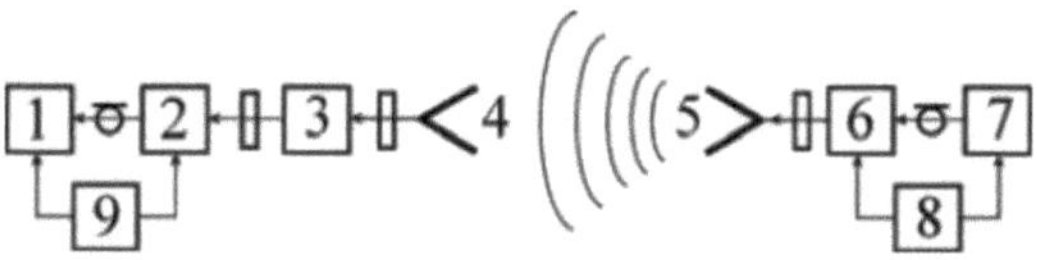

Fig. 5.4 Block diagram of the device for experimental determination of the orientation diagram: 1-indicator device; 2-receiver block; 3-adjustable attenuator; 4-receiver mouthpiece; 5-the researched transmitting mouthpiece; 6-generator block; 7-modulating block; 8, 9-food block

The reflection coefficient Q is determined using the following formula:

$$Q = \frac{Q_{dd} - 1}{Q_{dd} + 1}. \tag{5.1}$$

The measurement was performed with the following methodology. A horn radiator is attached to the output of the $P1$-28 waveguide measurement line. A microwave signal of length$\lambda = 3$cm is supplied from the generator to the input of the $P1$-28 type measuring line. The millivoltmeter is connected to the measuring line through a Hann diode. With the help of this millivoltmeter, its $\alpha_{\max}$– maximum and $\alpha_{\min}$– minimum values are determined. Then the standing wave factor Q_{dd} is determined according to the voltage according to the following formula:

$$Q_{dd} = \sqrt{\frac{\alpha_{\max}}{\alpha_{\min}}}. \tag{5.2}$$

After that, the reflection coefficient Q is determined based on the formula (5.1). The experimental determination of the directional diagram was performed in a closed room with dimensions 4 × 6 m with the help of a specially developed measuring device. The structural scheme of the measuring device is shown in Fig. 5.4, and its spatial model is shown in Fig. 5.5.

The measurement was made with the following methodology. The microwave klystron generator 6 is attached to the body along with the horn radiator 5. The receiving loudspeaker is placed at least 3 m away from the transmitting loudspeaker (far zone). Transmitting and receiving mouthpieces are placed along one axis. The signal from the output of the receiver mouthpiece is fed to the millivoltmeter device of the indicator block. By discretely changing the angle $\theta 1°$ in the vertical plane, the voltage U (mV) at the output of the receiving device is measured.

Then normalized orientation diagrams of the field intensity and power are constructed. The width of the orientation diagram at the 0.707 and 0.5 levels of intensity and power, respectively, is determined by the graphical method. The orientation diagram normalized by the intensity of the electromagnetic field is shown by curve 1 in Fig. 5.6.

Measurements were made in the interval from −90° to +90° of θ with respect to the axis of the mouthpiece device.

Fig. 5.5 Spatial model of the device for experimental determination of the directional diagram

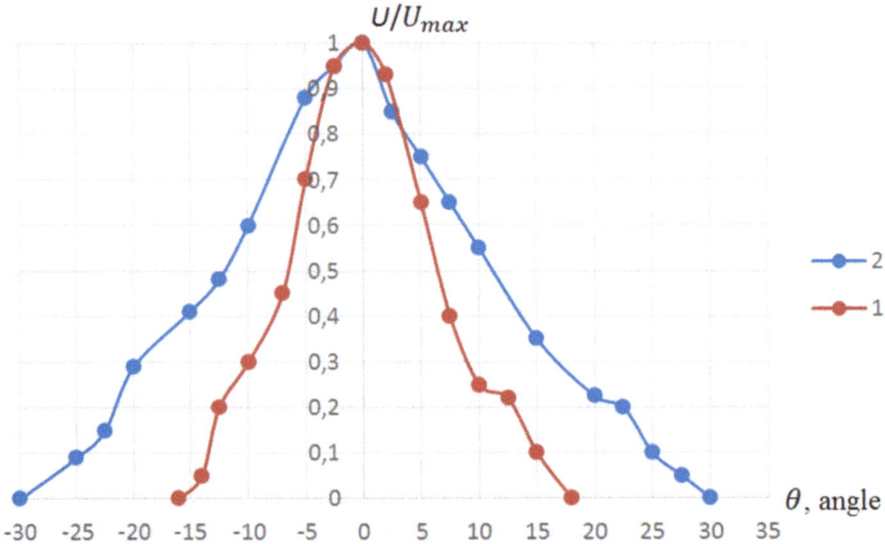

Fig. 5.6 Orientation diagrams of the loudspeaker radiator according to field intensity: 1 – vertical plane; 2 – horizontal plane

Schemes for measuring the modules of transmission and reflection coefficients are shown in Figs. 5.7 and 5.8, and the curves of these parameters obtained on the computer are shown in Fig. 5.9.

Standing wave coefficient $Q_{dd} = 1.42$ and reflection coefficient $Q = 0.17$ at $\lambda = 3$ cm wavelengths were determined. The standing wave coefficient of the studied loudspeaker radiator is $Q_{dd} = 1.96$, and the reflection coefficient is $Q = 0.32$.

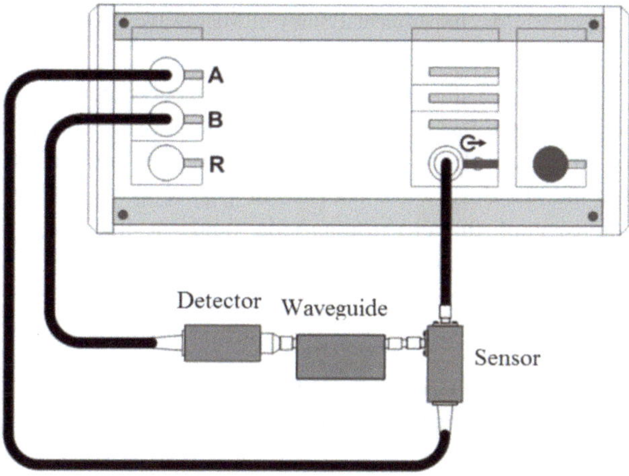

Fig. 5.7 Transmission and reflection coefficients module measurement scheme

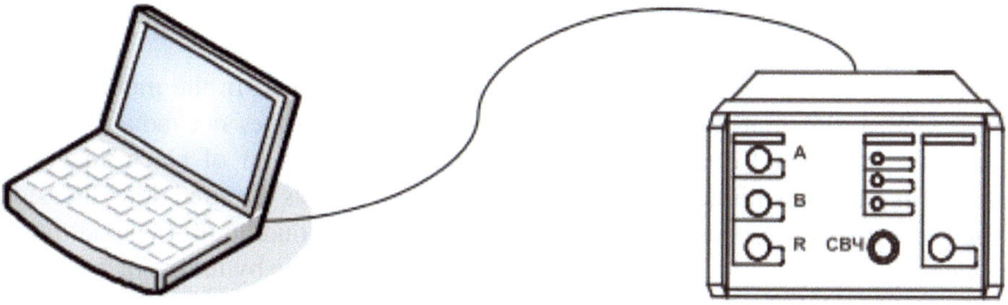

Fig. 5.8 Connecting the scheme for measuring the modules of transmission and reflection coefficients to a computer

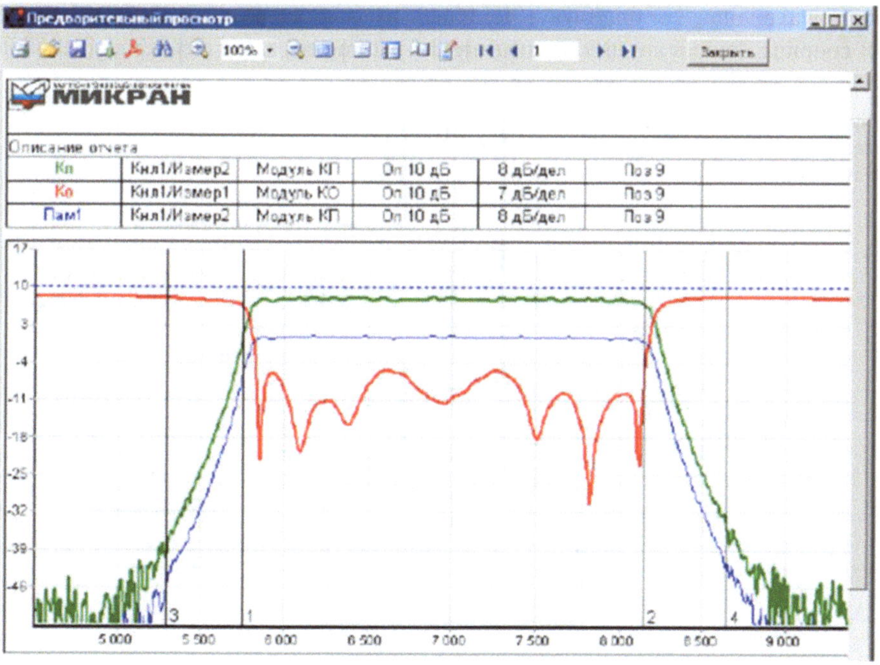

Fig. 5.9 Computer generated results of the modules of transmission and reflection coefficients

The large value of the standing wave coefficient Q_{dd} of the loudspeaker radiator is related to the large geometric dimensions of its output. For this purpose, smoothing of this part of the horn device was performed and the following results were obtained for the standing wave coefficient Q_{dd}: $Q_{dd} = 1.32$, $Q = 0.14$. According to the intensity of the electromagnetic field, at the level of 0.707, the width of the directional diagram of the horn radiator is $2\theta = 8.8°$.

5.2 Experimental Investigation of the Electromagnetic Field of Microwave Waveguides

Let's use the waveguide measuring line to experimentally determine the parameters of the electromagnetic field of waveguides in the microwave range. The scheme of the waveguide measuring line is shown in Fig. 5.10.

The base of the transducer measuring line is 1 transducer. In the middle of the large wall of this transducer, 2 slits are opened. This gap does not radiate electromagnetic waves and does not interfere with the propagation of electromagnetic waves in the transmission line. A lowered metal wire is inserted into the slot. These are called 3 probes. The probe passes through the middle of 4 coaxial lines. Six microwave detectors are coordinated with the coaxial line by means of 5 moving pistons. Tuning the transducer gauge line to the frequency of the wave is achieved by moving the piston. This is continued until the maximum current is obtained on the 8 indicator. The indicator is connected to the circuit through capacitor 7. By moving the probe with the detector, the rectified current of the detector is measured. Here, microwave voltage or microwave field intensity is not measured directly. A number of coupling devices are used to measure the parameters of the electromagnetic field

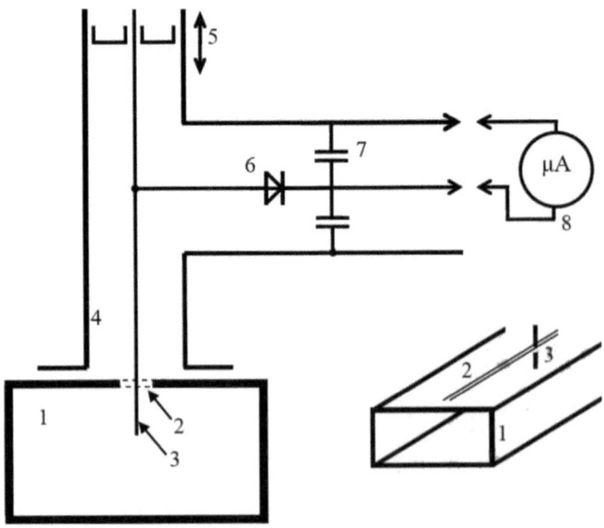

Fig. 5.10 Schematic of the measuring line of the waveguide

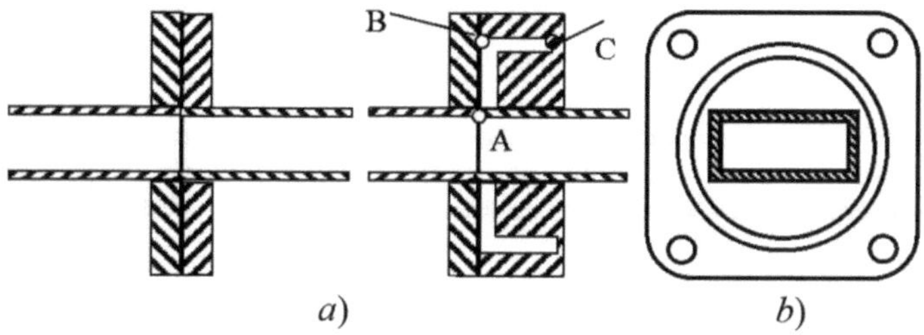

Fig. 5.11 Transducer connecting devices

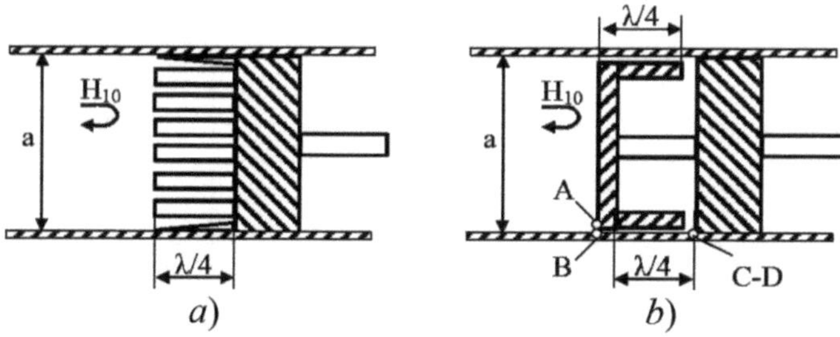

Fig. 5.12 Pistons in the waveguide

in extremely high frequency waveguides. They are mainly of two types: contact connections (Fig. 5.11a) and throttle connections (Fig. 5.11b).

Contact connections provide good electrical contact over a wide frequency range. However, it cannot be considered convenient during measurements. But good contact between the waveguides during throttle connections is ensured with the help of transmission lines. The part *AB* contains the radial line, where the *TEM* wave propagates. The *BC* section is a coaxial H_{11} wave short-circuited waveguide. Part *AB* is assumed to be $\lambda/4$ in length. *BC* part is accepted as $\lambda_{H11}/4$ in length.

The function of the piston in the scheme is to ensure that the reflection coefficient approaches unity. This is also possible when good electrical contact is ensured between the piston and the wall of the transducer. In order to ensure good electrical contact between the piston and the wall of the waveguide, it is important to place thin petals with a length of $\lambda/4$ on it (Fig. 5.12a). To ensure maximum electrical contact between the piston and the wall of the waveguide, throttle pistons should be applied (Fig. 5.12b).

In microwave waveguides, the matching load (Fig. 5.13) is also used when measuring the parameters of the electromagnetic field. The harmonic load is the end device of the waveguide tract. This device ensures complete extinction of the power of the electromagnetic field along the entire tract. Depending on the value of the dissipated power, matching loads are divided into two groups: matching load of low power level (dissipating power $P \leq 1$Wt), matching load of high power level (dissipating power $P \geq 1$Wt).

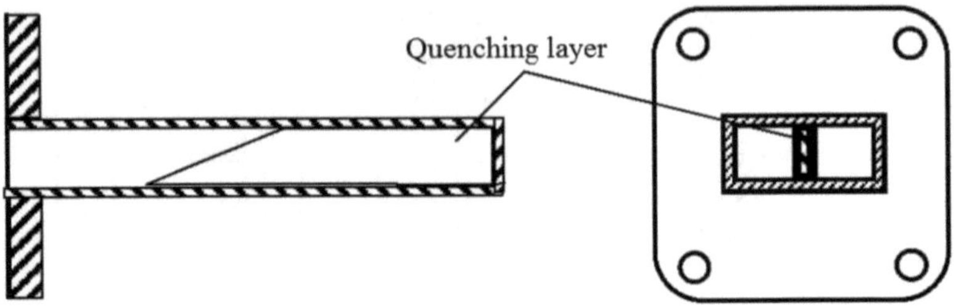

Fig. 5.13 Conciliatory loads

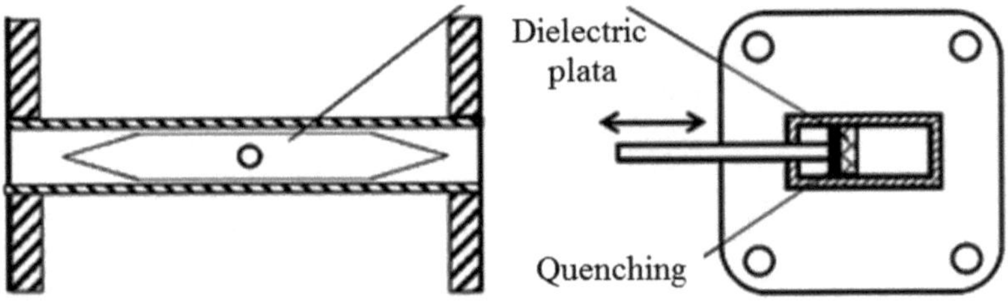

Fig. 5.14 Quenching attenuator

As can be seen from Fig. 5.13, small power-level matching loads are usually made in the form of a short-circuited part of the waveguide. A dielectric plate is inserted in the direction of its axis. This plate is covered with a quenching layer (graphite, lead, etc.). It is necessary to give a special shape to the board in order to make it agree with the waveguide. In high-power loads, the quencher is made in the form of a wedge. The standing wave factor for the used matching load is in the range of 1.02–1.05. This in turn ensures good coordination.

An attenuator (attenuator) was also used to conduct the experiments, which in turn is in the form of a double-shouldered junction and serves to weaken the power of the working wave.

Attenuators generally used in microwave techniques are divided into two groups: coupled and uncoupled.

According to the attenuation method, attenuators are also divided into two groups: quenching and polarized. At the same time fixed, adjustable, calibrated and non-calibrated attenuators, etc. is available.

A small power-level attenuator was used here (Fig. 5.14). This attenuator is made from the form of a small part of a waveguide, a dielectric plate is placed along its axis, and a quenching layer is applied to the plate, as in a matching load.

Attenuation adjustment is achieved by moving the plate inside the waveguide from its small wall to the middle of its large wall. The largest extinction is noted when the plate is in the middle of the waveguide. At this time, the electric field takes its maximum value.

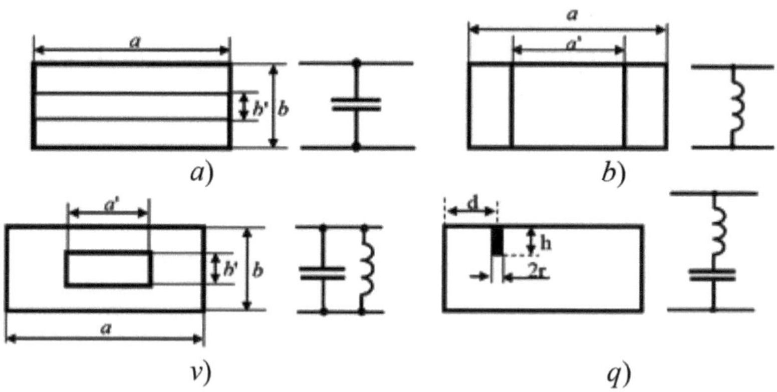

Fig. 5.15 Reactive elements in a rectangular waveguide

The value of the attenuation coefficient of the used attenuator varies from 0 to 30–40 dB. Thus, the standing wave coefficient is in the range of 1.1–1.2. Two quenching plates are used to obtain large attenuation.

At the same time, a phase shifter is used to conduct experiments. A phase shifter is quadrupole and undergoes a certain phase shift when a wave passes through it. Phase shifters, in turn, are divided into two groups: adjustable and non-adjustable. Adjustable phase shifters determine the phase shift between the output and input waveforms.

One of the most common phase shifters is dielectric phase shifters. In such phase shifters, the dielectric plate is placed along the length of the waveguide. This, in turn, means that the dielectric plate waveguide is placed parallel to the E vector of the H_{10} wave.

The construction of the tunable dielectric phase shifter for the H_{10}-wave rectangular waveguide is similar to the construction shown in Fig. 5.14. But here there is no extinguishing layer on the board.

Types of diaphragms in microwave rectangular waveguides and their equivalent circuits are shown in Fig. 5.15, *a*, *b*, *v*, *q*.

A waveguide diaphragm is also used to perform measurements. The waveguide diaphragm is placed in the cross-sectional area of the waveguide and reduces the linear dimensions of its cross-section. The diaphragm is a reactive element. So, this device creates almost no additional loss in the waveguide tract. The type of reactivity and its price depends on its size and shape.

5.3 Development of a Measuring Device for the Experimental Study of the Electromagnetic Field of Microwave Waveguides

The structural scheme and general appearance of the proposed measuring device for the investigation of the electromagnetic field in the waveguide are shown in Fig. 5.16a, b, respectively.

All nodes included in the waveguide tract of the measuring device are made on the basis of a microwave rectangular waveguide with a cross-sectional area of 40 × 20 mm. The electromagnetic wave is supplied to the line through a microwave with a Hann diode. This generator is fed through a power supply unit. A ferrite valve is placed at the input part of the measuring device to agree the input with the load. After the attenuator, which is used to adjust the power level of the wave, the measuring line of the waveguide is placed. With the help of the measurement line, standing wave coefficient and distribution of the electromagnetic field in the load, as well as the values of the electric and magnetic field intensity vector of the waveguide are determined. The measuring device connected to the output of the measuring line detector is a constant voltage microvoltmeter.

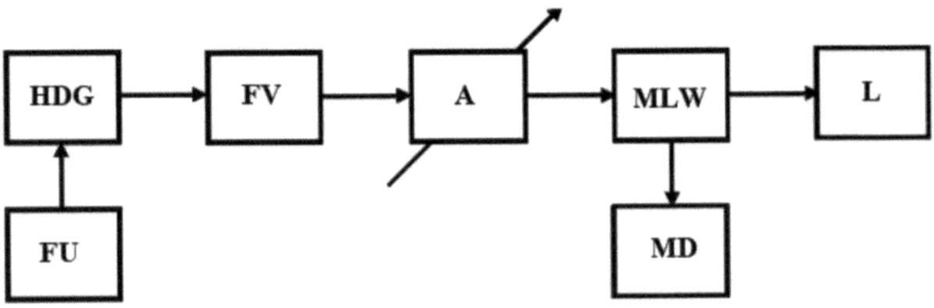

a)

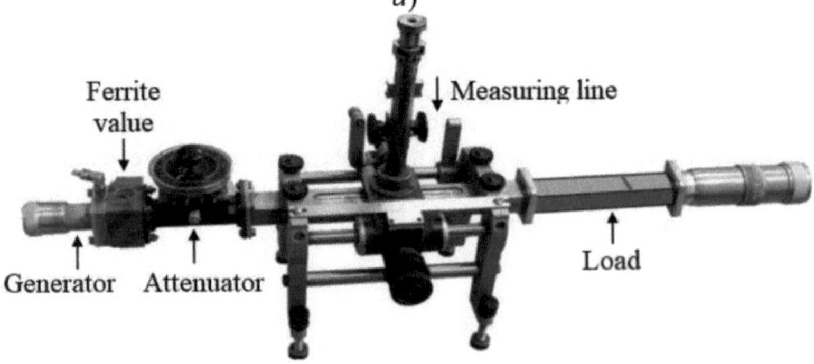

b)

Fig. 5.16 Measurement device for the study of the electromagnetic field in the microwave waveguide: (**a**) structural scheme; (**b**) general appearance: *HDG* Hann diode generator, *FV* ferrite valve, *A* attenuator, *MLW* measuring line of waveguide, *L* load, *FU* food unit, *MD* measuring device

Table 5.1 Experimental values of the electric field intensity of a microwave waveguide operating in the frequency range of 4.9–7.05 GHz

	For *E*-type wave		For *H*-type wave	
Number of elemen-tary fields	E_x, (10^5) *V/m*	E_y, (10^5) *V/m*	E_x, (10^5) *V/m*	E_y, (10^5) *V/m*
1	2.25	4.51	4.52	2.55
2	3.52	5.18	1.59	2.61
3	4.10	5.30	1.38	3.42
4	4.50	5.29	5.50	4.22
5	4.41	5.41	5.49	3.36
6	4.42	6.11	4.81	3.44
7	4.49	6.51	2.09	2.39
8	2.91	4.66	1.40	1.42
9	3.78	4.19	4.19	1.21
10	4.22	5.51	5.29	1.26
11	4.13	6.11	6.61	1.62
12	3.17	6.30	4.22	1.57
13	2.69	6.18	4.21	1.81
14	2.62	6.32	7.26	2.25
15	2.50	5.63	7.62	3.71
16	2.35	5.09	1.53	1.59
17	2.88	5.19	1.44	1.21
18	3.23	1.21	2.48	4.20
19	4.50	4.42	3.21	5.52
20	5.31	4.81	4.51	6.51

As a result of the measurements, the experimental values of the intensities of the electric (Table 5.1) and magnetic (Table 5.2) fields of the microwave rectangular waveguide operating in the frequency range of 4.9–7.05 GHz, microwave circular waveguide operating at the frequency of 9 GHz the experimental values of the intensities of the electric (Table 5.3) and magnetic (Table 5.4) fields of the waveguide were determined.

Based on the experimental values obtained in Tables 5.1 to 5.4, the electric power of the microwave rectangular waveguide operating in the frequency range of 4.9–7.05 GHz in Fig. 5.17, 4.9–7.05 GHz in Fig. 5.18. The dependencies of the experimental values of the magnetic field intensities of the microwave rectangular waveguide operating in the frequency range of on the length of its cross-section, and in Fig. 5.19 the electrical current of the microwave circular waveguide operating in the frequency range of 9 GHz, and in Fig. 5.20 in the frequency of 9 GHz dependencies of the experimental values of the magnetic field intensities of the working microwave circular waveguide on the length of its cross-section were established.

In order to more clearly describe the physical processes taking place in a two-mode microwave waveguide, let's show the distribution of several thresholds in the form of a function depending on the *X* and *Z* coordinates. Figure 5.21 shows the amplitude distribution of the electric field intensity of H_{10} and H_{20} waves on the

Table 5.2 Experimental values of the magnetic field intensity of a microwave waveguide operating in the frequency range of 4.9–7.05 GHz

	For *E*-type wave		For *H*-type wave	
Number of elemen-tary fields	H_x, *A/m*	H_y, *A/m*	H_x, *A/m*	H_y, *A/m*
1	51.82	24.31	54.02	28.30
2	67.84	43.24	25.43	35.63
3	44.53	25.63	35.82	77.12
4	63.72	35.42	46.22	45.74
5	11.64	85.81	66.14	62.71
6	52.61	43.70	34.53	64.74
7	54.74	38.92	76.32	22.62
8	61.23	32.13	24.31	34.73
9	63.41	25.24	12.11	37.72
10	53.20	44.34	32.63	11.73
11	52.92	45.93	36.22	33.71
12	43.54	47.11	74.72	35.44
13	34.73	52.53	36.14	20.60
14	55.72	66.90	66.23	41.91
15	47.52	31.72	34.33	42.53
16	59.20	43.71	34.10	29.52
17	79.34	62.94	64.32	10.23
18	61.71	44.43	75.83	37.71
19	61.82	44.21	41.94	29.54
20	49.51	25.83	41.11	21.63

Table 5.3 Experimental values of the electric field intensity of a microwave waveguide operating in the frequency range of 9 GHz

	For *E*-type wave		For *H*-type wave	
Number of elemen-tary fields	E_r, *V/m* (10^5)	E_z, *V/m* (10^5)	E_r, *V/m* (10^5)	E_z, *V/m* (10^5)
1	10.61	12.50	44.81	80.57
2	12.56	19.55	42.73	85.65
3	13.77	25.65	37.56	90.74
4	75.70	72.91	49.74	27.56
5	34.45	10.66	58.66	19.61
6	21.77	10.36	93.71	20.42
7	14.86	18.70	27.54	25.65
8	12.67	17.36	33.46	28.32
9	14.78	25.56	42.71	34.69
10	16.46	14.65	45.65	10.85
11	18.50	9.29	58.32	26.57
12	26.75	8.70	65.05	40.56
13	33.06	16.16	90.88	68.81
14	70.31	18.76	33.71	76.20
15	11.40	24.51	28.25	80.10
16	10.87	32.85	26.78	71.16
17	19.99	45.23	16.46	39.44
18	22.56	57.71	45.34	35.11
19	39.59	23.46	49.76	19.57
20	87.21	12.87	68.30	13.66

Table 5.4 Experimental values of the magnetic field intensity of a microwave waveguide operating in the frequency range of 9 GHz

	For *E*-type wave		For *H*-type wave	
Number of elemen-tary fields	H_r, *A/m*	H_z, *A/m*	H_r, *A/m*	H_z, *A/m*
1	50.56	14.46	33.47	30.16
2	53.36	38.34	29.51	40.14
3	52.12	30.35	36.24	87.67
4	55.18	29.24	50.54	35.26
5	16.53	76.33	59.32	75.14
6	60.43	54.18	29.42	75.11
7	41.56	29.16	73.32	34.86
8	57.32	39.41	30.14	42.89
9	51.57	32.33	18.42	44.36
10	61.14	39.46	41.17	18.22
11	45.46	57.15	24.21	48.37
12	14.31	50.54	75.27	25.39
13	40.18	54.56	14.34	29.26
14	39.34	56.75	70.14	38.83
15	52.13	29.08	43.29	53.12
16	49.14	41.32	25.27	33.84
17	66.44	63.32	69.13	17.56
18	52.42	47.15	63.83	37.71
19	59.14	48.17	53.16	18.31
20	33.88	30.19	45.26	30.19

large wall of the microwave waveguide [1]. Figure 5.22 shows the amplitude distribution of the final intensity of the electric field in the large wall of the microwave waveguide. Figure 5.23 shows the distribution of the electric field on the large wall of the microwave waveguide in the form of contour lines. The light zones shown in Figs. 5.21, 5.22, and 5.23 characterize larger values of the field intensity amplitude, and the dark zones characterize smaller values of the field intensity amplitude. The distributions shown in Fig. 5.22 and Fig. 5.23 characterize both incident and reflected waves for both types of waves in a microwave waveguide.

If only any type of wave (H_{10} or H_{20}) propagates along a microwave waveguide, then the power flux density (Poynting vector) distribution will not depend on the longitudinal coordinates of the waveguide. In this case, the value of the Poynting vector will have the same value across the cross section of the microwave waveguide.

If both H_{10} and H_{20} waves propagate along a microwave waveguide, then the power flux density (Poynting vector) distribution will depend on the longitudinal coordinates of the waveguide. In this case, the value of the Poynting vector will have a different value across the cross section of the microwave transducer. Such distribution of Poynting vector is shown in Fig. 5.24a.

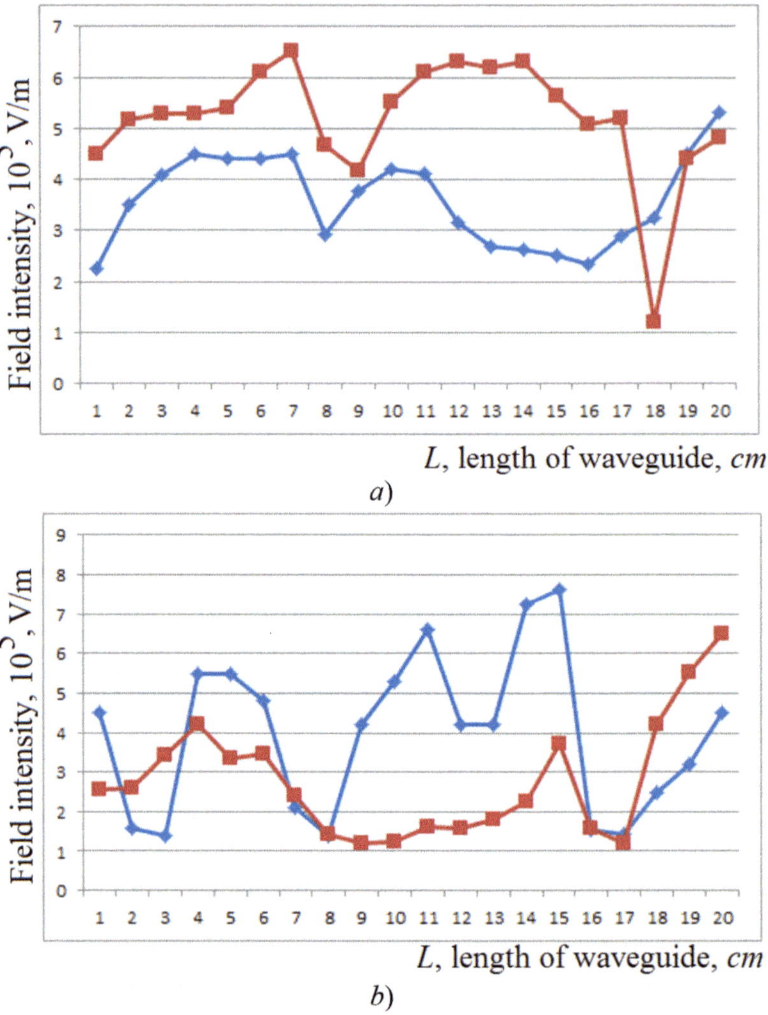

Fig. 5.17 For *E*-type (**a**) and *H*-type (**b**) waves, dependencies of the experimental values of the electric field intensities of the microwave rectangular waveguide operating in the frequency range of 4.9–7.05 GHz on the length of its cross-section: E_x – blue color, E_y – red color

If, in addition to the incident wave, the reflected wave manifests itself, then a more complex wave will be created, which is manifested in the form of the sum of both waves. The Poynting vector distribution for this case is shown in Fig. 5.24b.

As a result of the conducted studies, the scheme of the measuring device was proposed as shown in Fig. 5.25.

The input of the measuring section (1) of the dual-mode microwave transducer, which consists of a galvanomagnetic transducer (2), is supplied with microwave power for each of the H_{10} and H_{20} type waves. The measuring section is connected to

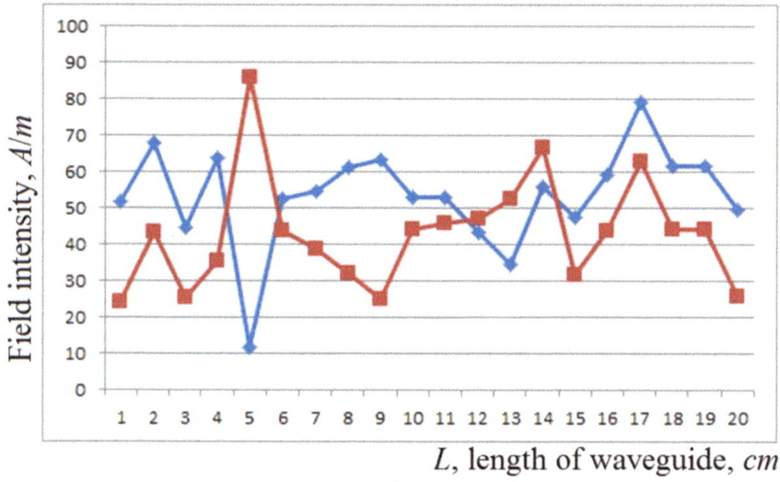

a)

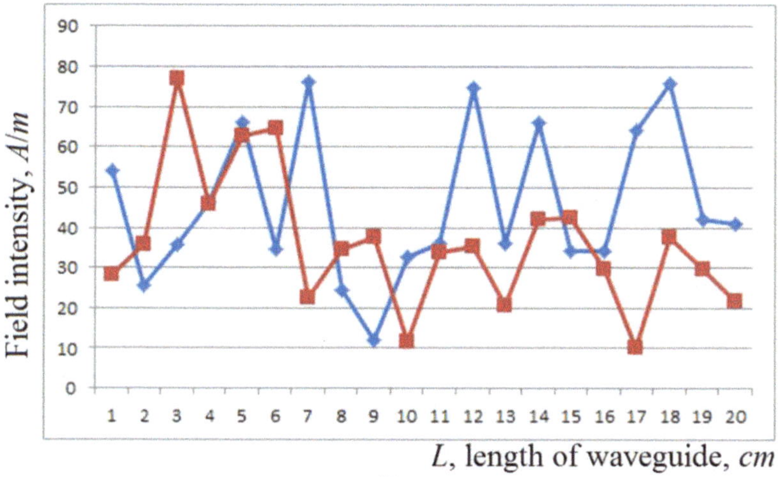

b)

Fig. 5.18 For *E*-type (**a**) and *H*-type (**b**) waves, dependencies of the experimental values of the magnetic field intensities of the microwave rectangular waveguide operating in the frequency range of 4.9–7.05 GHz on the length of its cross-section: H_x – blue color, H_y – red color

the load (3). At this time, the field in the gauge section can be described as a superposition of incident and reflected waves. The voltage is measured with the help of a galvanomagnetic transducer. This measured voltage is in turn proportional to the power flux density of the final electromagnetic field. The signal passes through the analog-to-digital converter (4) and enters the computer (5) for processing.

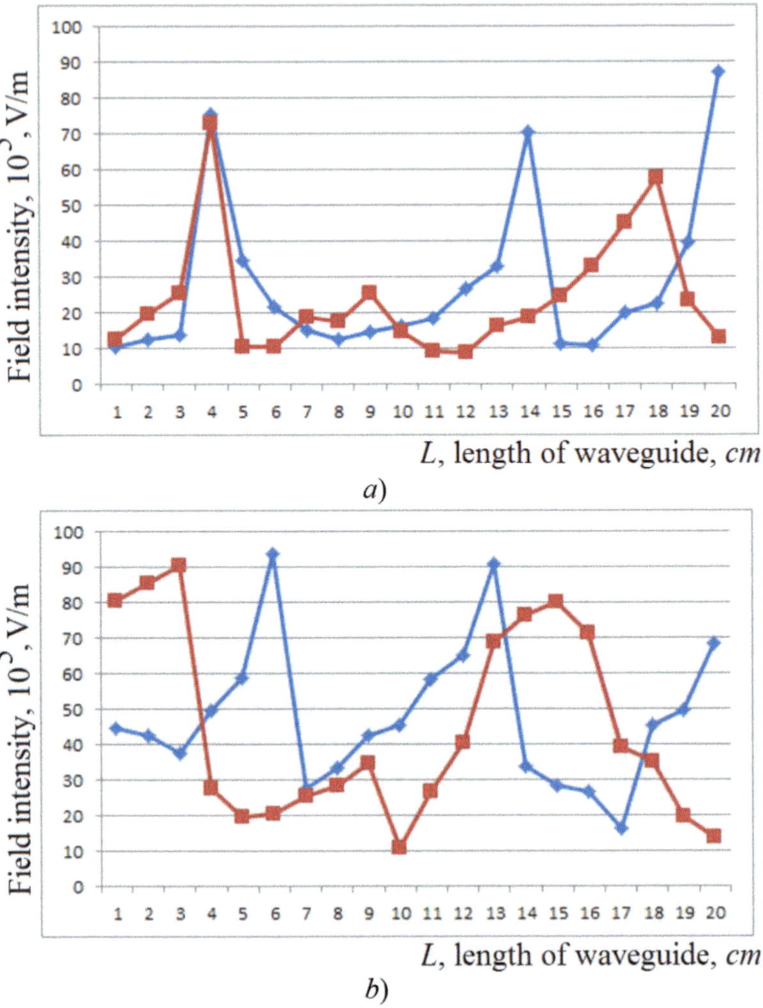

Fig. 5.19 For E-type (**a**) and H-type (**b**) waves, dependencies of the experimental values of the electric field intensities of the microwave circular waveguide operating in the frequency range of 9 GHz on the length of its cross-section: E_x – blue color, E_y – red color

5.4 Measurement of Dispersion and Attenuation in the Microwave Tract of a Rectangular Waveguide

In the microwave range, energy transfer through a waveguide can be directly characterized by the value of current and voltage. When waves are propagated through a waveguide in the microwave range, the value of current and voltage depends on the measurement points. However, there is currently no reliable method for measuring the absolute values of current and voltage at these frequencies. At this

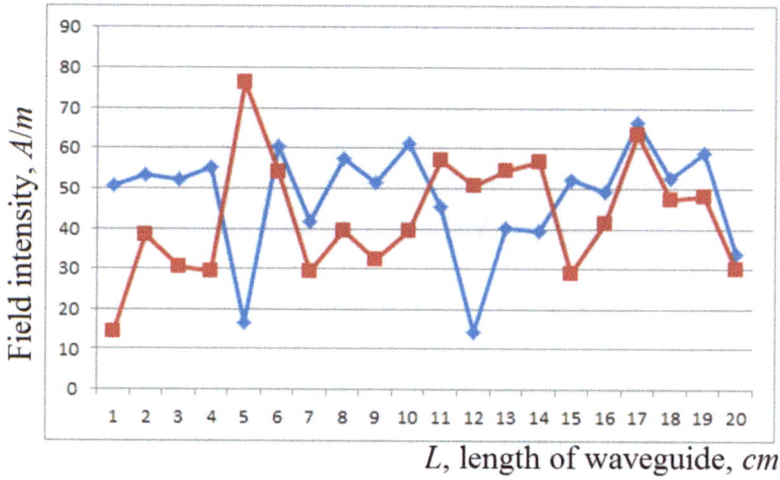

a)

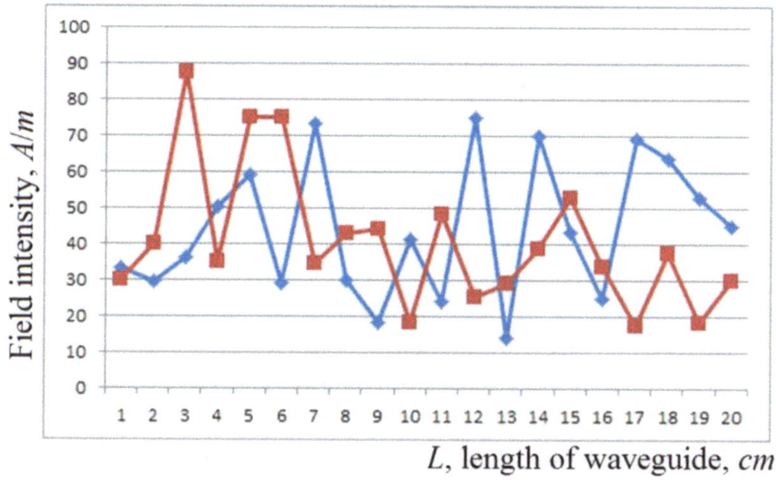

b)

Fig. 5.20 For *E*-type (**a**) and *H*-type (**b**) waves, dependencies of the experimental values of the magnetic field intensities of the microwave circular waveguide operating in the frequency range of 9 GHz on the length of its cross-section: H_x – blue color, H_y – red color

time, it is more appropriate to measure the electric and magnetic field intensity when the waves are propagated through the waveguide in the microwave range. During the transmission of electromagnetic waves through microwave transmitters, the most important characteristics are the measurement of the power, dispersion and attenuation carried by the waves. Another key parameter of the transmission line is the wave resistance (total resistance or impedance). This resistance is determined by the ratio of the total width of the electric and magnetic field intensities. For a

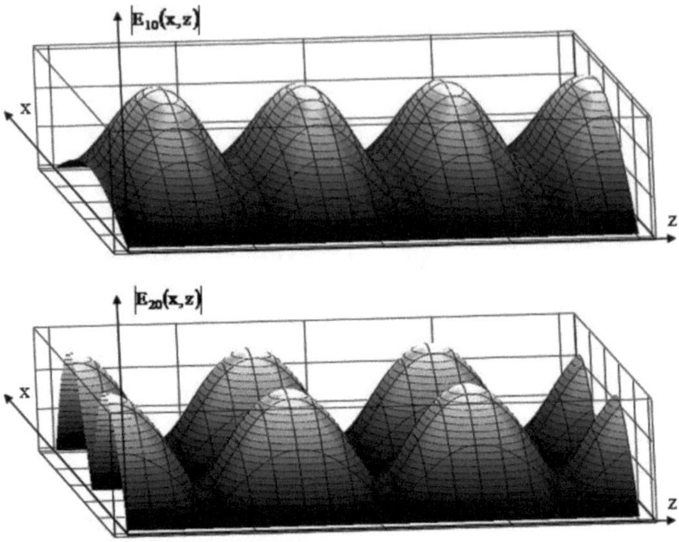

Fig. 5.21 Amplitude distribution of electric field intensity of H_{10} and H_{20} waves in the large wall of the microwave waveguide

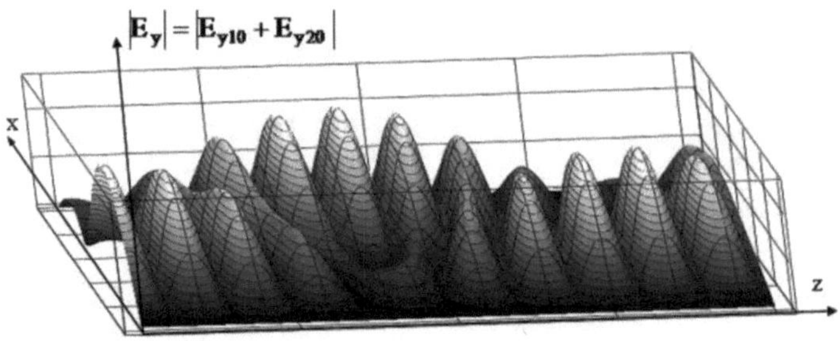

Fig. 5.22 Amplitude distribution of the final intensity of the electric field in the large wall of a microwave waveguide

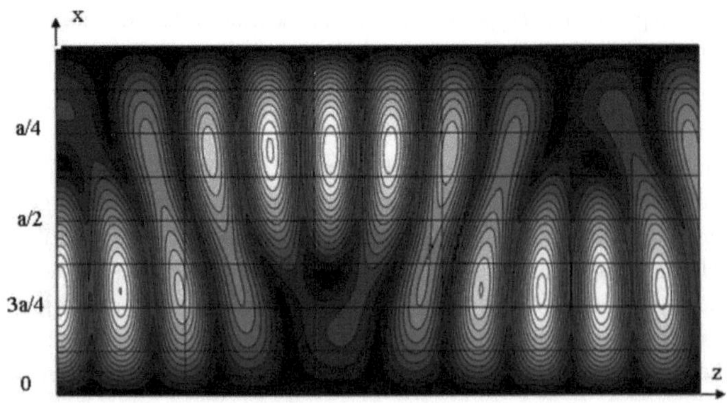

Fig. 5.23 The shape of the contour lines of the amplitude distribution of the E_y collected in the large wall of the microwave waveguide

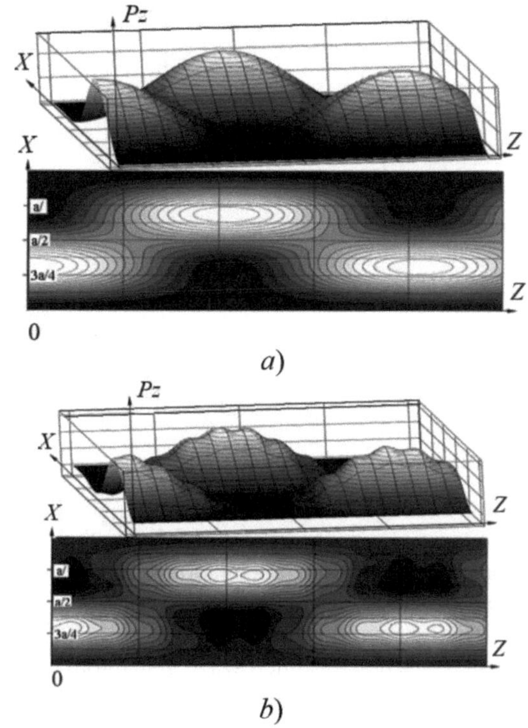

Fig. 5.24 Power flux density distribution of H_{10} and H_{20} waves in the large wall of the microwave waveguide: (**a**) in case of non-agreed cargo; (**b**) in case of agreed cargo

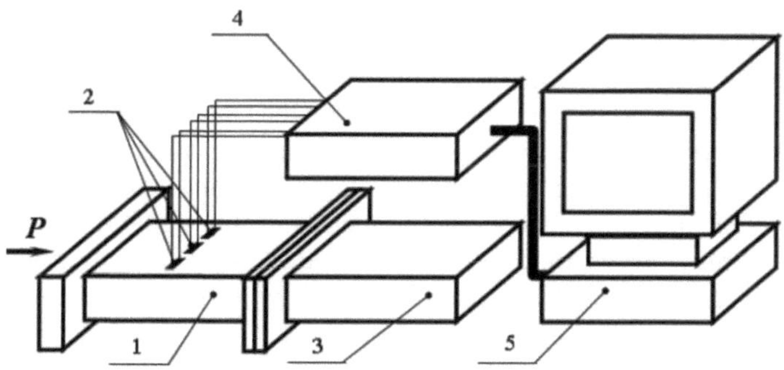

Fig. 5.25 Schematic diagram of a device for measuring power at both waveforms in a dual-mode microwave waveguide

non-homogeneous tract, the resistance has the same value along the entire cross-section of the waveguide. The ratio of resistances of different elements of the tract shows the efficiency of energy transfer. Thus, maximum power transmission occurs when the wave resistances of all elements are equal. The fact that wave resistances have different values leads to reflection. Thus, wave resistance is generally frequency dependent. In turn, reflection from different elements can be characterized by

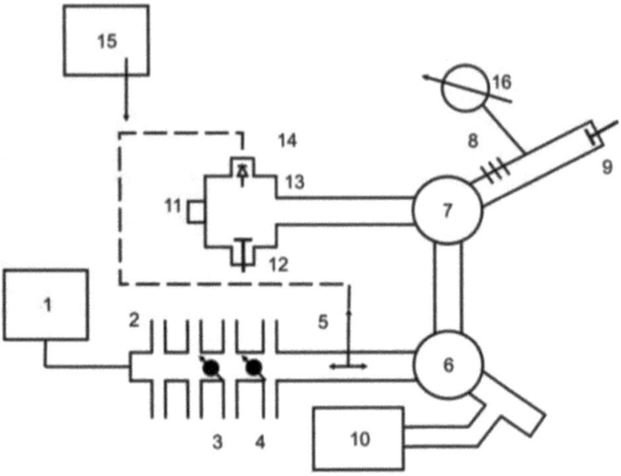

Fig. 5.26 Functional scheme of the experimental setup

frequency. From the above, it follows that the main parameters of the microwave tract are dispersion, attenuation and impedance. Therefore, it is appropriate to measure these parameters.

The functional scheme of the experimental unit is shown in Fig. 5.26.

In this scheme, a microwave generator (1) operates in the centimeter wave range. The source of radiation in this generator is a reflecting klystron. In the reflecting klystron, the generation of the frequency is determined by means of a digital indicator.

Attenuators (3 and 4) are used when measuring attenuation and power level. The measuring line (5) is used to measure the wavelength and the standing wave factor in the waveguide. Since the signal picked up by the probe is small, the measurement of the standing wave coefficient can be measured in the amplitude modulation mode of the microwave generator. The voltage ratio of alternating currents is determined by a special device (15). During measurement line operation, it must be tuned to the frequency of the microwave modulation signal. The input signal is measured using a voltmeter. Through waveguide converters (6 and 7), the path away from the generator is either short-circuited (to create a standing wave mode in the main path) or connected to various other elements of the device.

A thermistor-type power meter (10) is used to measure the power through the main tract. Power is measured in continuous generation mode. The detector section (9) is used for determining the relative power level during attenuator adjustment, including when determining the characteristics of the detector. A microammeter (16) is placed at the output of the detector. A full resistance transformer (8) is connected to the track before the detector section and serves to coordinate it with the track.

A double *T*-bridge (11) is used to measure the reactive loads on the comparison shoulders. Piston (12) is used to move the transducer. A short form of transducer (13) was used to make the measurements efficient.

A measuring line is used to measure wavelength in a transducer. By moving the measuring line along the transducer, the voltage values are recorded.

The extinction event in the transmission line is characterized by an exponential decrease in the amplitude of the field of electromagnetic waves.

The transmission coefficient for a tract of length l is defined as:

$$\tau = e^{-\alpha l}, \quad (5.3)$$

where l is the length of the microwave tract, the extinction coefficient per α unit length.

To determine the extinction coefficient according to the transmission coefficient, it is necessary to use the following formula:

$$\alpha \cdot l = \text{arctg} 1/k, \quad (5.4)$$

where k is the measured value of standing wave coefficient.

The following formulas are used to conduct the experiment:

$$l_{1,2} = \frac{z_1 + z_2}{2}, \quad (5.5)$$

where l_{12} – the corresponding lengths of the carriage, Z – the transverse distance between the holes drilled on the waveguide.

$$\lambda = \frac{c}{f}, \quad (5.6)$$

where λ– wavelength, $c = \frac{1}{\sqrt{\varepsilon_a \mu_a}}$ – is the speed of the wave.

$$\Delta = 2(l_{2\min} - l_{1\min}), \quad (5.7)$$

where Δ– is the pitch of the holes drilled on the waveguide.

$$E_{AB} = \frac{\pi \cdot \Delta z}{\Delta}, \quad (5.8)$$

where E_{AB}– it is the intensity between points A and B.

Based on the obtained results, dispersion (Fig. 5.27) and dependencies of extinction in the microwave tract (Fig. 5.28) were established.

The scheme of the experimental setup for measuring the module of the reflection coefficient and the standing wave coefficient in the load is shown in Fig. 5.29.

The result obtained from the measurement of the standing wave coefficient in the load is shown in Fig. 5.30, the result obtained from the measurement of the modulus of the reflection coefficient is shown in Fig. 5.31, and the result obtained from the

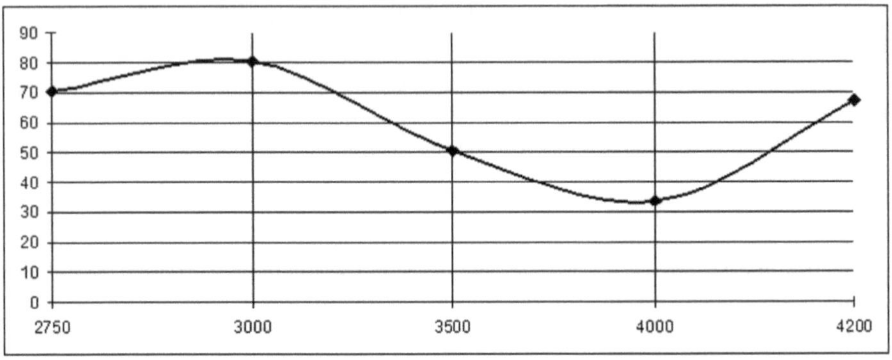

Fig. 5.27 Dispersion dependence

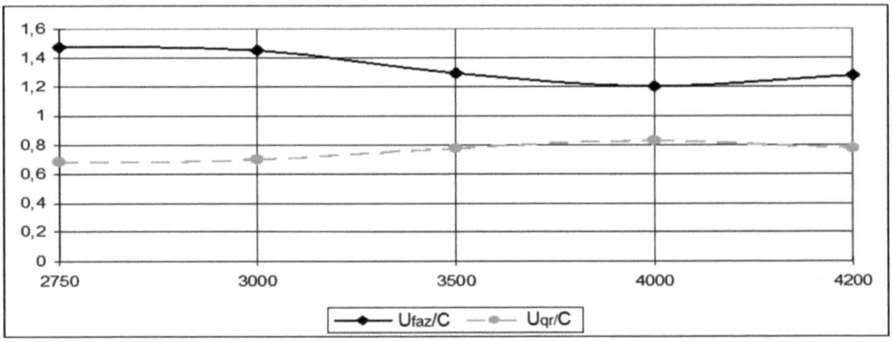

Fig. 5.28 Dependence of extinction in the microwave tract

Fig. 5.29 Modulus of reflection coefficient and in load schematic of the experimental setup for measuring the standing wave coefficient

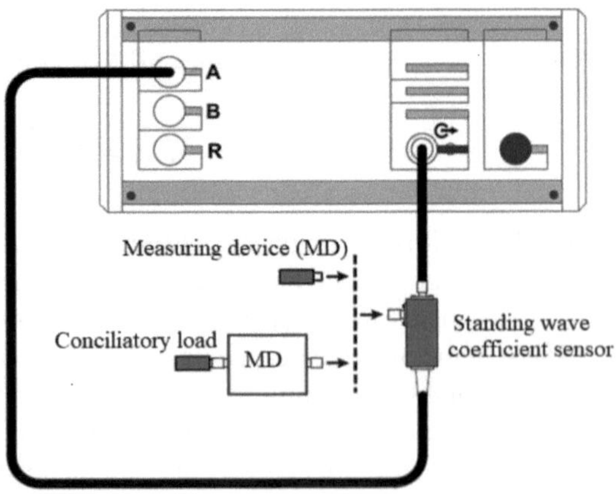

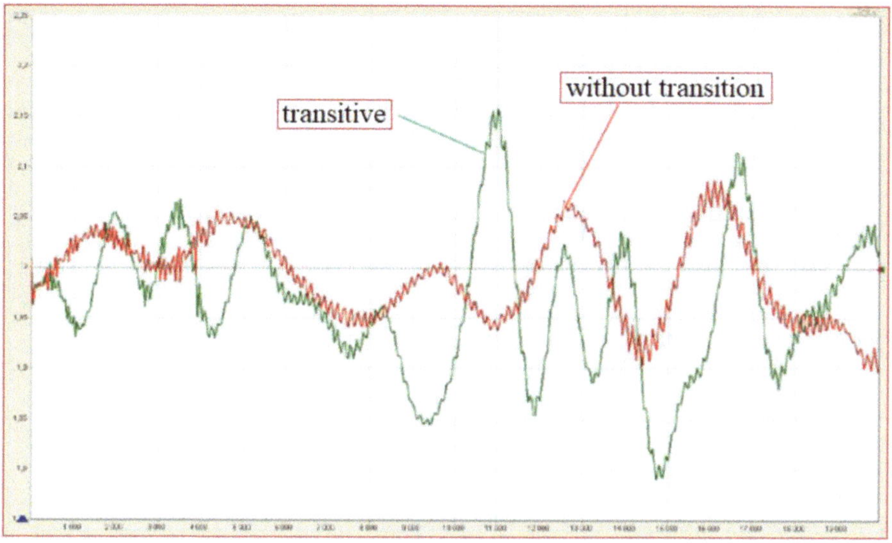

Fig. 5.30 Result from load standing wave factor measurement (load standing wave factor = 2.0)

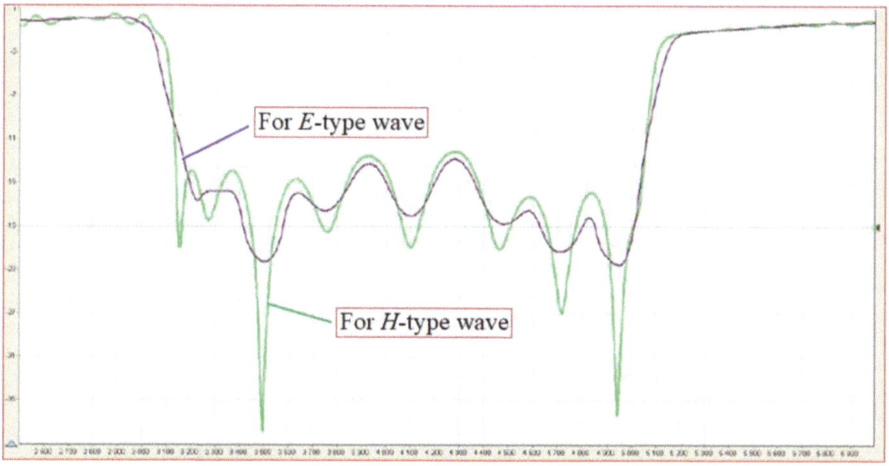

Fig. 5.31 The result obtained from the measurement of the modulus of the reflection coefficient

measurement of the standing wave coefficient in the load of the waveguide type switch is shown in Fig. 5.32.

Full resistance measurement in the microwave range is performed using a measuring line. The total resistance, that is, the complex reflection coefficient R measured by the vector analyzer of the impedance circuit [2] is calculated with the help. So, in the microwave range, the length of the transducer should not exceed 20 mm. Singh et al. [3] proposed a waveguide method for impedance measurement using a

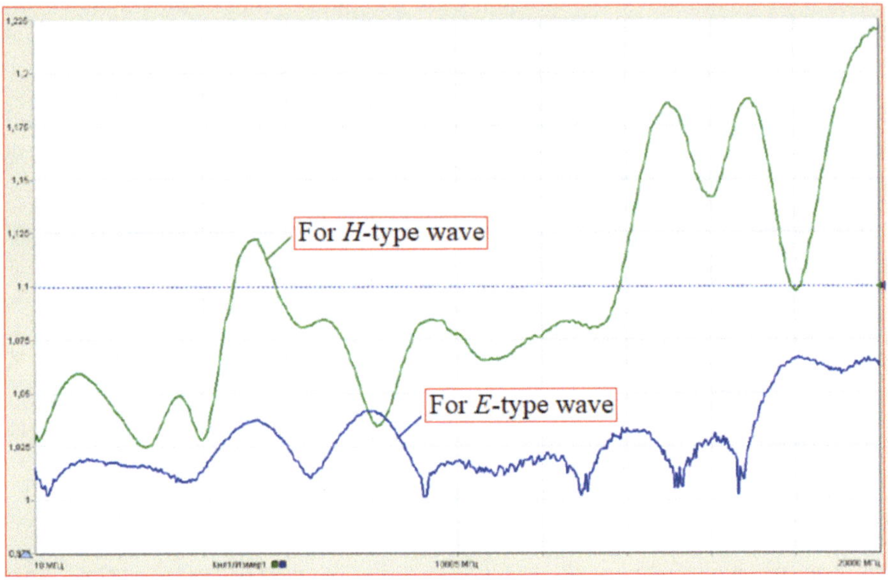

Fig. 5.32 The result obtained from the measurement of the standing wave coefficient of the waveguide-type switch in the load

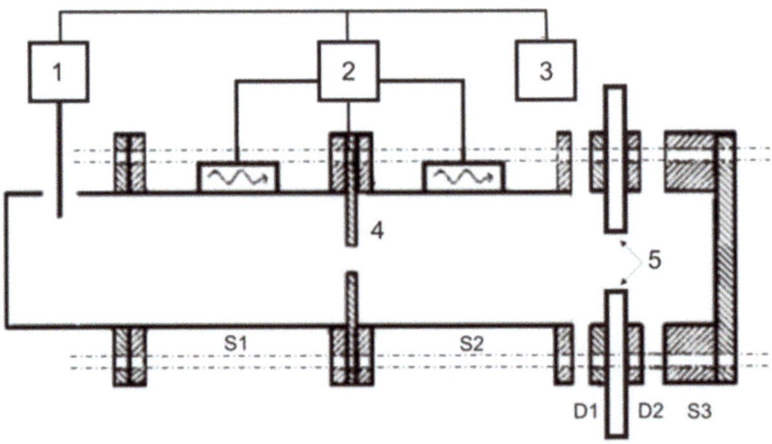

Fig. 5.33 Device for measuring impedance: 1 – frequency generator; circuit scalar analyzer; 3 – computer; 4 – diaphragm; 5 – waveguide

circuit scalar analyzer. This method is performed on the measuring device as shown in Fig. 5.33.

The experimental setup consists of a frequency excitation generator, two identical waveguide sections $S1$, $S2$ with an area of 40 × 20 mm and a length of 0.425 m, a symmetric inductive diaphragm with a thickness of 0.1 mm (a window with a length

Table 5.5 A rectangular waveguide operating at H_{10} wave values of experimentally determined parameters of the microwave tract

Limit power, P_d, Wt	4010.5
Releasable power, P_{bur}, Wt	1211
Extinction coefficient, α, dB/m	0.0201
Standing wave coefficient, K_{dd}	1021
Refractive coefficient, Q	0.0098
Special quality, Q_0	4497.2
The phase of the reflection coefficient, φ, rad/m	39.9π
Characteristic resistance, Z_H, Ohm	376.6

of 9 mm is opened on the diaphragm), consisting of a holding sample consisting of waveguide sections $D1$, $D2$ with lengths of 4.06 and 4.76 mm, respectively, and a shortened section $S3$ with a length of 10.02 mm.

Thus, the proposed waveguide method for measuring impedance in the microwave range is simpler and the results obtained are more accurate than those obtained by other methods. Table 5.5 lists the values of the experimental parameters for the tract of the microwave rectangular waveguide operating at the H_{10} wave.

5.5 With the Results Obtained from the Experimental Study of Microwave Rectangular and Circular Waveguides Comparative Analysis of the Results Obtained During Calculation by Numerical Methods and Error Estimation

During the theoretical studies conducted in Chaps. 2–4, the field intensities, which are the main parameters of the electromagnetic field of extremely high-frequency rectangular and circular waveguides in a nonlinear environment, were calculated using the finite difference method (finite difference method) and the finite element method (finite element method). At the same time, these parameters were experimentally measured in Chap. 5. The relative errors between theoretical and experimental results and their evaluation are given in Tables 5.6, 5.7, 5.8, 5.9, 5.10, 5.11, 5.12 and 5.13. Comparative analysis and error estimation of the results given in these tables show that the relative errors between the theoretical and experimental results do not exceed 4%. This is also considered satisfactory for the extremely high frequency range. The relative errors between theoretical and experimental results for these parameters are given in Table 5.14. Apparently, these relative errors are also satisfactory.

Table 5.6 Relative error between theoretical and experimental results for the electric field of a microwave rectangular waveguide

	Experimental values of electric field intensity in elementary regions, (10^5), V/m				The values obtained from the calculation of the electric field intensity in elementary regions by the finite difference method, (10^5), V/m				Relative error between theoretical and experimental results $\lvert X_{eksp.} - X_{hesab.}\rvert / \lvert X_{eksp.}\rvert \cdot 100\%$			
	E-wave		*H*-wave		*E*-wave		*H*-wave		*E*-wave		*H*-wave	
Number of elementary regions	*Ex*	*Ey*	*Ex*	*Ey*	*Ex*	*Ey*	*Ex*	*Ey*	*Ex*	*Ey*	*Ex*	*Ey*
1	2.25	4.51	4.52	2.55	2.18	4.45	4.45	2.45	3.11	1.33	1.54	3.92
2	3.52	5.18	1.59	2.61	3.48	5.24	1.6	2.59	1.13	1.15	0.62	0.76
3	4.10	5.30	1.38	3.42	4.07	5.28	1.43	3.39	0.73	0.37	3.62	0.87
4	4.50	5.29	5.50	4.22	4.47	5.29	5.42	4.13	0.66	0	1.45	2.13
5	4.41	5.41	5.49	3.36	4.44	5.36	5.51	3.44	0.68	0.92	0.36	2.38
6	4.42	6.11	4.81	3.44	4.43	6.07	4.78	3.37	0.22	0.65	0.62	2.03
7	4.49	6.51	2.09	2.39	4.43	6.42	2.11	2.42	1.33	1.38	0.95	1.25
8	2.91	4.66	1.40	1.42	2.83	4.75	1.37	1.40	2.74	1.93	2.14	1.40
9	3.78	4.19	4.19	1.21	3.85	4.27	4.21	1.19	1.85	1.90	0.47	1.65
10	4.22	5.51	5.29	1.26	4.15	5.47	5.21	1.23	1.65	0.72	1.51	2.38
11	4.13	6.11	6.61	1.62	4.19	6.03	6.52	1.60	1.45	1.30	1.36	1.23
12	3.17	6.30	4.22	1.57	3.16	6.25	4.12	1.54	0.31	0.79	2.36	1.91
13	2.69	6.18	4.21	1.81	2.67	6.26	4.11	1.79	0.74	1.29	2.37	1.10
14	2.62	6.32	7.26	2.25	2.55	6.29	7.31	2.19	2.67	0.47	0.68	2.66
15	2.50	5.63	7.62	3.71	2.43	5.59	7.52	3.61	2.8	0.71	1.31	2.69
16	2.35	5.09	1.53	1.59	2.27	5.14	1.49	1.57	3.4	0.98	2.61	1.25
17	2.88	5.19	1.44	1.21	2.91	5.12	1.42	1.19	1.04	1.37	1.38	1.65
18	3.23	1.21	2.48	4.20	3.16	1.20	2.51	4.1	2.16	0.82	1.20	2.35
19	4.50	4.42	3.21	5.52	4.37	4.37	3.19	5.49	2.88	1.13	0.62	0.54
20	5.31	4.81	4.51	6.51	5.24	4.78	4.42	6.49	1.31	0.62	1.99	0.30

Table 5.7 Relative error between theoretical and experimental results for the electric field of a microwave rectangular waveguide

	Experimental values of electric field intensity in elementary regions, (10^5), V/m				The values obtained from the calculation of the electric field intensity in elementary regions by the finite elements method, (10^5), V/m				Relative error between theoretical and experimental results $\lvert X_{eksp.} - X_{hesab.}\rvert / \lvert X_{eksp.}\rvert \cdot 100\%$			
	E-wave		*H*-wave		*E*-wave		*H*-wave		*E*-wave		*H*-wave	
Number of elementary regions	*Ex*	*Ey*	*Ex*	*Ey*	*Ex*	*Ey*	*Ex*	*Ey*	*Ex*	*Ey*	*Ex*	*Ey*
1	2.25	4.51	4.52	2.55	2.23	4.43	4.35	2.50	0.88	1.77	3.76	1.96
2	3.52	5.18	1.59	2.61	3.44	5.25	1.65	2.55	2.27	1.35	3.77	2.29
3	4.10	5.30	1.38	3.42	4.18	5.30	1.50	3.30	1.95	0	8.69	3.50
4	4.50	5.29	5.50	4.22	4.34	5.33	5.47	4.10	3.55	0.75	0.05	2.84
5	4.41	5.41	5.49	3.36	4.37	5.42	5.61	3.47	0.90	0.18	2.18	0.29
6	4.42	6.11	4.81	3.44	4.48	6.24	4.72	3.35	1.35	2.12	1.87	2.61
7	4.49	6.51	2.09	2.39	4.40	6.45	2.14	2.40	2.00	0.92	2.39	0.41
8	2.91	4.66	1.40	1.42	2.88	4.72	1.39	1.37	1.03	1.28	0.71	3.52
9	3.78	4.19	4.19	1.21	3.75	4.25	4.25	1.19	0.79	1.43	1.43	1.65
10	4.22	5.51	5.29	1.26	4.10	5.42	5.26	1.30	2.84	1.63	0.56	3.17
11	4.13	6.11	6.61	1.62	4.30	6.09	6.50	1.59	4.11	0.32	1.66	1.85
12	3.17	6.30	4.22	1.57	3.12	6.22	4.15	1.60	1.57	3.01	1.65	1.91
13	2.69	6.18	4.21	1.81	2.67	6.21	4.16	1.78	0.74	0.48	1.18	1.65
14	2.62	6.32	7.26	2.25	2.46	6.21	7.33	2.25	6.10	1.74	0.96	0
15	2.50	5.63	7.62	3.71	2.40	5.54	7.45	3.68	4	1.59	2.23	0.80
16	2.35	5.09	1.53	1.59	2.25	5.14	1.50	1.59	4.25	0.98	1.96	0
17	2.88	5.19	1.44	1.21	2.94	5.14	1.40	1.26	2.08	0.96	2.77	4.13
18	3.23	1.21	2.48	4.20	3.15	1.15	2.50	4.16	2.47	4.95	0.80	0.95
19	4.50	4.42	3.21	5.52	4.30	4.39	3.14	5.48	4.44	0.67	2.18	0.72
20	5.31	4.81	4.51	6.51	5.22	4.76	4.49	6.47	1.69	1.03	0.44	0.61

Table 5.8 Relative error between theoretical and experimental results for the magnetic field of a microwave rectangular waveguide

Number of elementary regions	Experimental values of magnetic field intensity in elementary regions, *A/m*				The values obtained from the calculation of the magnetic field intensity in elementary regions by the finite difference method, *A/m*				Relative error between theoretical and experimental results $\lvert X_{eksp.} - X_{hesab.}\rvert / \lvert X_{eksp.}\rvert \cdot 100\%$			
	E-wave		*H*-wave		*E*-wave		*H*-wave		*E*-wave		*H*-wave	
	Hx	*Hy*	*Hx*	*Hy*	*Hx*	*Hy*	*Hx*	*Hy*	*Hx*	*Hy*	*Hx*	*Hy*
1	51.82	24.31	54.02	28.30	51.86	24.2	53.6	27.94	0.08	0.45	0.78	1.27
2	67.84	43.24	25.43	35.63	67.11	44.25	24.5	36.3	1.08	2.34	3.66	1.88
3	44.53	25.63	35.82	77.12	43.56	25.43	36.57	76.23	2.18	0.78	2.09	1.15
4	63.72	35.42	46.22	45.74	64.16	34.98	44.89	46.98	0.69	1.22	2.88	2.71
5	11.64	85.81	66.14	62.71	11.34	86.78	65.79	63.39	2.58	1.13	0.53	1.08
6	52.61	43.70	34.53	64.74	53.12	44.9	34.79	65.15	0.97	2.75	0.75	0.63
7	54.74	38.92	76.32	22.62	55.56	39.19	74.76	22.37	1.50	0.69	2.04	1.11
8	61.23	32.13	24.31	34.73	60.47	33.18	24.26	33.71	1.34	3.27	0.21	2.94
9	63.41	25.24	12.11	37.72	62.12	25.17	11.98	38.18	2.03	0.28	1.07	1.22
10	53.20	44.34	32.63	11.73	51.45	43.45	33.76	11.56	3.29	2.01	3.46	1.45
11	52.92	45.93	36.22	33.71	51.9	44.86	35.69	32.96	1.93	2.33	1.46	2.22
12	43.54	47.11	74.72	35.44	42.45	46.58	73.72	34.48	2.39	1.13	1.34	2.71
13	34.73	52.53	36.14	20.60	35.19	53.56	35.43	20.11	1.32	1.96	1.96	2.38
14	55.72	66.90	66.23	41.91	56.61	67.5	67.31	40.81	1.6	0.9	1.63	2.62
15	47.52	31.72	34.33	42.53	48.13	31.34	34.53	43.12	1.28	1.2	0.58	1.39
16	59.20	43.71	34.10	29.52	60.16	42.95	33.99	30.15	1.62	1.74	0.32	2.13
17	79.34	62.94	64.32	10.23	80.78	61.89	64.44	10.35	1.81	1.67	0.19	1.17
18	61.71	44.43	75.83	37.71	62.34	43.56	76.45	38.48	1.02	1.96	0.82	2.04
19	61.82	44.21	41.94	29.54	60.45	43.56	41.25	30.05	2.22	1.47	1.65	1.73
20	49.51	25.83	41.11	21.63	50.23	26.21	40.38	22.38	1.45	1.47	1.78	3.47

Table 5.9 Relative error between theoretical and experimental results for the magnetic field of a microwave rectangular waveguide

Number of elementary regions	Experimental values of magnetic field intensity in elementary regions, *A/m*				The values obtained from the calculation of the magnetic field intensity in elementary regions by the finite elements method, *A/m*				Relative error between theoretical and experimental results $\lvert X_{eksp.} - X_{hesab.} \rvert / \lvert X_{eksp.} \rvert \cdot 100\%$			
	E-wave		*H*-wave		*E*-wave		*H*-wave		*E*-wave		*H*-wave	
	Hx	*Hy*	*Hx*	*Hy*	*Hx*	*Hy*	*Hx*	*Hy*	*Hx*	*Hy*	*Hx*	*Hy*
1	51.82	24.31	54.02	28.30	50.78	24.79	53.67	28.72	2	1.97	0.65	1.48
2	67.84	43.24	25.43	35.63	66.45	42.87	25.78	36.46	2.05	0.86	1.38	2.33
3	44.53	25.63	35.82	77.12	43.79	25.45	34.87	76.54	1.66	0.7	2.65	0.75
4	63.72	35.42	46.22	45.74	62.17	34.78	47.34	46.38	2.43	1.81	2.42	1.4
5	11.64	85.81	66.14	62.71	11.45	83.49	67.81	61.56	1.63	2.7	2.52	1.83
6	52.61	43.70	34.53	64.74	51.93	44.12	34.13	65.45	1.29	0.96	1.16	1.1
7	54.74	38.92	76.32	22.62	53.58	39.23	75.76	22.34	2.12	0.8	0.73	1.24
8	61.23	32.13	24.31	34.73	60.52	33.11	24.18	35.12	1.16	3.05	0.53	1.12
9	63.41	25.24	12.11	37.72	64.14	25.29	12.36	38.89	1.15	0.2	2.06	3.1
10	53.20	44.34	32.63	11.73	52.95	43.97	31.57	11.48	0.47	0.83	3.25	2.13
11	52.92	45.93	36.22	33.71	53.12	46.76	35.34	32.92	0.38	1.81	2.43	2.34
12	43.54	47.11	74.72	35.44	44.14	47.56	75.65	34.76	1.38	0.96	1.24	1.92
13	34.73	52.53	36.14	20.60	35.11	51.65	35.93	20.23	1.09	1.68	0.58	1.8
14	55.72	66.90	66.23	41.91	54.78	65.97	67.14	40.99	1.69	1.39	1.37	2.2
15	47.52	31.72	34.33	42.53	48.34	30.87	33.12	43.28	1.73	2.68	3.52	1.76
16	59.20	43.71	34.10	29.52	60.31	42.78	33.85	30.11	1.88	2.13	0.73	1.99
17	79.34	62.94	64.32	10.23	80.14	63.21	63.45	10.47	1	0.43	1.35	2.35
18	61.71	44.43	75.83	37.71	60.71	45.41	76.54	36.76	1.62	2.44	0.94	2.52
19	61.82	44.21	41.94	29.54	62.36	45.87	42.32	30.05	0.87	3.75	0.91	1.73
20	49.51	25.83	41.11	21.63	50.12	26.13	40.59	21.22	1.23	1.16	1.26	1.9

Table 5.10 Relative error between theoretical and experimental results for the electric field of a microwave circular waveguide

Number of elementary regions	Experimental values of electric field intensity in elementary regions, (10^5), *V/m*				The values obtained from the calculation of the electric field intensity in elementary regions by the finite difference method, (10^5), *V/m*				Relative error between theoretical and experimental results $\lvert X_{eksp.} - X_{hesab.}\rvert / \lvert X_{eksp.}\rvert \cdot 100\%$			
	E-wave		*H*-wave		*E*-wave		*H*-wave		*E*-wave		*H*-wave	
	Ex	*Ey*	*Ex*	*Ey*	*Ex*	*Ey*	*Ex*	*Ey*	*Ex*	*Ey*	*Ex*	*Ey*
1	10.61	12.50	44.81	80.57	10.5	12.4	44.7	80.6	1.04	0.01	0.25	0.04
2	12.56	19.55	42.73	85.65	12.6	19.6	42.8	85.7	0.32	0.26	0.16	0.06
3	13.77	25.65	37.56	90.74	13.8	25.7	37.6	90.8	0.21	0.19	0.10	0.07
4	75.70	72.91	49.74	27.56	75.6	72.8	49.8	27.6	0.13	0.15	0.12	0.15
5	34.45	10.66	58.66	19.61	34.5	10.7	58.7	19.5	0.15	0.38	0.07	0.57
6	21.77	10.36	93.71	20.42	21.8	10.4	93.6	20.3	0.14	0.39	0.12	0.59
7	14.86	18.70	27.54	25.65	14.9	18.13	27.6	25.7	0.27	3.00	0.22	0.20
8	12.67	17.36	33.46	28.32	12.6	17.4	33.5	28.2	0.55	0.23	0.12	0.42
9	14.78	25.56	42.71	34.69	14.8	25.6	42.6	34.6	0.14	0.16	0.26	0.26
10	16.46	14.65	45.65	10.85	16.5	14.7	45.7	10.9	0.66	0.34	0.11	0.46
11	18.50	9.29	58.32	26.57	18.4	9.2	58.2	26.6	0.54	0.97	0.21	0.11
12	26.75	8.70	65.05	40.56	26.8	8.6	65.1	40.6	0.19	1.15	0.08	0.10
13	33.06	16.16	90.88	68.81	33.1	16.2	90.9	68.7	0.12	0.25	0.02	0.16
14	70.31	18.76	33.71	76.20	70.2	18.8	33.6	76.1	0.16	0.21	0.33	0.13
15	11.40	24.51	28.25	80.10	11.3	24.4	28.3	80.0	0.88	0.45	0.18	0.12
16	10.87	32.85	26.78	71.16	10.9	32.9	26.7	71.2	0.28	0.15	0.30	0.06
17	19.99	45.23	16.46	39.44	19.9	45.3	16.5	39.5	0.45	0.15	0.24	0.15
18	22.56	57.71	45.34	35.11	22.6	57.6	45.4	35.0	0.18	0.19	0.13	0.31
19	39.59	23.46	49.76	19.57	39.5	23.5	49.8	19.6	0.23	0.17	0.08	0.15
20	87.21	12.87	68.30	13.66	87.1	12.9	68.2	13.7	0.13	0.23	0.15	0.30

Table 5.11 Relative error between theoretical and experimental results for the electric field of a microwave circular waveguide

	Experimental values of electric field intensity in elementary regions, (10^5), *V/m*				The values obtained from the calculation of the electric field intensity in elementary regions by the finite elements method, (10^5), *V/m*				Relative error between theoretical and experimental results $\lvert X_{eksp.} - X_{hesab.}\rvert / \lvert X_{eksp.}\rvert \cdot 100\%$			
	E-wave		*H*-wave		*E*-wave		*H*-wave		*E*-wave		*H*-wave	
Number of elementary regions	*Ex*	*Ey*	*Ex*	*Ey*	*Ex*	*Ey*	*Ex*	*Ey*	*Ex*	*Ey*	*Ex*	*Ey*
1	10.61	12.50	44.81	80.57	10.57	12.44	44.74	80.66	0.37	0.48	0.15	0.11
2	12.56	19.55	42.73	85.65	12.64	19.63	42.89	85.75	0.63	0.40	0.37	0.11
3	13.77	25.65	37.56	90.74	13.89	25.75	37.63	90.86	0.87	0.38	0.18	0.13
4	75.70	72.91	49.74	27.56	75.61	72.87	49.85	27.63	0.11	0.05	0.22	0.25
5	34.45	10.66	58.66	19.61	34.53	10.73	58.72	19.55	0.23`	0.65	0.10	0.30
6	21.77	10.36	93.71	20.42	21.84	10.45	93.64	20.34	0.32	0.86	0.07	0.39
7	14.86	18.70	27.54	25.65	14.96	18.66	27.63	25.75	0.67	0.21	0.32	0.38
8	12.67	17.36	33.46	28.32	12.69	17.49	33.57	28.22	0.15	0.74	0.32	0.35
9	14.78	25.56	42.71	34.69	14.82	25.65	42.63	34.63	0.27	0.35	0.18	0.17
10	16.46	14.65	45.65	10.85	16.54	14.73	45.77	10.99	0.48	0.54	0.26	1.29
11	18.50	9.29	58.32	26.57	18.43	9.25	58.23	26.64	0.37	0.43	0.15	0.26
12	26.75	8.70	65.05	40.56	26.85	8.62	65.12	40.62	0.37	0.91	0.10	0.14
13	33.06	16.16	90.88	68.81	33.18	16.21	90.94	68.77	0.36	0.30	0.06	0.05
14	70.31	18.76	33.71	76.20	70.29	18.83	33.66	76.14	0.02	0.37	0.14	0.07
15	11.40	24.51	28.25	80.10	11.32	24.42	28.32	80.01	0.70	0.36	0.24	0.11
16	10.87	32.85	26.78	71.16	10.97	32.95	26.75	71.23	0.91	0.30	0.11	0.09
17	19.99	45.23	16.46	39.44	19.91	45.39	16.58	39.53	0.40	0.35	0.72	0.22
18	22.56	57.71	45.34	35.11	22.62	57.66	45.44	35.02	0.26	0.08	0.22	0.25
19	39.59	23.46	49.76	19.57	39.54	23.54	49.81	18.78	0.12	0.34	0.10	4.00
20	87.21	12.87	68.30	13.66	87.16	12.97	68.20	13.9	0.05	0.77	0.14	1.75

Table 5.12 Relative error between theoretical and experimental results for the magnetic field of a microwave circular waveguide

Number of elementary regions	Experimental values of magnetic field intensity in elementary regions, *A/m*				The values obtained from the calculation of the magnetic field intensity in elementary regions by the finite difference method, *A/m*				Relative error between theoretical and experimental results $\lvert X_{eksp.} - X_{hesab.}\rvert / \lvert X_{eksp.}\rvert \cdot 100\%$			
	E-wave		*H*-wave		*E*-wave		*H*-wave		*E*-wave		*H*-wave	
	Hx	*Hy*	*Hx*	*Hy*	*Hx*	*Hy*	*Hx*	*Hy*	*Hx*	*Hy*	*Hx*	*Hy*
1	50.56	14.46	33.47	30.16	50.14	14.13	33.12	29.99	0.83	2.28	1.05	0.56
2	53.36	38.34	29.51	40.14	53.13	38.12	29.24	39.58	0.43	1.85	0.91	1.40
3	52.12	30.35	36.24	87.67	52.34	30.59	36.56	86.85	0.42	0.79	0.88	0.94
4	55.18	29.24	50.54	35.26	54.91	29.35	50.14	35.45	0.49	0.38	0.79	0.54
5	16.53	76.33	59.32	75.14	16.43	77.06	60.11	74.87	0.60	0.96	1.33	0.36
6	60.43	54.18	29.42	75.11	60.14	53.86	29.18	74.78	0.48	0.59	0.81	0.44
7	41.56	29.16	73.32	34.86	41.17	29.43	72.82	34.55	0.94	0.93	0.68	0.89
8	57.32	39.41	30.14	42.89	57.17	40.03	30.48	42.57	0.26	1.57	1.13	0.75
9	51.57	32.33	18.42	44.36	52.23	32.86	18.23	43.83	1.28	1.64	1.03	1.19
10	61.14	39.46	41.17	18.22	60.74	40.01	41.41	18.37	0.65	1.39	0.58	0.82
11	45.46	57.15	24.21	48.37	45.19	56.78	24.36	47.81	0.59	0.65	0.62	1.16
12	14.31	50.54	75.27	25.39	14.61	51.07	74.49	25.76	2.10	1.05	1.04	1.46
13	40.18	54.56	14.34	29.26	39.79	54.89	14.42	29.51	0.97	0.60	0.56	0.85
14	39.34	56.75	70.14	38.83	39.84	57.18	69.75	39.16	1.27	0.76	0.51	0.85
15	52.13	29.08	43.29	53.12	51.95	29.27	43.88	53.75	0.35	0.65	1.36	1.19
16	49.14	41.32	25.27	33.84	49.34	41.73	25.46	34.15	0.41	0.99	0.75	0.92
17	66.44	63.32	69.13	17.56	65.84	63.89	68.56	17.41	0.90	0.90	0.82	0.85
18	52.42	47.15	63.83	37.71	52.76	47.43	63.46	37.31	0.65	0.59	0.58	1.06
19	59.14	48.17	53.16	18.31	58.78	47.89	53.56	18.75	0.61	0.58	0.75	2.40
20	33.88	30.19	45.26	30.19	33.31	30.92	44.79	30.58	1.68	2.42	1.04	1.29

Table 5.13 Relative error between theoretical and experimental results for the magnetic field of a microwave circular waveguide

Number of elementary regions	Experimental values of magnetic field intensity in elementary regions, *A/m*				The values obtained from the calculation of the magnetic field intensity in elementary regions by the finite elements method, *A/m*				Relative error between theoretical and experimental results $\lvert X_{eksp.} - X_{hesab.} \rvert / \lvert X_{eksp.} \rvert \cdot 100\%$			
	E-wave		*H*-wave		*E*-wave		*H*-wave		*E*-wave		*H*-wave	
	Hx	*Hy*	*Hx*	*Hy*	*Hx*	*Hy*	*Hx*	*Hy*	*Hx*	*Hy*	*Hx*	*Hy*
1	50.56	14.46	33.47	30.16	50.39	14.35	32.95	29.98	0.34	0.76	1.55	0.60
2	53.36	38.34	29.51	40.14	54.15	39.12	29.35	41.17	1.48	2.03	0.54	2.57
3	52.12	30.35	36.24	87.67	51.75	30.54	35.89	86.73	0.71	0.63	0.97	1.19
4	55.18	29.24	50.54	35.26	54.78	29.14	51.16	35.69	0.72	0.34	1.23	1.22
5	16.53	76.33	59.32	75.14	16.33	77.34	60.13	75.78	1.20	1.32	1.37	0.85
6	60.43	54.18	29.42	75.11	61.16	53.87	29.22	74.68	1.21	0.57	0.68	0.57
7	41.56	29.16	73.32	34.86	40.69	30.05	74.14	35.21	2.09	3.05	1.12	1.00
8	57.32	39.41	30.14	42.89	56.76	40.34	30.33	43.32	0.98	1.85	0.63	1.00
9	51.57	32.33	18.42	44.36	51.11	31.93	18.34	44.13	0.89	1.24	0.43	0.52
10	61.14	39.46	41.17	18.22	60.87	40.04	40.85	18.01	0.44	1.47	0.78	1.15
11	45.46	57.15	24.21	48.37	45.31	56.76	24.11	47.78	0.33	0.68	0.41	1.22
12	14.31	50.54	75.27	25.39	14.21	49.75	76.46	24.94	0.70	1.56	1.58	1.77
13	40.18	54.56	14.34	29.26	39.72	54.15	14.22	28.97	1.14	0.75	0.84	0.99
14	39.34	56.75	70.14	38.83	38.94	56.14	69.68	38.35	1.02	1.07	0.66	1.24
15	52.13	29.08	43.29	53.12	51.86	29.39	43.14	52.89	0.52	1.07	0.35	0.43
16	49.14	41.32	25.27	33.84	48.81	41.54	25.18	33.03	0.67	0.53	0.36	2.39
17	66.44	63.32	69.13	17.56	66.25	62.76	70.02	17.36	0.29	0.88	1.29	1.14
18	52.42	47.15	63.83	37.71	52.13	47.35	64.16	37.17	0.55	0.42	0.52	1.43
19	59.14	48.17	53.16	18.31	58.79	48.57	52.88	18.12	0.59	0.83	0.53	1.04
20	33.88	30.19	45.26	30.19	33.46	30.64	44.92	30.43	1.24	1.49	0.75	0.79

Table 5.14 Relative error between the theoretical and experimental results of the parameters of the microwave tract of the waveguide

Name of the microwave tract parameter	Experimental value of the parameter of the microwave tract	Theoretical value of parameter of microwave tract	Relative error between theoretical and experimental results
Limit power, P_d, Wt	4010.5	4000	0.26
Releasable power, P_{bur}, Wt	1211	1200	0.91
Extinction coefficient, α, dB/m	0.0201	0.0203	0.99
Standing wave coefficient, K_{dd}	1021	1.02	0.1
Refractive coefficient, Q	0.0098	0.0099	1.02
Special quality, Q_0	4497.2	4490	0.16
The phase of the reflection coefficient, φ, rad/m	39.9π	39.8π	0.25
Characteristic resistance, Z_H, Ohm	376.7	376.6	0.03

5.6 Conclusions

1. The reflection coefficient, transmission coefficient and directivity diagram of the microwave rectangular waveguide-to-sector horn antenna converter were determined experimentally. In a microwave rectangular waveguide that transmits electromagnetic waves, the type wave is determined to be the fundamental type wave.
2. The main parameters of the microwave rectangular waveguide measuring 40 × 20 mm were calculated by Galerkin and Ritz methods. The theoretical and experimental results were compared and the error between them was determined to be satisfactory.
3. A method was developed to experimentally measure the level of transmitted power in microwave rectangular and circular waveguides. The difference of this method from other existing methods is the use of a galvanomagnetic converter. At the same time, one of the advantages of the developed method is the use of a small number of probes.
4. A functional scheme for measuring dispersion and attenuation in the microwave tract of a rectangular waveguide was proposed, and based on this, the dependence of dispersion and the dependence of attenuation in the microwave tract were established.
5. A measuring device was developed for the experimental study of the electromagnetic field of microwave rectangular and circular waveguides. The electric and

magnetic fields of the rectangular waveguide operating in the frequency range of 4.9–7.05 GHz for *E*-type and *H*-type waves and the electric and magnetic fields of the circular waveguide operating at the frequency of 9 GHz for *E*-type and *H*-type waves on the developed device the dependence of the intensity values obtained from the experiment on the length of its cross-section was established.
6. The scheme of the experimental device for measuring the module of the reflection coefficient and the standing wave coefficient in the load is proposed. On the basis of this scheme, the module of the reflection coefficient of the standing wave coefficient in the load, the standing wave coefficient in the load of the waveguide type switch were determined experimentally.
7. An experimental device was developed for measuring total resistance (impedance) in the microwave range.
8. The field intensities, which are the main parameters of the electromagnetic field of microwave rectangular and circular waveguides in a nonlinear environment, were experimentally measured, the relative errors between the theoretical and experimental results were evaluated and determined to be up to 4%.
9. Parameters such as limit power, output power, extinction coefficient, standing wave coefficient, reflection coefficient, specific quality, phase of the reflection coefficient, characteristic resistance of the rectangular waveguide are determined experimentally. The relative errors between the theoretical and experimental results for these parameters were determined to be up to 1%. This proves the adequacy of the obtained results.

References

1. Shu, C. C., & Weng, C. C. (2017). Electromagnetic theory with discrete exterior calculus. *Progress In Electromagnetics Research, 159*, 59–78.
2. Singh, H., et al. (2022). Synthesis of non-uniform circular antenna array for low side lobe level and high directivity using self-adaptive cuckoo search algorithm. *Arabian Journal for Science and Engineering, 47*, 3105–3118.
3. Singh, H., et al. (2021). Designing and performance evaluation of metamaterial inspired antenna for 4G and 5G happlications. *International Journal of Electronics, 108*, 1035–1057.

Chapter 6
Optimal Synthesis of Microwave Waveguide Systems of Complex Structure in Nonlinear Media

6.1 Setting the Problem of Optimal Synthesis of Devices with a Microwave Range: Structural Optimization

The development process of devices with a microwave range can be divided into stages that are conditionally related to each other. The block diagram of the synthesis process of devices with a microwave range is shown in Fig. 6.1 [1].

It should be noted that the realization of the synthesis process is constantly developing with the development of new effective mathematical methods.

In the process of optimal synthesis, the first step is to set the problem (Fig. 6.1) and it is brought to the demand for providing the technical-economical and operational indicators of the microwave unit. Technical conditions – constructive features of devices, mechanical effects, etc. – requirements can be attributed. Only after fulfilling all the conditions imposed on the device does it become possible to proceed to the second stage of the process – structural optimization.

The process of choosing the optimal structure of microwave devices is an extremely complex issue. Structural optimization is often heuristic in nature and a number of factors need to be taken into account. Structural optimization of low-frequency telecommunication devices has been studied in sufficient detail, but in the microwave range, this issue has hardly been studied. Therefore, a subjective factor can play a role in the structural optimization of microwave devices.

6.2 Parametric Optimization of Microwave Range Devices

Parametric optimization is one of the stages of the synthesis process. Parametric optimization includes four stages. The block diagram of parametric optimization is shown in Fig. 6.2 [2].

I. Islamov, *Radio Engineering and Telecommunications Waveguide Systems in the Microwave Range*, https://doi.org/10.1007/978-3-031-37916-1_6

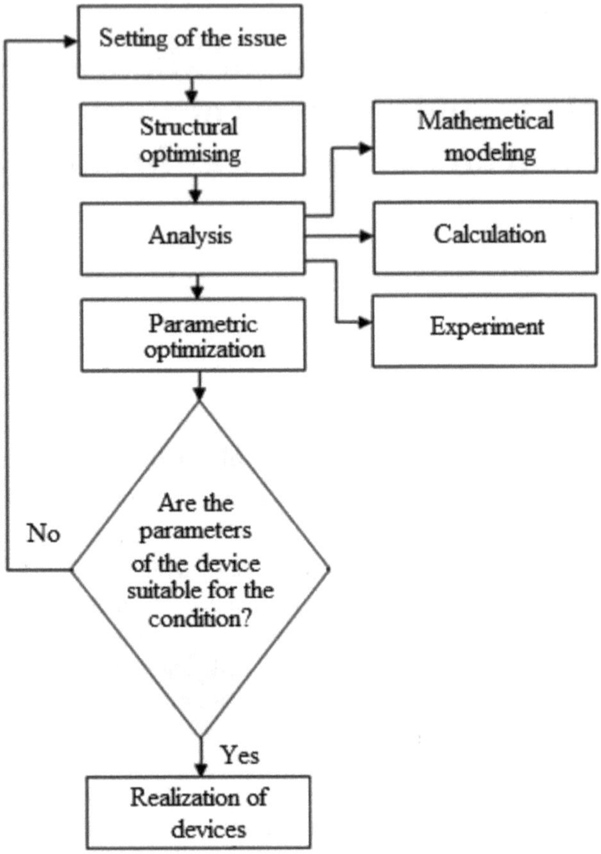

Fig. 6.1 Block diagram of the synthesis process of microwave range devices

A designed microwave range device must meet many conflicting requirements. The following can be attributed to them: it should have stable electrical parameters; should have high reliability suitable for operation, etc. Among the various technical and economic indicators characterizing the device, it is possible to distinguish indicators-criteria related to the quality of its activity [3]. For example, a microwave device that requires little energy and has high reliability can be considered optimal. The specified criteria are performed based on the analysis of the technical task for the design of the device. Let's assume that the criterion is characterized from a numerical point of view. Let's denote the criteria vector by g.

Let's choose the V parameters from the parameters of the mathematical models of the device. The restriction on the variation of the components of the vector V can be indicated by a number of factors. Thus, consideration of the requirements for technical and economic indicators of the device is closely related to the limitation imposed on it. The most common restriction is the restriction on the variation of the individual components of the vector V. Let's assume that the value of the components of the vector V is set to be smaller if the criteria improves the performance of

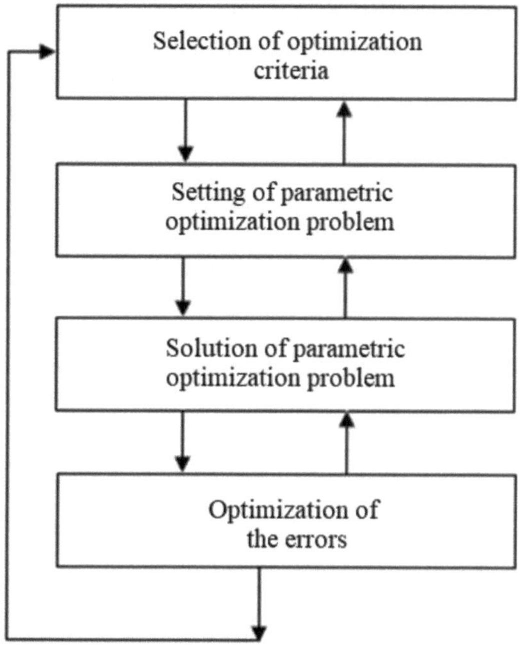

Fig. 6.2 Block diagram of parametric optimization

the device enough. In this case, the parametric optimization problem is reduced to the minimization of the g component, that is, the multi-criteria optimization problem. In the general case, g is a function of the vector P, q, r of the mathematical model, and the optimization of the device leads to the optimal selection of the components of the vectors P, q. The following features that complicate the solution of this issue can be noted: (*a*) multi-criteria character. In this case, the developed device must meet a number of criteria; (*b*) multiextremity of criteria quality; (*v*) nonlinear dependence of criteria quality and limitation on V vector; (*q*) in the general case, the vector V is infinite-dimensional.

These features require different approaches to solving the optimization problem. Basically, the following methods are more widespread: analytical and numerical optimal control methods; calculus of variations; mathematical programming. The last method can be considered more universal and effective. This method can be used subject to restrictions on the V vector. Factors showing the effectiveness of this method [4] is shown clearly.

Optimization of devices with a microwave range is primarily performed based on the characterization of the electromagnetic field by the elements of that device. The influence of the electromagnetic field of the environment is determined by the internal geometric dimensions of the device. Thus, the problem of optimization is brought to the search for the optimal function described by the law of variation of the geometric dimensions of the device and the electrophysical parameters of its elements. This type of problem can be related to the problem of optimization of optimal control systems with distributed parameters [5]. In this case, the optimized function (control function) is considered to be the elements of the infinite-dimensional Hilbert space, and thus the problem of parametric optimization of the device is, in principle, infinite-dimensional. It should be noted that in practice it is more appropriate to use devices with a simple control function. Optimization of the control function is carried out by performing logical and numerical operations. This is possible if the control function V is characterized by a set of finite numbers. The form of the dependence of the control function on V depends on the chosen parameterization method. The method of parameterization is determined by the composition of optimization algorithms, speed of assembly, stability.

Mathematical modeling of microwave range devices using telegraph equations allows the control function to be viewed as a one-dimensional function of $h(z)$. Z is the spatial coordinate. Geometric dimensions of the device as a function of $h(z)$ and wave resistance, coupling coefficient, etc. an auxiliary function characterized by its change is understood. Accordingly, solving the problem of parametric optimization of microwave devices can be divided into two approaches: in the first approach, auxiliary functions can be taken as an optimization function, and in the second approach, it can be taken as a function that includes the internal geometric dimensions of the device.

The first approach is a traditional approach and is mainly used for parametric optimization of T-type wave devices. This approach is distinguished by its universality and simplicity.

The second approach has less universality and complexity. However, this approach does not allow to take into account constructive and technological limitations. It is more appropriate to apply this approach in the development of the optimal construction during mass production.

Technical, economic and operating parameters of microwave range devices are considered its indicators. Adequacy of these indicators allows to assess the quality of the unit's work. Numerical indicators criteria related to the quality of the device's activity [6] is called. For microwave device, etc. can be. Choosing the right criteria is very important. Thus, the selection of the criterion allows to compare different manufacturing options of the device.

It is more complicated to give the system of indicators and criteria of a complex device. The following factors should be taken as a basis for solving this issue. First, the system of indicators should reflect all important aspects of the device. From this point of view, the system should be complete. The completeness of the system of indicators is important for the correct formulation of the optimization problem. Second, the system of indicators should be minimal. Increasing the number of indicators allows for more accurate characterization of the device. However, the excessive number of indicators and criteria complicates the issue of optimization.

Therefore, it is necessary to limit the number of indicators and criteria. On the one hand, this creates the conditions to reflect the main indicators of the composition of the device, and on the other hand, to have more information. The third factor is related to information. It is clear that many indicators can be functionally dependent on each other (for example, the mass of the device can depend on its geometric dimensions). In such cases, it is appropriate to consider functionally dependent indicators when setting the optimization issue.

6.3 Multi-criteria Optimization of Microwave Range Devices

The performance quality indicators (monotonic (unequivocal) dependence on the change of criteria $g_i, i = \overline{1, m}$) of the microwave range devices allows to formulate the parametric optimization problem as follows:

$$\min_{v \in V} g_i, i = 1, \overline{m}. \tag{6.1}$$

In writing the Eq. 6.1, the following is assumed: (*a*) the functional dependence of criteria $g_i = g_i(v)$ on the parameters of the V vector is determined with the help of the mathematical model of the microwave range device; (*b*) the criterion is written in such a way that a small value of $g_i(v) \geq 0, i = \overline{1, m}$ and $g_i(v)$ corresponds to the best quality of operation of the unit with the the microwave range; (v) The constraints on the parameters of the vector V are joint and the domain V is not empty. If the constraints on the maximum value of $g_i(v)$ are not taken into account in the formulation of the problem (6.1), then this leads to a formal solution of the equation, which may be unsatisfactory from a practical point of view. Therefore, we will accept the following constraint:

$$g_i(v) \leq g_i^+. \tag{6.2}$$

The value of the threshold g_i^+ is given in the requirements for the technical and economic indicators of the microwave range device. But in some cases, choosing g_i^+ can be difficult. Thus, if the threshold g_i^+ is small enough, then the region V may be empty. In [7], the method of optimal selection of threshold g_i^+ was considered [8], the algorithm used is based on the results obtained from the numerical modeling of a microwave range device.

The solution (6.1), as a result of solving the multi-criteria optimization problem, the optimal vector V, is determined. Unlike the one-criteria problem ($m = 1$), the solution of (6.1) is more complicated in the multi-criteria problem. The main problem here is to define the concept of optimization of vector V. In the case of a single-criteria optimization problem, the advantage of one objective function of the vector V_1 over the other objective function for the vector V_2 is clear:

$$g_1(v_1) \leq g_1(v_2). \tag{6.3}$$

In the case of multi-criteria optimization, this factor is more complicated:

$$([g_2(v_1) > g_2(v_2)]).$$

Let's assume that.

$$g_i(v_1) \leq g_i(v_2), i = \overline{1, m}\,, \tag{6.4}$$

$$g_j(v_1) < g_j(v_2). \tag{6.5}$$

In this case, the option corresponding to the values of the parameters of the V_1 vector of the microwave range device will definitely be better than the second option. At the same time point V_1 will be better than point V_2. Let's look at point $V_{\mathrm{eff}} \in V$. In this case, it is impossible to find a better point at point $v \in V$ than at point V_{eff}. This point is called the V_{eff} effective point. The effective point itself is not unique, that is, there is a set of V_{eff} points.

Let G denote the set of points of vector g. The flat areas of the boundary of G correspond to the $g_1(v) \leq g_1^+; g_2(v) \leq g_2^+$ constraints. The main indication that the vector V belongs to the V_{eff} set is that the point $g(v)$ is located in the lower left part of the G set. Such V-points facilitate the determination of effective points. Let's denote the criteria space of the set of effective points $g(V_{\mathrm{eff}})$ by G_{eff}. In two-category ($m = 2$) problems, the G_{eff} set forms a curved line, and this is called a compromise option.

The study of effective points and the composition of sets V_{eff}, G_{eff} facilitates the solution of (6.1) and allows the construction of methods for finding such points.

Based on the step of finding the set of effective points for the solution of (6.1), let's see their numerical determination methods [9]. The simplest of them are methods based on minimization within the $v \in V$ condition. Where

$$g(v) = \sum_{i=1}^{m} \lambda_i g_i(v), \tag{6.6}$$

where $\lambda_i \geq 0$ and $\sum_{i=1}^{m} \lambda_i = 1$ are the weight limits.

The assignment of λ_i in (6.6) makes it possible to determine all points of the V_{eff} set in many real cases. (6.6) is useful when the function $g_i(v)$ is a differentiable function of v. In this case, the minimization of (6.6) can be performed with the help of simple minimization methods for differentiable functions.

The effective points of the set V can also be determined by minimizing the criterion $g_i(v)$. In this case, the parameters of the vector v, which is the solution of the problem, belong to the set of points V_{eff} in all cases:

$$\min_{v \in V} g_i(v); g_i(v) \leq g_j^0, j = \overline{1, m}\,, j \neq i. \tag{6.7}$$

Giving different g_j^0 in (6.7) allows finding effective points. Note that the constraint in the form of the inequality in (6.7) can be replaced by the constraint in the form of the following equality:

$$g_j(v) \leq g_j^0.$$

In this case, the point v found as a result of solving (6.7) will be an effective point. A more common method of finding efficient points is based on the minimization of the maximum function

$$g(v) = \max_{i=\overline{1,m}} \lambda_i g_i(v), \tag{6.8}$$

where $\lambda_i \geq 0$ are weight strokes.

It should be noted that (6.8) is non-differentiable in the general case even when $g_c(v)$ is differentiable.

6.4 Approximation

The realization of the given frequency (dispersion) characteristic of the microwave range devices is the most important indicator of its effective operation. Therefore, optimization can be viewed as a matter of approximation of given characteristics. As a result of the approximation, a microwave band device should be created that simultaneously has several optimal dispersion characteristics. Therefore, in general, this issue is multi-criteria. Its writing in the form (6.1) is possible only after determining $g_i(v)$ criteria. In this case, the $g_i(v)$ criteria should reflect the characteristics of the dispersion characteristic of the device with microwave range.

The criteria are given on the basis of the requirements imposed on the $f(v, \theta)$ characteristics of the microwave range devices and the dependence of the E_θ set on the θ. At this time, the function $f(v, \theta)$ determines the output reflection coefficient, extinction, intensities of electric and magnetic fields, etc. can describe. θ can describe frequency, phase shift. E_θ can describe the sum of continuous intervals and discrete points in general. Let's look at the criteria set for the $f(v, \theta)$characteristics of a microwave range device. If the given function F_θ is approximated to the function $f(v, \theta)$ in the set E_θ, then the following equality is fulfilled as the $g(v)$ criterion [10]:

$$g(v) = \|f(v, \theta) - F(\theta)\|_{E_\theta}. \tag{6.9}$$

In many cases, equal-sized norms are used in (6.9). If the parameter E_θ falls into the interval $[\theta_1, \theta_2]$, then the criterion can be written as follows:

$$g(v) = \left(\int_{\theta_1}^{\theta_2} \lambda(\theta) |f(v, \theta) - F(\theta)|^P d\theta \right)^{1/P}, \tag{6.10}$$

$$g(v) = \max_{\theta \in [\theta_1, \theta_2]} \lambda(\theta) |f(v, \theta) - F(\theta)|. \tag{6.11}$$

The weight function $\lambda(\theta)$ in expressions (6.10) and (6.11) is given under condition $\lambda(\theta) > 0$. P is constant. If the set E_θ consists of a finite sum of points $\theta_i\left(i=\overline{1,m}\right)$, then criteria similar to (6.10) and (6.11) can be written as follows:

$$g(v)=\left(\sum_{i-1}^{m}\lambda_i|f(v,\theta_i)-F(\theta_i)|^P\right)^{1/P}, \tag{6.12}$$

$$g(v)=\max_{i=\overline{1,m}}\lambda_i|f(v,\theta_i)-F(\theta_i)|, \tag{6.13}$$

where $\lambda_i > 0$ are weight strokes.

Criteria (6.12), (6.13) can be obtained as a result of approximation of integrals (6.10). The weight function $\lambda(\theta)$ in the expressions (6.10) to (6.13) and the multiplication of λ_i are different from 1 if the θ at different points meet the requirement for the accuracy of $F(\theta)$. Increasing $\lambda(\theta)$ and λ_i leads to a decrease in the absolute value of the difference between $f(v,\theta)$ and $F(\theta)$ at points θ. For multi-criteria optimization as in (6.7) and (6.8), the use of (6.9) to (6.13) in finding the optimal values of $\lambda(\theta)$, λ_i limits during approximation should be performed by applying the iteration process. In the case of a set of F_θ, criteria, (6.9) to (6.13) can be generalized.

Suppose that E_θ consists of a point θ_0 and it is required to add m special derivatives of $f(v,\theta)$, $F(\theta)$ and θ to it. In this case, the following criteria can be given:

$$g(v)=\left(\sum_{i=0}^{m}\lambda_i\left|f^{(i)}(v,\theta_0)-F^{(i)}(\theta_0)\right|^P\right)^{1/P}, \tag{6.14}$$

$$g(v)=\max_{i=\overline{0,m}}\lambda_i\left|f^{(i)}(v,\theta_0)-F^{(i)}(\theta_0)\right|, \tag{6.15}$$

where the index i is the i-th derivative of θ.

In this case, λ_i weight multipliers allow adjusting the accuracy of the value of the function and its derivative during minimization. In practice, it is necessary to have characteristics of $f(v,\theta)$not lower than or not higher than $F(\theta)$.

Thus, the realization of the function ($f(v,\theta) \le F(\theta)$) in the interval $[\theta_1,\theta_2]$ is possible after the minimization of the following criteria:

$$g(v)=\left(\int_{\theta_1}^{\theta_2}\lambda(\theta)[f(v,\theta)-F(\theta)]^P+d\theta\right)^{1/P}, \tag{6.16}$$

$$g(v)=\max_{\theta\in[\theta_1,\theta_2]}\lambda(\theta)[f(v,\theta)-F(\theta)]_+, \tag{6.17}$$

where $[x]_+ = \max(0, x)$ is the intercept function.

$f(v, \theta)$ exact satisfaction of the imposed demand corresponds to the condition $g(v) = 0$. A widespread variant of the approximation problem corresponds to the requirement to minimize or maximize the function $f(v, \theta)$ within the condition $\theta \in E_0$. If it is $E_\theta = [\theta_1, \theta_2]$, this situation can be changed to what was mentioned earlier. At this time, $F(\theta) = 0$ is taken in the expression (6.16), (6.17). In the case of $f(v, \theta)$ maximization, it is necessary to replace function $f(v, \theta)$ with function $1/f(v, \theta)$ in expressions (6.16) and (6.17). The form of the approximation corresponding to $g(v)$ in (6.10), (6.11), (6.14) and (6.15) is called top, Chebyshev (equal) and maximal plane, respectively. Let's assume that criteria (6.12), (6.16) are met by the average approximation and (6.11)-Chebyshev approximation. In the expressions (6.10), (6.12), (6.16), the case $p = 2$ corresponds to the more common mean square approximation criterion. The above forms of approximation are widely used in the synthesis of devices with a microwave range. The choice of one or another form of approximation is determined by various factors.

Note that the average criterion is more favorable. So, in this case (especially at small values of p) $f(v, \theta)$ may have significant local deviations from $F(\theta)$. Maximal plane convergence criterion allows controlling only the local composition of the function $f(v, \theta)$. Therefore, the use of this criterion is only used to achieve convergence of $f(v, \theta)$ and $F(\theta)$ in a small interval of variation of θ. In the case of $p > 1$, the upper mean criterion is a differentiable function of v. This allows us to use the methods of minimization of the differentiable function to solve the approximation problem. The Chebyshev criterion is a non-differentiable function of v. Therefore, in order to solve extreme problems, it is necessary to use special methods that are more complex in this case. When using the maximal plane criterion, problems arise in calculating the high-order derivatives of function $f(v, \theta)$ with respect to θ.

From the point of view of the machine time consumption of the above-mentioned criteria, it can be concluded that the maximal plane, then Chebyshev and average criteria are more favorable.

Depending on the complexity of the mathematical model of the microwave range device, this difference increases even more. Let's mark the maximum value of criteria (6.16), (6.17) with δ. It is clear that $\delta \geq 0$. This case $\delta = 0$ corresponds to the identity of $f(v, \theta) \equiv F(\theta)$. If the parameters of the vector v are satisfactory, then $F(\theta)$ fulfills the condition of physical realization of a microwave range device.

6.5 Mathematical Programming Methods for the Optimization of Microwave Range Devices

Consider an extremal problem of the following form:

$$\min_{v \in V} g(v). \tag{6.18}$$

As a result of solving (6.18), the vector V^* corresponding to the minimum value of $g(v)$ in the set V is determined. Various aspects of the problem (6.18) and the problems arising from its solution have been considered in several works [11].$V \in E_n$ (E_n is an n-dimensional Euclidean space).

$$v = \lambda v_1 + (1 - \lambda)v_2 (0 \leq \lambda \leq 1)$$

When the condition is paid, it becomes sediment.

The function $g(v)$ defined by E_n becomes convex if arbitrary.

$$V_1, V_2 \text{ and } 0 \leq \lambda \leq 1$$

for

$$g(\lambda v_1 + (1 - \lambda)v_2) \leq \lambda g(v_1) + (1 - \lambda)g(v_2)$$

to be paid conditionally. If the function $[-g(v)]$ is convex, then $g(v)$ is convex. If the condition $g(v^*) \leq g(v)$ is satisfied for any $v \in V$, then the global minimum point of $g(v)$ from the set V is $v^* \in V$.

The point $v^* \in V$ is a local minimum of $g(v)$ when the inequality $g(v^*) < g(v)$ is satisfied for any $v \in V_\varepsilon \cap V$. If the vector $g^1(v_0)$ exists, then the function $g(v)$ is a function differentiable to V_0.

$$g(v) = g(v_0) + g^1(v_0)\Delta + 0(\|\Delta v\|), \Delta v = v - v_0, \tag{6.19}$$

where $0(x)$ is an infinitesimal quantity.

If $g(v)$ is twice continuously differentiable at v_0, then we get

$$g(v) = g(v_0) + g^1(v_0)\Delta V + 0,5v^T H(v_0)\Delta v + 0\left(\|\Delta v\|^2\right), \tag{6.20}$$

where the $H(v_0)$-second formulation is the matrix of special derivatives.

Derivative of function $g(v)$ with respect to$P(\|P\| = 1)$ directions

$$\frac{\partial g(v_0)}{\partial P} = \lim_{\alpha \to 0} \frac{g(v_0 + \alpha P) - g_0(v_0)}{\alpha}.$$

If $g(v)$ is – degenerate, then the derivative

$$\frac{\partial g(v_0)}{\partial P} = \max_{g^1(v_0)} g^1(v_0)P. \tag{6.21}$$

$g(v) \max_{i \in I} f_i(v)$ the derivative in the direction of the maximum function is defined analogously to (6.21):

$$\frac{\partial g(v_0)}{\partial P} = \max_{i \in I_0} f_i^1(v_0)P, \tag{6.22}$$

where I – a finite set of indices; $f_i(v)$ – soft function of v; $I_0 - f_i(v)$ is the set of indices of the function (in the case of $v = v_0$, this function is equal to $g(v_0)$, i.e., $I_0\{i \in I, f_i(v_0) = g(v_0)\}$). In (6.18), the set of v is given as systems of equality and inequality

$$f_{0j}(v) = 0, j \in J_0; f_j(v) \leq 0, j \in J, \tag{6.23}$$

where J_0, J – finite sets of indices; $f_{0j}(v)$, $f_j(v)$ are soft functions.

$g(v)$, $f_{0j}(v)$, $f_j(v)$, depending on the functions, mathematical programming problems are divided into separate classes: quadratic, geometric, convex, integer, etc. It is more appropriate to use nested programming methods in the optimization of devices in the microwave range. Therefore, let's focus on the programming method. Let's assume that the functions included in (6.18) are soft and exactly computable. Let's be satisfied with deterministic methods for solving (6.18). In this case, it is necessary to use logical and numerical operations to solve (6.18).

The minimization algorithm V_k is an iterative process consisting of constructing successive points, and this process is defined as:

$$V_{k+1} = V_k + \alpha P_k, \tag{6.24}$$

where the vector P_k indicates the displacement of the point V_k.

In general, it should be noted that the more effective algorithm for solving the optimization problem is the quasi-Newton minimization algorithm. At this point, the following equation holds.

$$P_k = -D_k g^1(V_k), \tag{6.25}$$

where D_k is a symmetric matrix.

There are different algorithms for constructing D_k. The most widespread of them are the following:

(a) Davidon-Fletcher-Powell algorithm [12].

$$D_{k+1} = D_k + \frac{\Delta V_k \Delta V_k^T}{\Delta V_k \Delta g_k^1} - \frac{D_k \Delta g_k^1 \left[D_k \Delta g_k^1\right]^T}{\Delta g_k^{1T} D_k \Delta g_k^1}; \tag{6.26}$$

(b) Broyden-Shenno-Fletcher algorithm [12].

$$D_{k+1} = D_k + \frac{\Delta V_k \Delta V_k^T}{\Delta V_k \Delta g_k^1} + \Delta g_k^{1T} D_k \Delta g_k^1 x_k x_k^T - - \frac{D_k \Delta g_k^1 \left[D_k \Delta g_k^1\right]^T}{\Delta g_k^{1T} D_k \Delta g_k^1}, \tag{6.27}$$

where $x_k = \Delta V_k / \Delta g_k^{1T} \Delta V_k - D_k \Delta g_k^1 / \left(\Delta g_k^{1T} D_k \Delta g_k^1 \right), D_0 = E. \Delta V_k = V_{k+1} - V_k;$ $\Delta g_{k1} = g^1(V_{k+1}) - g^1(V_k)$; T – imported operator; E – is a uniform matrix.

It is determined as a result of solving the one-dimensional minimization problem in direction α_k, P_k in quasi-Newton algorithms:

$$\alpha_k = \arg \min_{\alpha > 0} g(V_k + \alpha P_k).$$

In some cases, the application of Newton's method is considered appropriate. In this case

$$P_k = -H^{-1}(V_k) g^1(V_k). \tag{6.28}$$

A general programming problem can be transformed into a sequential unconditional optimization problem. These algorithms can be performed with the help of the method of centers, the replacement of unrelated variables, and the application of the Lagrange function.

The methods for searching for global extremums are more detailed [13]. Note that these specified methods can be applied when ($n \leq 5$) parameters smaller than five are included in the vector v.

6.6 Minimization of the Maximum Function

Let's look at the methods of minimizing a function of the following form:

$$g(v) = \max_{i \in I} f_i(v), \tag{6.29}$$

where I – set of finite indices; $f_i(v)$ is a continuously differentiable function.

The function (6.29) can arise during the solution of the multi-criteria optimization problem, Chebyshev approximation, system of nonlinear equations [14]., a method of converting a general mathematical programming problem to an unconditional minimization problem of the form (6.29) is given. The complexity of minimizing the maximum function (6.29) is that the function $g(v)$ is not differentiable. Therefore, the above methods cannot be used here. Let's look at the construction of algorithms for the minimization of the maximum function. For the minimization of $g(v)$ [15], general algorithms can be used. The simplest method for software implementation is the generalized gradient descent method. The disadvantage of these methods is that the function $g(v)$ is convex. In practical matters, this condition is rarely fulfilled. Other minimization methods are based on the piecewise-linear approximation of the function $g(v)$.

$$V^* = \lim_{P \to \infty} V_P^*, \tag{6.30}$$

where $V_P^* - g_P(v)$ is the minimum solution.

Let's show the simpler form of $g_P(v)$ as follows:

$$g_P(v) = \left(\sum_{i \in J} f_i^P(v) \right)^{1/P}. \tag{6.31}$$

Important results for the form of $g_P(v)$ in (6.31) [16]. Difficulties arise in minimizing (6.29) from large values of P parameters. Nevertheless, this approach is widely used in the development of devices with a microwave range. Introducing additional variables to minimize problem (6.29) turns it into a nonlinear programming problem.

$$\min_{v,\beta} \beta, f_i(v) - \beta \leq 0, i \in I. \tag{6.32}$$

Analogously, we get (6.29) for the minimization of the maximum function

$$\min_{v,\beta} \beta, f_i(v) - \beta \leq 0, i \in I; f_j(v) \leq 0, j \in J. \tag{6.33}$$

Thus, the minimization of the maximum function was brought to its equivalent forms (6.32), (6.33),

In solving some practical problems, it is appropriate to use the rule (6.23) to construct V_k. In this case, the vector P_k is determined at each step of the iteration process:

$$\begin{cases} \min_{P\beta} \left(\beta + 0,5\|P\|^2 \right); \\ f_i^1(v)P + f_i(v_k) - \beta \leq 0; i \in I_k; I_k = \{ i \in I, f_i(v_k) \geq g(v_k) - \delta \}, \\ \delta \geq 0; g(v_k) = \max_{i \in I} f_i(v_k), \end{cases} \tag{6.34}$$

where $\alpha_k = (0,5)^L$, $L = 0, 1, 2, \ldots$ is the first index and the following inequality is satisfied for it:

$$g\left(v_k + P_k/2^L\right) \leq g(v_k) - \varepsilon \|P_k\|^2 / 2^L, 0 \leq \varepsilon \leq 1. \tag{6.35}$$

The generalized form of the above algorithm [17] is given. This time

$$\min_{v,\beta}\left(\beta+0,5\|P\|^2\right);$$

$$f_i^1(v_k)P+f_i(v_k)-\beta\le 0,\, i\in I_k; f_j^1(v_k)P+f_j(v_k)\le 0,$$
$$j\in J_k; I_k=\{i\in I, f_i(v_k)\ge g(v_k)-\delta\},\delta\ge 0; \qquad (6.36)$$
$$J_k=\left\{j=J, f_j(v_k)\ge F(v_k)-\delta^1\right\},\delta^1\ge 0;$$
$$g(v_k)=\max_{i\in I} f_i(v_k); F(v_k)=\max_{j\in J}\left\{0, f_j(v_k)\right\},$$

where $\alpha_k=(0,5)^L$, $L=0, 1, 2, \ldots$ is the first index and the following condition is fulfilled for it:

$$g\left(v_k+P_k/2^L\right)+NF\left(v_k+P_k/2^L\right)\le g(v_k)+NF(v_k)-\varepsilon\|P_k\|^2/2^L, 0\le\varepsilon\le 1. \qquad (6.37)$$

$f_i^1(v)$ in (6.34) and (6.36) – gradient of functions in (6.37), N is a sufficiently large number. During the implementation of algorithms (6.34) to (6.37), additional problems of quadratic programming cause difficulties. In these problems, the objective function $\beta+0,5\|P\|^2$ is not completely collapsed, which complicates the solution of the problem. Therefore, let's generalize the algorithms (6.34) to (6.37) to overcome the above problem. Determining the vector P_k to minimize the unconstrained maximum function is considered as a solution to the problem.

$$\min_{P,\beta}\beta+0,5\left(\|P\|^2+\gamma\beta^2\right); f_i^1(v_k)P+f_i(v_k)-\beta\le 0, i\in I_k, \qquad (6.38)$$

where $\alpha_k=(0,5)^L$, $L=0, 1, 2, \ldots$ is the first index and the following inequality is satisfied for it:

$$g\left(v_k+P_k/2^L\right)\le g(v_k)-\varepsilon\|P\|^2/2L,$$
$$0\le\varepsilon\le(1+R_1)^{-1}; R_1=\max_{k=\overline{1,\infty}}\gamma\beta_k>-1. \qquad (6.39)$$

Algorithm (6.24) is analogous to algorithms (6.36), (6.37). Where P_k is the solution of the problem.

$$\min_{P,\beta}\beta+0,5\left(\|P\|^2+\gamma\beta^2\right); f^1(v_k)P+f_i(v_k)-\beta\le 0,$$
$$i\in I_k; f_j^1(v_k)P+f_j(v_k)\le 0, j\in J_k, \qquad (6.40)$$

where $\alpha_k=(0,5)^L$, $L=0, 1, 2, \ldots$ is the first index and the following condition is fulfilled for it:

$$g(v_k + P_k/2^L) + NF(v_k + P_k/2^L) \leq g(v_k) + \; + NF(v_k) - \varepsilon \|P_k\|^2/2^L, 0 < \varepsilon \angle (1 + R_1)^{-1};$$

$$R_1 = \max_{k=\overline{1,\infty}} \gamma\beta_k > -1. \quad (6.41)$$

The indices of the function I_k, C_k, $F(v)$ in (6.38), (6.40) are determined as in (6.34), (6.36). The objective functions in (6.38) and (6.20) are fully convex.

In some cases, the heuristic algorithm [18] is effective in minimizing the maximum function. In this case, P is considered as a solution to the problem.

$$\min_{P,\beta} \|P\|^2 + \gamma\beta^2; f_i^1(v_k)P + f_i(v_k) - \beta = 0, \quad (6.42)$$

where $\alpha_k = (0,5)^L$, $L = 0, 1, 2, \ldots$ is the first index and the following condition is fulfilled for it:

$$g\left(v_k + P_k/2^L\right) < g(v_k). \quad (6.43)$$

In certain cases, the advantage of algorithms (6.42), (6.43) over algorithms (6.34), (6.36) and (6.43), (6.38) is that the auxiliary problem in (6.42) can be explained by a very simple solution.

6.7 An Experimental Calculation Method for the Optimization of Microwave Range Devices

Devices with a microwave range must be developed several times during the research process. Each development project allows to develop a more improved microwave range device, to adjust its electrical, structural, operational parameters. Thus, the developed physical model is universal and its characteristics are more stable. Let's show the results of the development of the method of parametric optimization of devices with a microwave range. The basis of this is the optimal use of experience. This method allows you to use an improved mathematical model and get an accurate result. At this time, let's use numerical methods and a computer to solve the problem. Let's assume that the physical realization of the device with a microwave range is not the last stage of the synthesis process, but the intermediate-fifth stage (Fig. 6.3). Let's take the obtained optimal solution of the optimization problem (stage 4) as the first approach. Then the following expression can be written:

$$\min_V g(v), g(v) = \max_{\theta \in [\theta_1 \theta_2]} |f(v,\theta) - F(\theta)|, \quad (6.44)$$

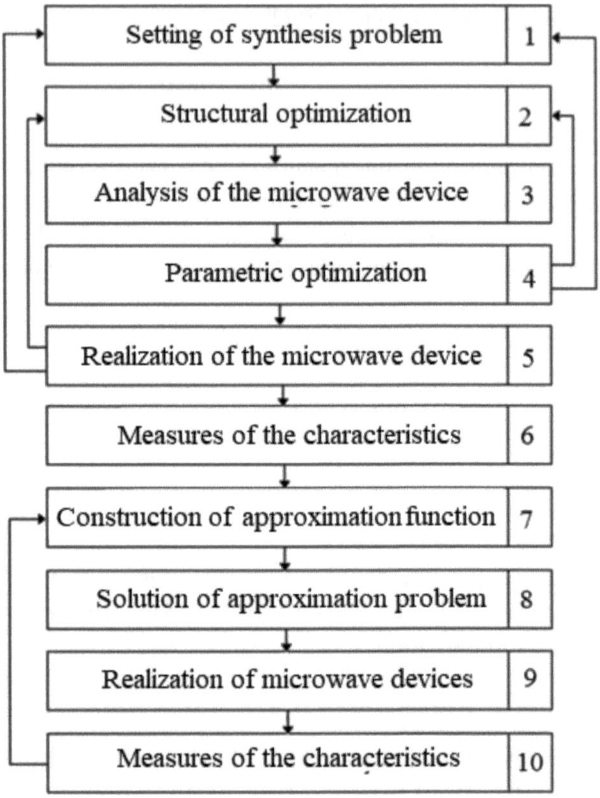

Fig. 6.3 Scheme of experimental-computational synthesis process of microwave range devices

where V is a vector of a set of parameters; θ_1, θ_2 – boundaries of the working frequency band; $F(\theta)$ – given approximated function; $f(v, \theta)$ is an approximation function.

After the fifth stage of the preparation of devices with a microwave range, its electrical characteristics are measured (sixth stage) and in the seventh stage, graphs of the following functions are constructed in the same coordinate system: $F(\theta)$ – approximated, $f(v, \theta)$ – calculated and $f_0^t(\theta)$ experimental. The working frequency band is divided into a grid consisting of m points so that it is possible to construct a new approximating function $F_1(\theta) = F(\theta) + \left[f\left(v_0^*, \theta\right) - f_0^t(\theta)\right]$ by finding $f\left(v_0^*, \theta\right) - f_0^t(\theta)$ differences at each point. At the eight stage, the following approximation problem is solved:

$$\min_V g_1(v), g_1(v) = \max_{\theta \in [\theta_1 \theta_2]} |f(v, \theta) - F(\theta)| \quad (6.45)$$

The optimal value of the vector v_1^* determined as a result of the solution of (6.45) is taken as the basis for building the natural model of the next microwave range device (stage 9). After that, an experimental analysis of its characteristics is carried out and new $f_1^t(\theta)$ characteristics are established (stage 10). If necessary, the entire experimental-calculation process can be repeated. However, studies show that an empirical calculation process can be sufficient. Thus, without knowing the exact mathematical model of the characteristic $f(v, \theta)$, it is possible to clarify how the function $f(v_0^*, \theta)$ changes as a result of the experiment. This change is determined by function $\Delta = f(v_0^*, \theta) - f_0^t(\theta)$, which is called the correction of the original function $f(\theta)$ approximating.

6.8 Optimizing Errors That Can Be Omitted in the Development of Microwave Range Devices

Designing microwave range devices with optimal remissible errors is an important issue in the synthesis process. Considering its difficulty, currently they are satisfied with the parametric optimization of devices in the microwave range. Therefore, sometimes this approach does not justify itself, and the issue of optimization of errors that can be omitted is relevant.

Assume that a microwave range device with a manufacturing cost of F has an allowable measurement error of ΔV. Where, $\Delta V = (\Delta V_1, \Delta V_2, \ldots \Delta V_n)$, ΔV_n is the allowable error of the i-th parameter.

At the same time, the allowable error of the device dimensions depends on the value of the parameters of the ΔV v vector. Therefore, the function F does not directly depend on these values.

Thus,

$$\min_{V, \Delta V} F(\Delta V), \tag{6.46}$$

find the limit.

Let's assume that during the development of a microwave range device, its parameters are realized independently of each other. The condition for the operation of the device is that the vector v is located in the region V. This is equivalent to fulfilling the following condition.

$$P_{v,\Delta v} \in V, \tag{6.47}$$

where $P_{v,\ \Delta v}$ is a parallelepiped with center point V and the length of its i-th edge is equal to $2\Delta v_i$.

Let's look at the possible solutions of (6.46) and (6.47). Let's denote the vector $\tau \in E_n$. So, its components $\tau_i = \Delta V_i / \|\Delta V\|$. $\|\|$ symbols show the Euclidean norm. For a fixed value of τ, let's accept the following norm:

$$\rho_\tau(v) = \max_{i=\overline{1,n}} \frac{|\Delta v_i|}{\tau_i}. \tag{6.48}$$

It is clear that the limit $\rho_\tau(v - y)$ is equal to half the length of the diagonal of the parallelepiped $P_v(y, \tau)$. At the same time, the maximum allowable error for the device parameters is as follows:

$$\Delta v_i = \tau_i \min_{y \in \partial v} \rho_\tau(v - y), i = \overline{1, n} . \tag{6.49}$$

6.9 Realization of Optimal Synthesis of Devices with Microwave Range

The combination of Empipe 3D software with High Frequency Structure Simulator (HFSS) software can be effectively applied in the optimal synthesis of devices in the microwave range. These applications include waveguide transformer, waveguide rejector filters, maximum waveguide transition, waveguide bridges, etc. used in synthesis. However, this program complex was not used in the process of optimal synthesis of microwave rectangular and circular waveguides. Therefore, this chapter deals with the issue of optimal synthesis of microwave rectangular and circular waveguides using Empipe 3D software and HFSS software. Empipe 3D allows you to enter structural and physical parameters of microwave rectangular and circular waveguides. Using HFSS built-in 3D plotting tool, microwave rectangular and circular waveguides are created in several variants (projects) with different parameters or dimensions. These parameters are determined by the increment of parameter changes in the structure. Empipe provides information about changes in 3D parameters by reading geometric data (Fig. 6.4).

Empipe 3D uses several optimizers. Among them, the least squares method, the minimax method, the quasi-Newton method, and the random search method can be shown as an example (Fig. 6.5).

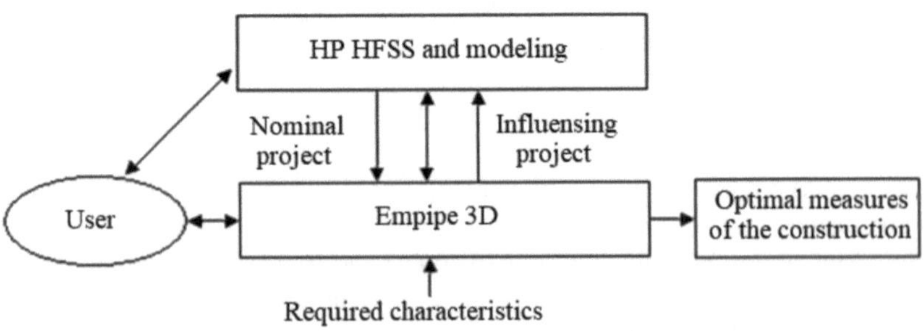

Fig. 6.4 Data flood in Empipe 3D system

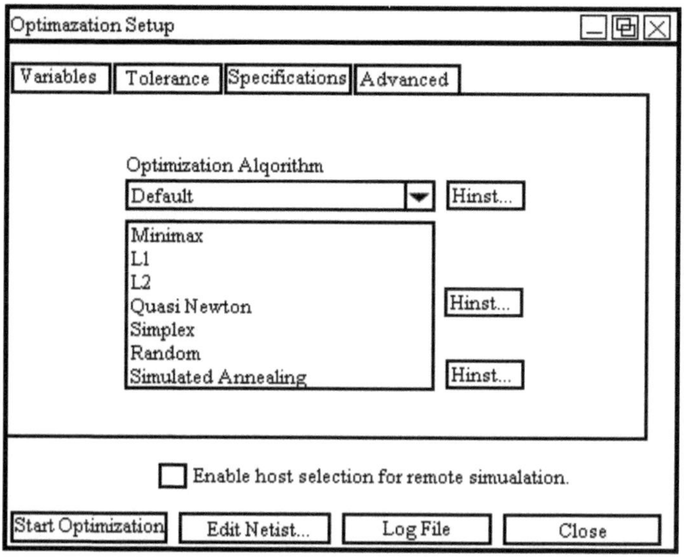

Fig. 6.5 Optimization method selection page in Empipe 3*D*

Empipe 3*D* software uses parameter discretization. This is only intended to speed up the optimal synthesis process. Modeling of the electromagnetic field of microwave and circular waveguides can be significantly simplified by applying interpolation. This is particularly important in choosing the gradient method of optimal synthesis (Fig. 6.5).

Linear or quadratic interpolation can be selected in the software complex.

6.10 Selection of Criteria in Optimal Synthesis of Microwave Rectangular and Circular Waveguides

Let's assume that the operator L is the operator of the analysis problem of microwave rectangular and circular waveguides. The operator L is characterized by the electric and magnetic field intensity of these devices E, H (a, b, r) and the wave propagation constant $\gamma(f)$.Where f is the normalized frequency of microwave rectangular and circular waveguides.

$f = 4.9–7.05$ GHz for the microwave rectangular waveguide we are looking at, and $f = 9$ GHz for the microwave circular waveguide.

$$\gamma(f) = L[E, H(a, b, r)]. \tag{6.50}$$

Let's denote the required dependence of the normalized propagation constant of waves by $\widetilde{\gamma}(f)$. The proximity of the functions $\gamma(f)$ and $\widetilde{\gamma}(f)$ is determined with the help of the evaluation function $F = \{L[E, H(a,b,r)], \widetilde{\gamma}\}$.

$$F = \{L[E, H(a,b,r)], \widetilde{\gamma}\} = \max_{f \in [f_1 f_2]} |L[E, H(a,b,r)] - \widetilde{\gamma}|, \tag{6.51}$$

or

$$F = \{L[E, H(a,b,r)], \widetilde{\gamma}\} = \int_{f_1}^{f_2} \sigma(f)(L[E, H(a,b,r)] - \widetilde{\gamma})^2 df, \tag{6.52}$$

where f_1, f_2 are the operating frequency range of microwave rectangular and circular waveguides. The inclusion of a weighting function $\sigma(f)$ allows the exact realization of $\widetilde{\gamma}(f)$ in separate subbands of the range f. Physical and structural parameters of devices should be taken into account when choosing the methods for solving the problem of optimal synthesis. So, the following conditions must be taken into account. For microwave rectangular and circular waveguides, these are:

$$0 < E_{\min}, H_{\min} \le E, H(a,b,r) \le E_{\max}, H_{\max},$$

$$E, H(a,b,r) \in M,$$

$$N^P[E, H(a,b,r)] = F\{L[E, H(a,b,r)] - \widetilde{\gamma}\} + \\ + pT[E, H, (a,b,r)], \tag{6.53}$$

$$F = \{L[E, H(a,b,r)], \widetilde{\gamma}\} \le \Delta. \tag{6.54}$$

Let's look at the optimal synthesis of a microwave rectangular waveguide in the Empipe 3*D* program under the specified conditions (Fig. 6.6).

In the process of solving the problem, the following should be considered:

- Empipe 3*D* intensity description.
- parameterization of geometric dimensions.
- providing the objective function.
- release of electromagnetic modeling of Empipe 3*D* program.
- performance of optimization.
- saving optimized solutions in memory.

Figure 6.6 shows the optimized parameters *a*, *b*. Optimization of these parameters is performed with a step of 1 GHz *in* the frequency range $f = 4.9$–7.05 GHz. Let's use the bend 1 function of the project. For this, let's use the box subroutine and the Sweep operation (Figs. 6.7 and 6.8).

Fig. 6.6 Optimized parameters of a microwave rectangular waveguide

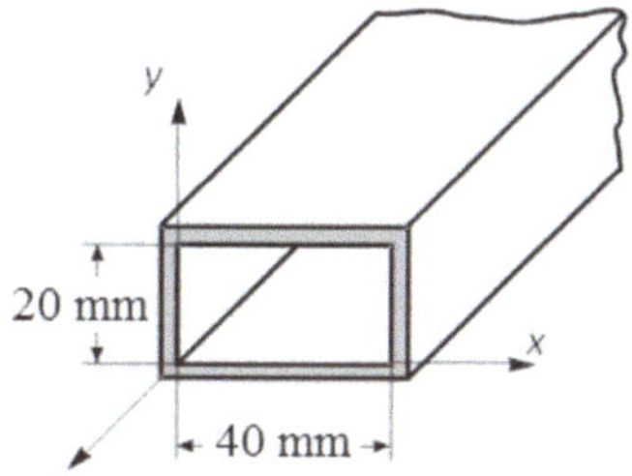

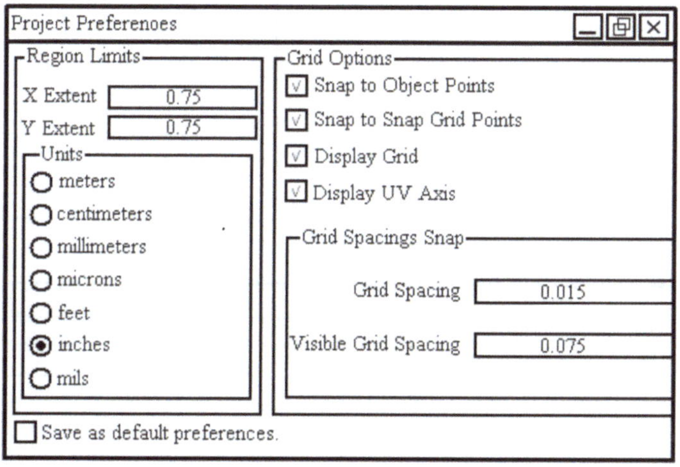

Fig. 6.7 Building the project

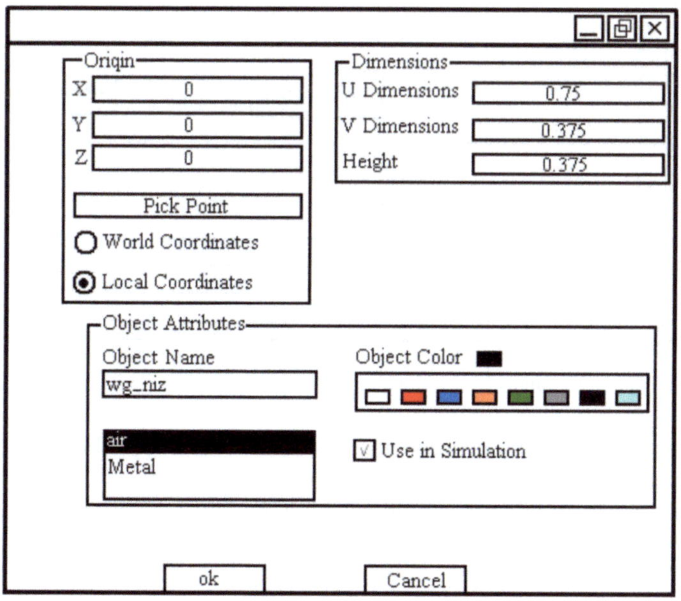

Fig. 6.8 Low microwave rectangular waveguide template

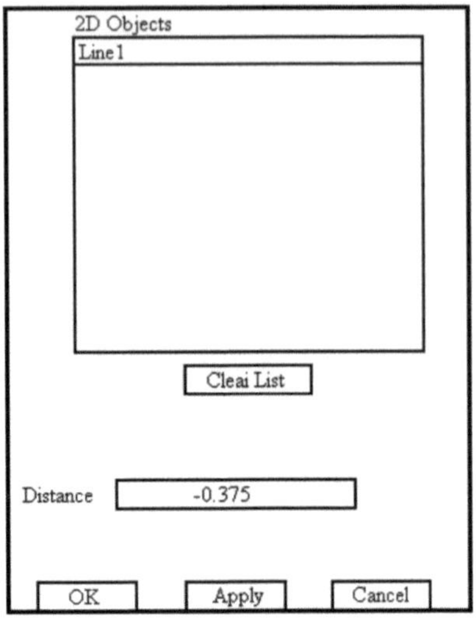

Fig. 6.9 The Sweep command creates a microwave rectangular waveguide of the desired shape

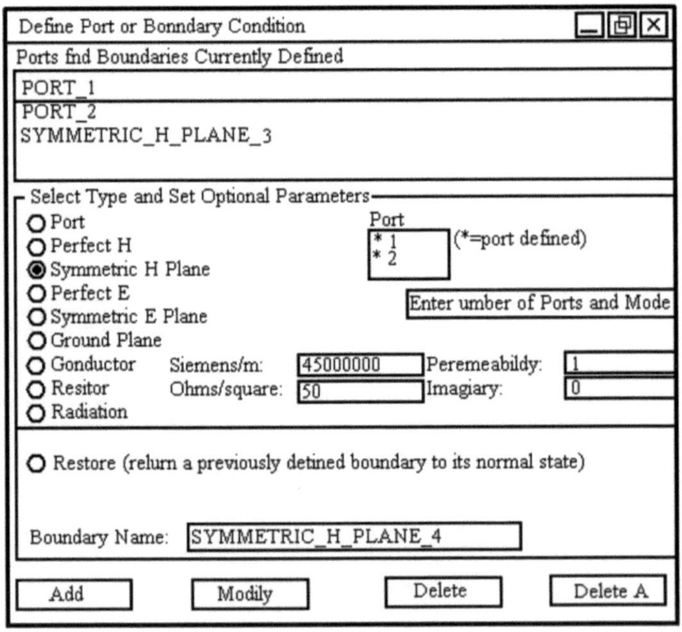

Fig. 6.10 Inclusion of boundary conditions

Let's perform a Sweep operation at a distance of 0.375 inch (Fig. 6.9) to draw an average microwave rectangular waveguide.

Let's show the boundary conditions of a microwave rectangular waveguide (Fig. 6.10).

The remainder of the boundary of a microwave rectangular waveguide is the OUTER boundary, which is equivalent to the metallization surface of the device.

After building the base project and calculating its characteristics (the frequency range of the analysis (4.9–7.05 GHz frequency range divided into 5 points)) the Empipe 3*D* program is released.

6.11 Release of Empipe 3*D*

Start>Programs>HP HFSS>Empipe 3*D* buttons are clicked and then the project manager Empipe 3*D* appears on the monitor. This dialog box (Fig. 6.11) is similar to the HP HFSS project dialog box.

Any directory listed in "Directoris" can be selected. The names displayed in "Procect Folders" are from the HP HFSS project. "Empipe 3*D* Procects" shows and optimizes the project.

Work is performed step by step with the helper (Procect Mizard). So, at this time, it is necessary to answer the assistant's questions and fill in the necessary windows. After pressing the New button, we get to the first step of the assistant (Fig. 6.12).

You can enter an arbitrary name in the window. Project optimization will include several projects. The first of them is the nominal project to be created by HFSS (Fig. 6.13).

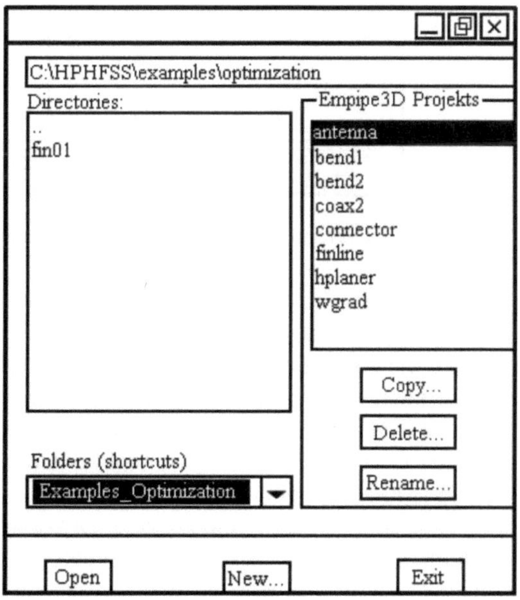

Fig. 6.11 Empipe 3*D* project manager window

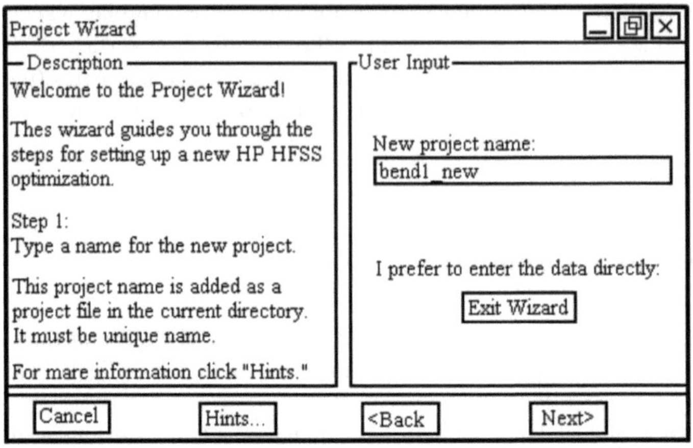

Fig. 6.12 Naming the optimized project in the first step

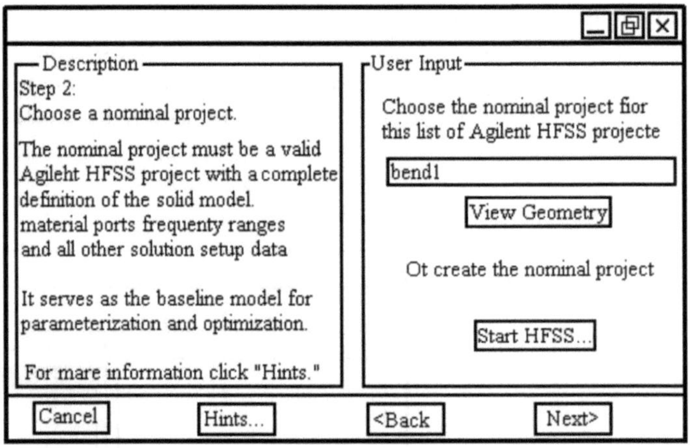

Fig. 6.13 In the second step, selecting a nominal project from the HP HFSS project list

If the "Open" button is pressed in the window shown in Fig. 6.11, then the view of the project already created for optimization is opened. If the "New..." button is clicked, the optimization process assistant will be launched.

The name of the nominal project already in the Empipe *3D* project folder is selected. Then you can view the geometric view of the nominal project by clicking the View Geometry button. For complete clarity, you can click on the "Hints" button (Fig. 6.13) in these and other steps of the optimization process. You can also click the "Start HFSS..." button to create a normal project.

In the third step, it is necessary to enter the geometric dimensions of the project (Fig. 6.14). These dimensions may change during the optimization process.

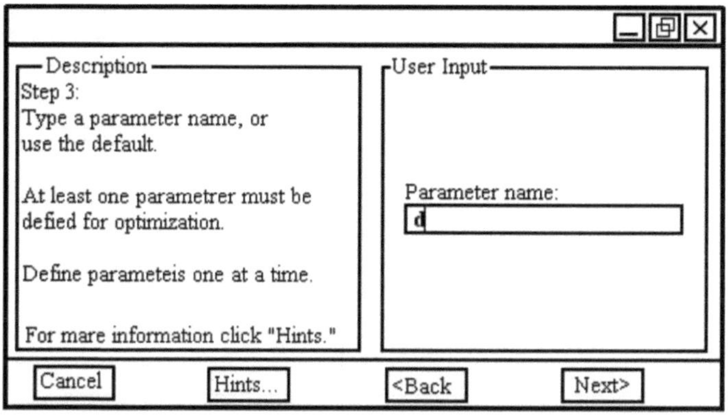

Fig. 6.14 In the third step, the selection of the optimization parameter

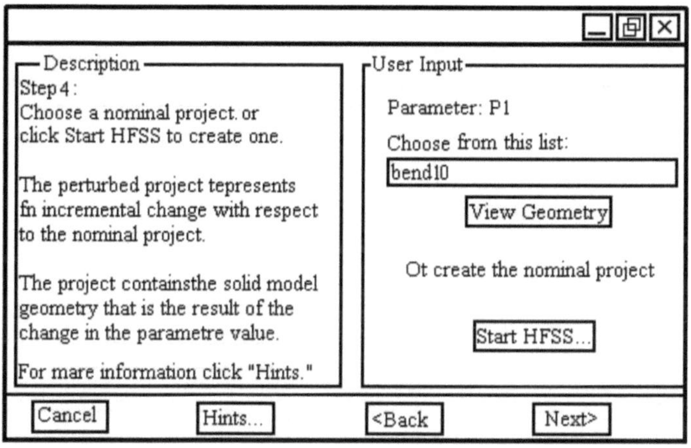

Fig. 6.15 In the fourth step, choosing a variable project

The parameter can be given an arbitrary name. By pressing the “Next” button, a dialog window appears and the variable project is created there. The installed variable project will exactly coincide with the nominal project except for the d parameter (Fig. 6.15).

A variable project can be set up by clicking the “Start HFSS...” button. Then the dialog shown in Fig. 6.16 is called by pressing the “Next” button. In this dialog window, the values $a = 40$, $b = 20$ for the nominal project, $a = 20$, $b = 10$ for the variable project are entered.

During the process of searching for the optimal structure of a microwave rectangular waveguide, the range of changes of parameters a, b is divided into N parts (Fig. 6.17).

Then let’s enter the lower and upper limits of the change of parameters a, b (Fig. 6.18). This will help you get the right physical model.

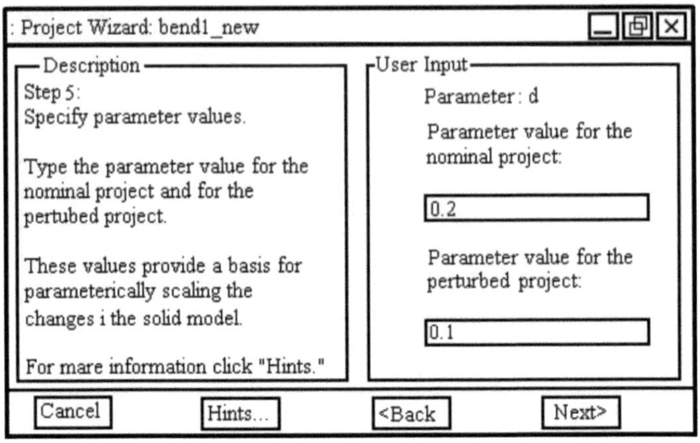

Fig. 6.16 Entering variable parameters in step five

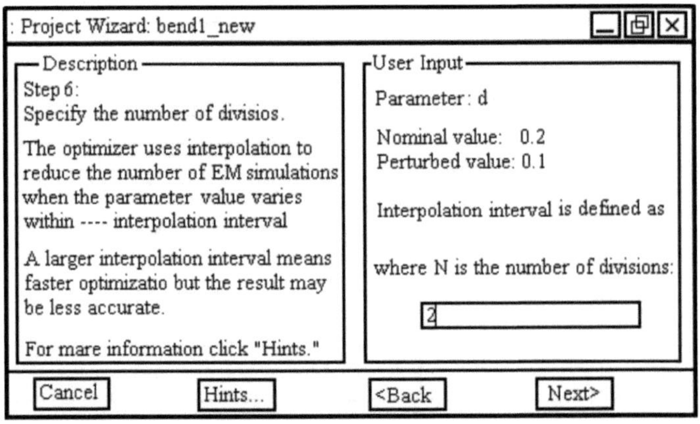

Fig. 6.17 In the sixth step, dividing the range of parameters into N parts

After that, the dialog shown in Fig. 6.19 appears. After entering all the parameters in the dialog window shown in Fig. 6.20, the "Done" button is pressed. This completes the project creation phase with the wizard and returns to the dialog shown in Fig. 6.21.

In the dialog window shown in Fig. 6.20, the "Simulate..." button is clicked and the frequency characteristic of the microwave rectangular waveguide is obtained.

It should be noted that the names of nominal and variable projects are known. Therefore, in the project manager dialog window, select optimize_examples, then bend 1 from Empipe 3*D*, then Open. With these, the main window of Empipe 3*D* shown in Fig. 6.21 appears. In Empipe 3*D*, the bend 1 project is used as a working project. That is, it is improved each time during the optimization process. The subdirectories used in the optimization process are shown in Table 6.1.

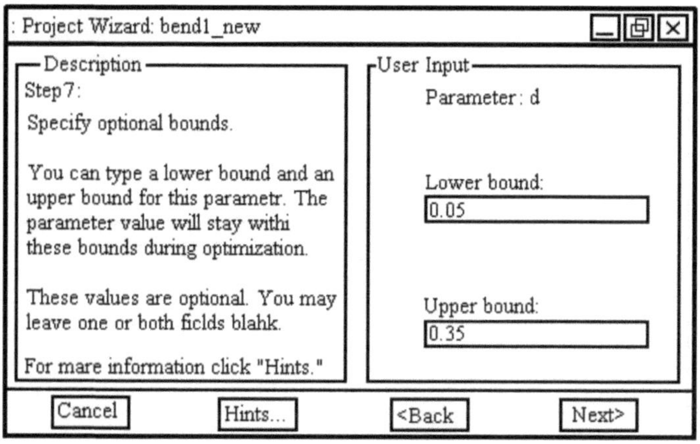

Fig. 6.18 Entering the lower and upper bounds of the variation of optimized parameters in the seventh step

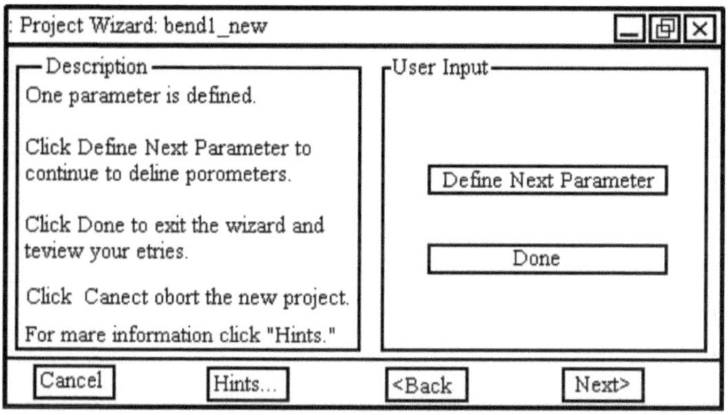

Fig. 6.19 Move to next parameter or end of optimization input

Where, bend 10 is a nominal project. It contains nominal (initial) geometric dimensions, boundaries, frequency range and other data of microwave rectangular and circular waveguides. It also serves as a base project for parameterization.

To visually see the spatial model of the nominal project, you need to press the View Geometry button (Fig. 6.22).

To exit this command, enter the command File>Return to Main. In HFSS, geometric symmetry can be used to reduce computation time. Therefore, the Perfect *H* Boundary and Perfect *E* Boundary commands must be included when analyzing a microwave rectangular waveguide with HFSS. The actual dimensions of the microwave waveguide are 40 × 20 mm.

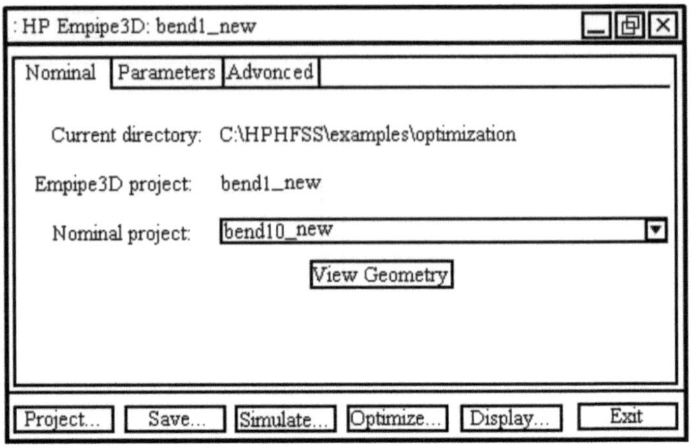

Fig. 6.20 Analysis and optimization management dialog window

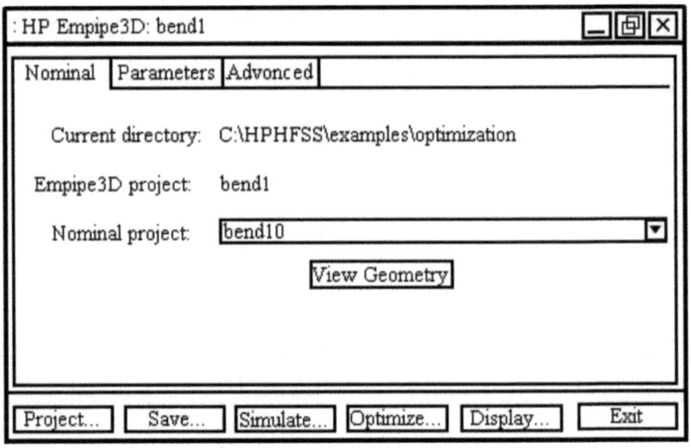

Fig. 6.21 Empipe 3*D* main window

Table 6.1 Subdirectories used in Empipe 3*D*

Bend 1	The name of the working project created by Empipe 3*D*
Bend 1_opt	Configuration and database files
Bend 10	Nominal project
Bend 11	Changing project

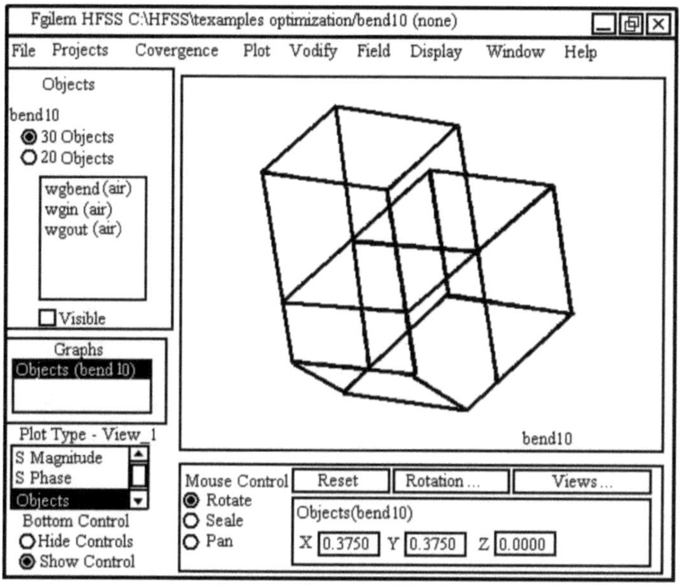

Fig. 6.22 Space model of the nominal design of a microwave rectangular waveguide

To prepare microwave rectangular and circular waveguides for optimization, it is necessary to build a new design for each parameter of these devices. By comparing the nominal and variable designs, Empipe 3*D* automatically records the information for the parameter change value corresponding to the spatial model.

Note that in bend 1 there are parameters *a*, *b*. The new project includes an increase in the change of parameters *a, b*. Two values of each parameter of nominal and variable projects are optimal. The two values of each parameter may not be their minimum or maximum values. They only indicate the direction of change. The step can be negative or positive. That is, the values of parameters *a*, *b* can both increase and decrease.

When preparing nominal and variable projects, all information about the composition, boundaries and parameters of materials in both projects should be entered into the program. In other words, both projects should be ready for optimization. In the main window of Empipe 3*D*, the Parameters sub-window is selected. Then the parametrized shape of the microwave rectangular waveguide can be seen (Fig. 6.23).

The List of Parameters field lists all the parameters. After selecting the parameters shown in the list, information about their composition appears. For the considered bend 1 project, parameters *a*, *b* are variable parameters. Information about the composition of parameters can be obtained with the help of the following subprograms.

- Parameter name.

It is an ASJII arbitrary string and cannot exceed 32 characters. Thus, a name was chosen for the parameters *a*, *b*.

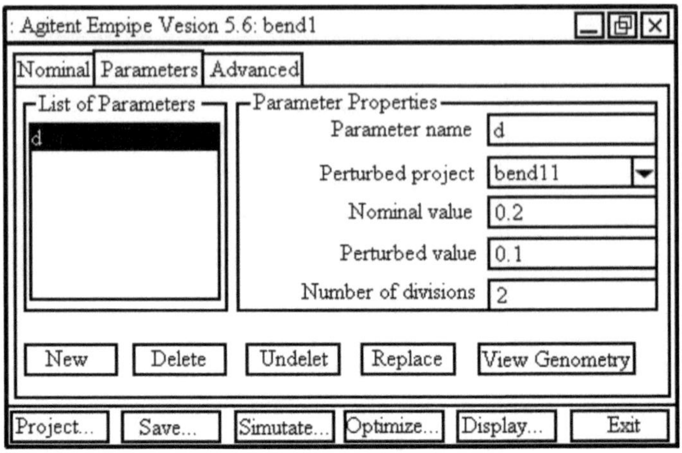

Fig. 6.23 Assigning parameterization

- Pertubed project.

This subroutine compares variable and nominal projects and shows its change growth. Bend 11 is the project consisting of the increase in the change of parameters *a*, *b*.

- Nominal value.

The nominal price refers to the parameters described in the nominal project. The nominal design values for microwave rectangular and circular waveguide are $a \times b = 40 \times 20$ mm and $d = 60$ mm, respectively. These values are entered into the program as simple numbers. Physical units are set in the HP HFSS program complex itself.

- Perurbated value.

This parameter is the parameter after the variable increment.

- Number of divisions.

This input parameter gives the interpolation interval as:

Interpolation Interval = |Perturbed Value – Nominal Value|/Number of Divs.

To reduce the amount of modeling, Empipe 3*D* uses interpolation. If parameter values vary within an interpolation interval, then Empipe 3*D* uses *S*-parameter interpolation in conjunction with HFSS. This is also an important feature when calculating the gradient. The number of sections for parameters *a*, *b*, *d* is equal to 2 (Fig. 6.23). So, since the change increment is equal to 60 mm/2 = 30 mm, the interpolation interval is 30 mm/2 = 15 mm. For $a \times b = 40 \times 20$ mm, the change increment is 20×10 mm, so the interpolation interval will be 10×5 mm. If, during optimization, *a*, *b* changes in the range of 20×10 mm to 25×15 mm, and for *d* parameter, it varies in the range of 30–35 mm. Then Empipe 3*D* calls HFSS for the

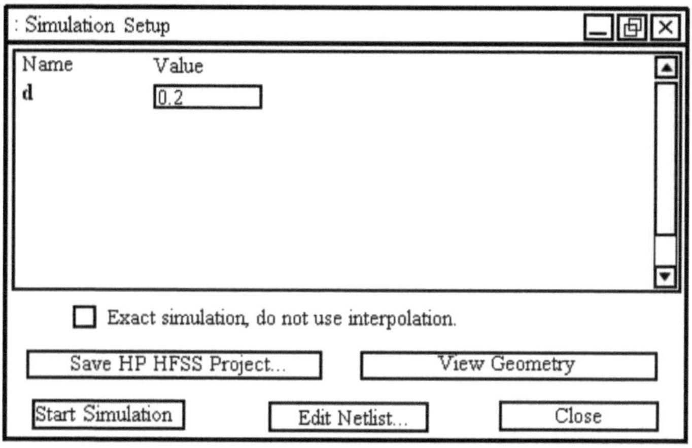

Fig. 6.24 The window that performs the modeling

following boundary points, respectively: $a \times b = 20 \times 10$ mm and $a \times b = 25 \times 15$ mm; $d = 30$ mm, $d = 35$ mm and then interpolation is performed. This is to obtain the *S*-parameters of other values of these parameters between the values of 20×10 mm and 25×15 mm for $a \times b$ and 30 mm and 35 mm for *d*. The smaller the number of partitions, the larger the interpolation interval or helps the optimization to complete faster. However, in this case, the obtained result may be inaccurate. On the other hand, if the interpolation interval is too small, the number of HFSS modeling increases, ultimately leading to an unsatisfactory use of interpolation. It should be noted that during the optimization process, it is necessary to start with a large interval and then gradually reduce the size of the interval according to the reduction of the objective function. Before optimization, modeling is performed and characteristics are checked according to technical requirements. Then the Simulate button is pressed in the main window of Empipe 3*D* and a dialogue window appears (Fig. 6.24). Then the Start Simulation button is pressed.

Note that the *S*-parameters of devices calculated with HP HFSS are already stored in the database. Therefore, the modeling will be completed very quickly. If there is no database, then Empipe 3*D* calls the HP HFSS program and a control window appears.

After the modeling is completed, return to the Empipe 3*D* main window and click the Display button. To exit this window, click File>Exit.

Then click the Optimize button in the Empipe 3*D* window and the Optimizaton Setup window will open. Variables enumerates each optimization parameter. If Varibles? flag is set, then the corresponding parameter is marked as a variable. That is, its value is given by the optimizer. Parameters not marked as variable parameters remain constant during optimization. Value – this value is used as an initial value during optimization (Fig. 6.25).

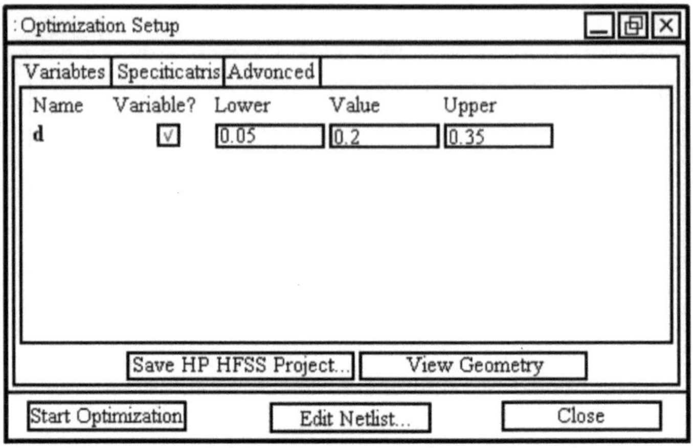

Fig. 6.25 The window for determining the change limit of variables *a*, *b*

If the parameter is selected as a variable, then Upper and Lower limits are set. This is to limit the ranges of parameter changes during optimization.

To determine the objective function, the following sequence should be followed. First select Specifications. In general, to determine the technical requirements, the following is necessary:

1. Selecting the frequency range.
2. Selecting the *S*-parameter or impedance characteristics.
3. Choosing the type of objective function.
4. Enter numerical values for optimization.
5. Enter the weight factor.
6. Press the Add button.

Let's set the technical demand for bend 1 as shown in Fig. 6.26.

Let's look at the selection of the frequency range. All technical requirements are given in the frequency range. The frequency range for the optimization is a set of discrete frequencies set only by the nominal HFSS project (with the Solve>Setup commands).

The following Response menu is used to select characteristics in Empipe 3*D*:

It is necessary to select two numbers of the port for giving *S*-parameters, and one number for the impedance characteristic.

The list below allows you to select a technical requirement.

The symbols $<$,$>$, and $=$ represent lowercase, uppercase, and equal, respectively.

The desired parameter is entered to determine the technical requirement. At the same time, the weight factor can be included. The weight factor must be a positive number.

That is, the Add button is pressed to set the technical requirements. This adds a new row to the Specifications Currently Defined list. This is done based on the required characteristics and weighting factor.

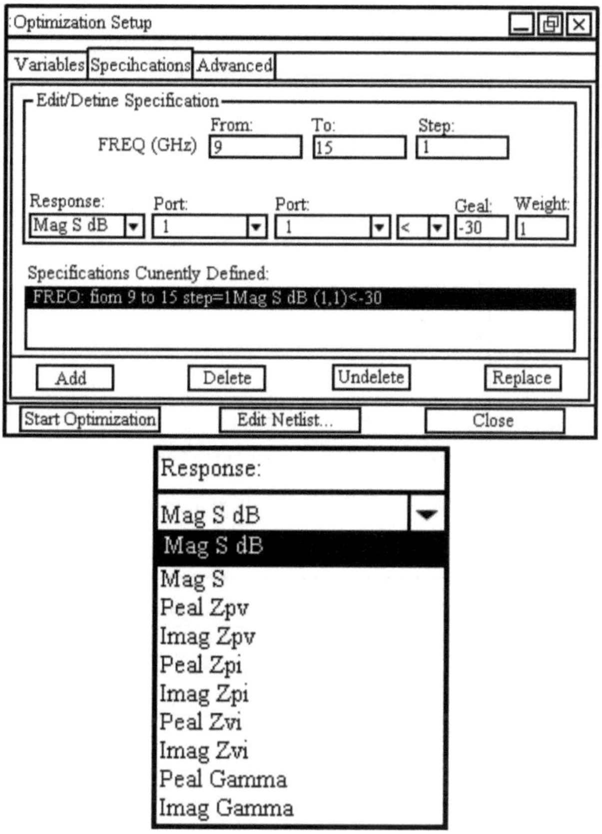

Fig. 6.26 A window that defines the type of the objective function

Then the optimization process starts. For this, the Start Optimization button is pressed and the optimization window appears on the monitor. This window shows the optimization process and the calculation of the objective function at each iteration (Fig. 6.27).

The results of the modeling using the HFSS software were stored in the database. Therefore, the optimization process will be performed quickly.

If there is no database, then Empipe 3*D* calls HFSS and the HFSS task execution window appears.

Empipe 3*D* software chooses the effective method needed for optimization.

For microwave rectangular and circular waveguides, the optimator minimax method is selected. In this case, Max Error indicates the largest deviation between the calculation and the required characteristic. Click Ok to close the Optimization Monitor window and go to the Optimization Setup window. After that, the value of the optimized parameter appears (Fig. 6.28).

Then the View Geometry button is clicked in the window shown in Fig. 6.30 and the HFSS program is called and the optimized structure of microwave rectangular and circular waveguides is displayed (Figs. 6.29 and 6.30).

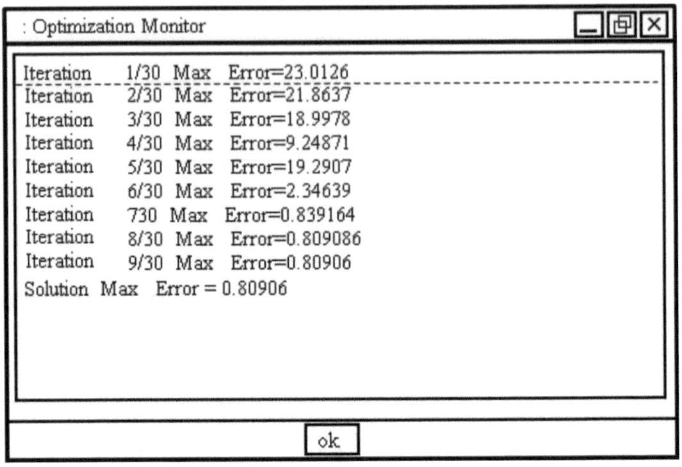

Fig. 6.27 Optimization window for entering iteration number and objective function value

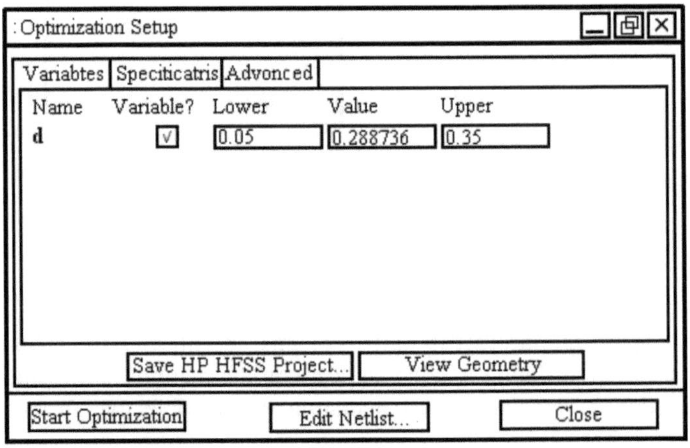

Fig. 6.28 Optimizing parameter value setting window

In the Optimization Setup window, click the Close button. Next, in the main window of Empipe 3*D* you need to press the Display button.

To get better characteristics of the designed construction (device), it is necessary to make it a little more complicated, and then a more accurate construction can be obtained (Fig. 6.30).

The frequency characteristic of the microwave rectangular waveguide with improved parameters is shown in Fig. 6.31, and the frequency characteristic of the circular waveguide is shown in Fig. 6.32.

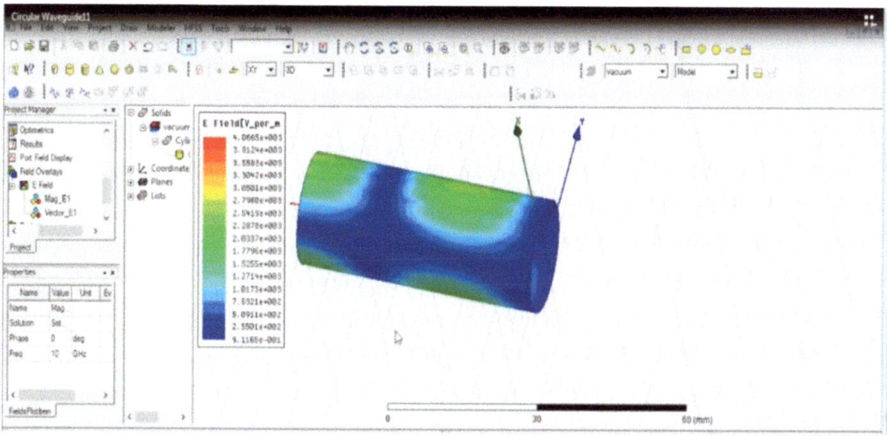

Fig. 6.29 View window of the space model of an optimized microwave circular waveguide

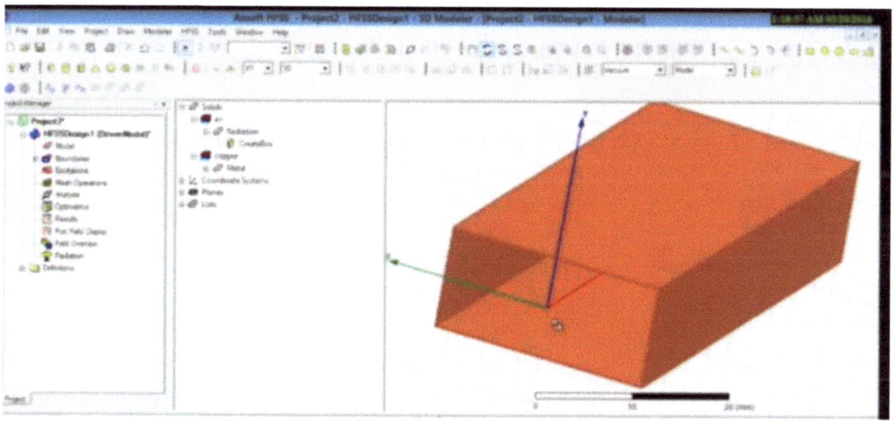

Fig. 6.30 View window of the space model of the optimized microwave rectangular waveguide

It should be noted that during the optimization of the structures of connected (connected) objects, the optimization process remains unchanged despite the fact that the geometric dimensions of each object are different. Therefore, such optimization can be called structural optimization. In modern optimization, the method that realizes the synthesis of the structure is also called the genetic optimization method. Thus, the microwave rectangular waveguide obtained after optimal synthesis is shown in Fig. 6.33.

Its dimensions after structural optimization are 47.55 × 22.15 mm. The frequency range corresponding to these measurements is $f = 4.9$–7.05 GHz. It can be seen that since the frequency range of the microwave rectangular waveguide is reduced after

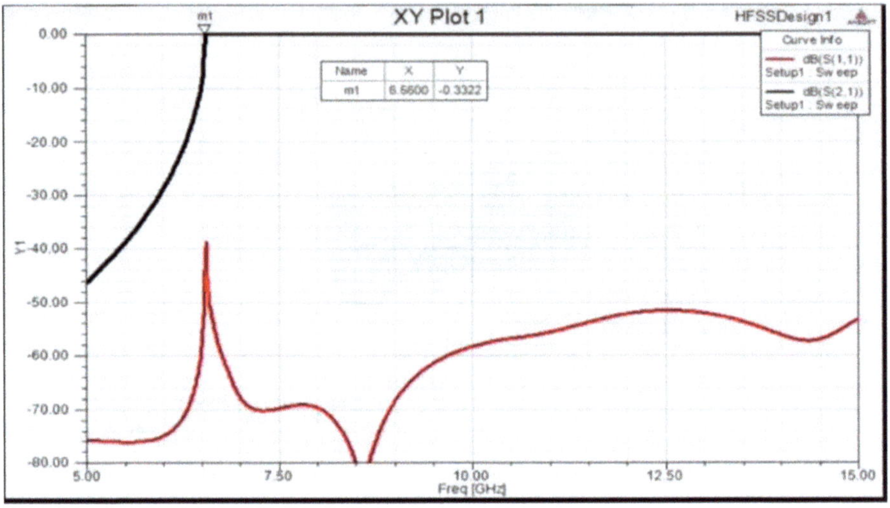

Fig. 6.31 Frequency characteristics of a microwave rectangular waveguide with improved parameters

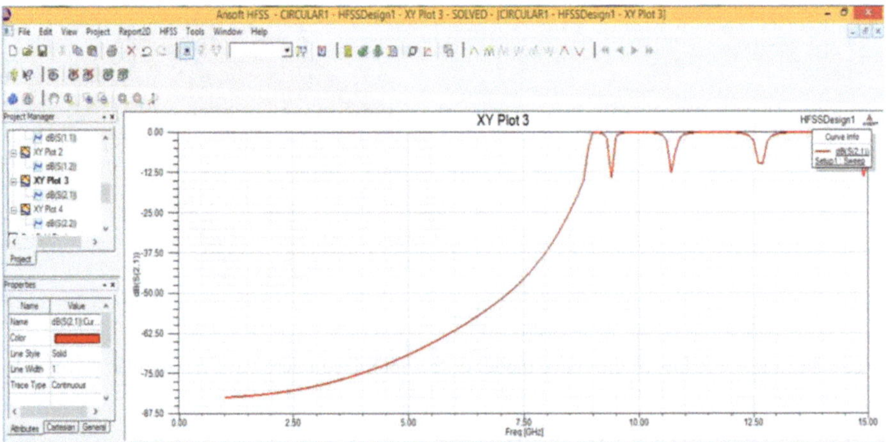

Fig. 6.32 Frequency characteristics of a microwave circular waveguide with improved parameters

optimization, the attenuation coefficient is reduced from 0.0431 to 0.0201 dB/m. And for the microwave circular waveguide, the operating frequency range has been reduced from 9 to 7.5 GHz. This, in turn, allowed the attenuation coefficient to decrease from 0.0527 to 0.0211 dB/m.

The parameters of the improved microwave rectangular and circular waveguides are shown in Tables 6.2 and 6.3.

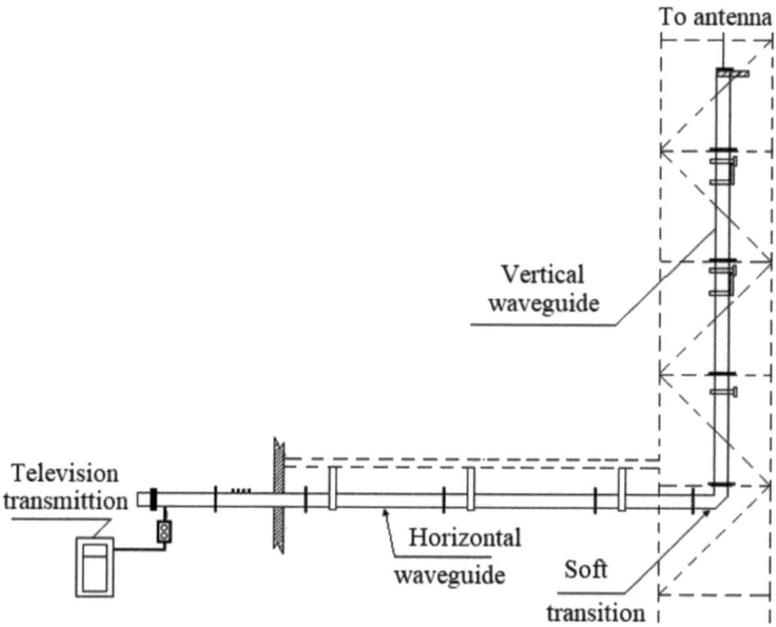

Fig. 6.33 A new waveguide tract system developed to connect the transmitter and antenna in TV towers

Table 6.2 Parameters of the microwave rectangular waveguide obtained after optimization

Parameters of the microwave rectangular waveguide before optimization			
Operating frequency range, *GHz*	Cross-sectional area, *mm*		Attenuation coefficient for brass transducer, dB/m
	A	*b*	
4.9–7.05	40	20	0.0431
Parameters of the microwave rectangular waveguide after optimization			
3.94–5.99	47.55	22.15	0.0201

Table 6.3 Parameters of the microwave circular waveguide obtained after optimization

Parameters of the microwave circular waveguide before optimization		
Operating frequency range, *GHz*	Diameter, *d*, *mm*	Attenuation coefficient for brass transducer, dB/m
9	50	0.0422
Parameters of the microwave circular waveguide after optimization		
7.5	60	0.0211

Thus, a waveguide tract system, shown in Fig. 6.33, connecting the transmitter to the antenna from a microwave rectangular waveguide, developed to improve the quality of broadcasts in television towers, was proposed.

6.12 Conclusions

1. It was determined that the optimal synthesis of devices with a microwave range is based on the search for the optimal structure of the electromagnetic field of these devices. The optimality of this structure depends on the selection of criteria, the nonlinearity of the environment, and the types of waves. Unsatisfactory selection of the optimal structure of the electromagnetic field of rectangular and circular waveguides in the microwave range cannot be improved by compensating in the parametric optimization process.
2. Performance of parametric optimization of *E*-type and *H*-type microwave range rectangular and circular waveguides is defined in two stages. In the first stage, a normal (initial given) rectangular and circular waveguide was created and the fulfillment of the technical requirements was assessed. The second stage is the verification stage, based on the higher-level model, and ensures the refinement of its parameters.
3. Optimization of *E*-type and *H*-type wave microwave range devices leads to computational generalizations. At the same time, using these generalizations several times during the optimization process saves time and resources.
4. One of the main stages of the process of optimal synthesis of *E*-type and *H*-type wave microwave range devices is the construction of their mathematical models. For the optimal organization of the synthesis process, the construction of at least two mathematical models of these devices – basic (approximate, single-wave) and auxiliary (refined) models is determined.
5. For the first time, parametric analysis and structural optimization of rectangular and circular waveguides from microwave range devices using HFSS and Empipe 3*D* software complex was carried out. According to the results obtained from the conducted studies, the structural dimensions of the specified devices have been significantly improved.
6. It is justified to use the *S*-parameters of the microwave band device as an objective function during the optimization process.
7. The problem of minimizing the maximum function is solved.
8. New algorithms for the optimal synthesis of *E*-type and *H*-type wave microwave range devices were given, their advantages were justified, and new rectangular and circular waveguides were developed based on these algorithms. The electrical, magnetic, structural, operational parameters and characteristics of these transducers have been significantly improved compared to the previous ones.
9. On the basis of the developed waveguides, a new waveguide tract system connecting the transmitter and the antenna was developed to improve the quality of transmission in TV towers.

References

1. Singh, R. R., & Priye, V. (2018). Numerical analysis of film-loaded silicon nanowire optical rectangular waveguide: An effective optical sensing. *Nano-Micro Letters, 13*(9), 1291–1295.
2. Singh, V., & Bhattacharyya, S. (2021). A free space frequency-time-domain technique for electromagnetic characterization of materials using reflection-based measurement. *International Journal of RF and Microwave Computer-Aided Engineering, 31*(1), 55–67.
3. Song, K., et al. (2019). High-isolation diplexer with high frequency selectivity using substrate integrated waveguide dual-mode resonator. *IEEE Access, 7*, 116676–116683.
4. Stratis, I. G., & Yannacopoulos, A. N. (2015). Some remarks on a class of inverse problems related to the parabolic approximation to the Maxwell equations: A controllability approach. *Mathematical Methods in the Applied Sciences, 38*(17), 3866–3878.
5. Sun, D. Q., & Xu, J. P. (2022). Real time rotatable waveguide twist using contactless stacked air-gapped waveguides. *IEEE Microwave and Wireless Components Letters, 27*(3), 215–217.
6. Sun, K., et al. (2016). Fields and wave modes analysis of rectangular waveguide filled with uniaxial medium. *IEEE Transactions on Microwave Theory and Techniques, 64*(11), 3429–3440.
7. Taghizadeh, H., et al. (2019). Grounded coplanar waveguide-fed compact MIMO antenna for wireless portable applications. *Radioengineering, 28*(3), 528–534.
8. Taisir, H. I., & Zoubir, M. H. (2010). Array pattern synthesis using digital phase control by quantized particle swarm optimization. *IEEE Transactions on Antennas and Propagation, 58*(6), 2142–2145.
9. Tian, D., & Chen, Y. (2021). Optical waveguides in organic crystals of polycyclic Arenes. *Advanced Optical Materials, 9*(23), 55–65.
10. Trujillo-Flores, J., et al. (2020). CPW-fed transparent antenna for vehicle communications. *Applied Sciences, 10*, 6001.
11. Tsuburaya, T., et al. (2017). Fast computation of linear systems based on parallelized preconditioned MRTR method supported by block-multicolor ordering in electromagnetic field analysis using edge-based finite element method. *Electronics and Communications in Japan, 100*(8), 59–70.
12. Vakili, B., et al. (2015). All-optical switching using a new photonic crystal directional coupler. *Advanced Electromagnetics, 4*(1), 63–67.
13. Van den Brande, Q., et al. (2018). Highly efficient impulse-radio ultra-wideband cavity-backed slot antenna in stacked air-filled substrate integrated waveguide technology. *IEEE Transactions on Antennas and Propagation, 66*(5), 2199–2209.
14. Varshney, P. K., & Akhtar, M. J. (2019). A compact planar cylindrical resonant RF sensor for the characterization of dielectric samples. *Journal of Electromagnetic Waves and Applications, 33*, 1700–1717.
15. Varshney, P. K., & Akhtar, M. J. (2021). Permittivity estimation of dielectric substrate materials via enhanced SIW sensors. *IEEE Sensors Journal, 21*, 12104–12112.
16. Wang, L. F., et al. (2005). Electromagnetic scattering model for rice canopy based on Monte Carlo simulation. *Progress In Electromagnetics Research, 52*, 153–171.
17. Wang, R. Q., & Jiao, Y. C. (2019). Synthesis of wideband rotationally symmetric sparse circular arrays with multiple constraints. *IEEE Antennas and Wireless Propagation Letters, 18*, 821–825.
18. Wang, W., et al. (2019). A waveguide slot filtering antenna with an embedded metamaterial structure. *IEEE Transactions on Antennas and Propagation, 67*(5), 2953–2960.

Chapter 7
Modeling of Microwave Waveguide Systems of a Special Design

7.1 Electrodynamic Characteristics of a Specially Designed Rectangular Waveguide

Currently, antenna cages are widely used to transmit and receive information. The most widespread of the antenna arrays is the waveguide-slit antenna arrays. Waveguide-slot antenna cages have a number of advantages, which include: no additional protrusions; the possibility of exciting several irradiators at the same time, etc.

Analysis of waveguide-slit antenna cages with analytical methods [1] can be carried out by numerical methods as well. The most efficient of these numerical methods is the Galerkin method [2]. In this case, the equivalent magnetic currents $J_n(s_n)$ in each N slot are searched in the form of a series, that is, in the form of a trigonometric linear independent basis function [3]:

$$J_n(s_n) = \sum_{p=1}^{P} J_{np} \sin \frac{p\pi(L_n + s_n)}{2L_n}, \tag{7.1}$$

where s_n – the local coordinates of the slit with length $2L_n$, P – the total number of basis functions, J_{np} – unknown coefficients.

During the implementation process of Galerkin method, it is necessary to numerically solve the system of $\{N \times P\}$-order linear algebraic equations. In this case, the calculation time increases proportionally to the limit of $\{(N \times P) \times N\}$ [4].

A number of authors are satisfied with performing the approximation of the current in the gap in the form of a function ((case $P = 1$ in expression 7.1)) [5].

However, such an approach is correct for the value of the ratio of λ wavelengths to the slit length equal to 0.5. For other values of ratio $2L_n/\lambda$, this approach is not justified. If the condition for the ratio $(2L_n/\lambda) \neq 0.5$ is fulfilled, including if the distance between adjacent slits is significantly smaller than the wavelength λ_g in the

I. Islamov, *Radio Engineering and Telecommunications Waveguide Systems in the Microwave Range*, https://doi.org/10.1007/978-3-031-37916-1_7

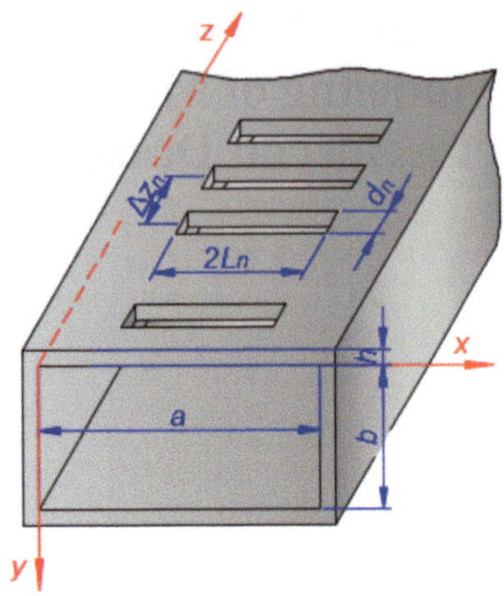

Fig. 7.1 Rectangular waveguide-slit system

waveguide, the waveguide-slit antenna lattice behaves like an antenna lattice radiating electromagnetic waves [6]. As a rule, during the analysis of such antennas, a number of conditions are accepted: the number of slits is infinite [7]; the thickness of the waveguide system is zero [8]; the length of the slit coincides with the dimensions of the large wall of the waveguide [9]; the electric field in the gap is in the form of a half-wave sinusoidal [10]. In a number of works [11] during the study of the characteristics of the system, the above conditions were not considered, and the issue was solved by applying Galerkin's method. However, due to the conditions mentioned above, the effectiveness of the considered solution decreases as the number of radiators increases during the numerical implementation.

Therefore, with the help of the approximation function, the issue of radiation of electromagnetic waves of the waveguide-slit system placed on the large wall of the rectangular waveguide for magnetic currents has been solved. It is taken into account that the system is sufficiently large and finite. In order to evaluate the effect of the parameters of the considered structure on its electrodynamic characteristics, the calculation of the radiation coefficient and the directivity diagram was performed. At the same time, the numerical results obtained by applying the Galerkin method were compared with the results obtained from the experiment.

The considered system consists of N number of thin linear slits

$$\{d_n/(2L_n)\} << 1, \ \ \{d_n/\lambda\} << 1,$$

where d_n – the width of the slit is the n-th slit. These slits are placed symmetrically to its longitudinal axis on the large wall of the rectangular waveguide with thickness h and cross-sectional area $a \times b = 20 \times 40$ mm (Fig. 7.1). Current in the gap

$$J_n(s_n) = J_{0n} f_n(s_n), \ \ f_n(\pm L_n) = 0,$$

where $f_n(s_n)$ is the given function.

When approximating this function in the form of using the boundary condition of the tangential summation of the magnetic field, we get a system of linear algebraic equations with respect to the amplitude constant of the current J_{0n} for each slot:

$$\sum_{m=1}^{N}\sum_{n=1}^{N} J_{0n} Y_{mn}^{Wg,R,Hs}(kL_m, kL_n) = -\frac{i\omega}{2k} M_m(kL_m), \tag{7.2}$$

where ω – circular frequency, $k = 2\pi/\lambda$,

$$Y_{mn}^{Wg,R,Hs}(kL_m, kL_n) = \\ = \frac{1}{2k}\int_{-L_m}^{L_m} f_m(s_m)\left[\begin{array}{l}\left(\frac{d^2}{ds_m^2}+k^2\right)\times \\ \times\int_{-L_n}^{L_n} f(s_n')G_{s_m}^{Wg,R,Hs}(s_m, s_n')ds_n'\end{array}\right]ds_m \tag{7.3}$$

– normalized ($m = n$) and specific ($m \neq n$) permeability of the slit;

$$M_m(kL_m) = \int_{-L_m}^{L_m} f_m(s_m) H_{0s_m}(s_m)\, ds_m; \tag{7.4}$$

– magnetic force; $G_{s_m}^{Wg,R,Hs}$ – Green's function for an infinite (semi-infinite) Wg rectangular waveguide, R rectangular resonator and H_s infinite screen, H_{0s_m} – the projection of the magnetic field of external sources on the axis of the slits.

Let's assume that the H_0 amplitude, H_{10} wave propagates in the region $z = -\infty$ in the system. Let's choose the function $f_n(s_n)$ for the slit located in the large wall of the rectangular waveguide:

$$f_n(s_n) = \cos ks_n \cos k_c L_n - \\ - \cos kL_n \cos k_c s_n, \tag{7.5}$$

where $k_c = 2\pi/\lambda_c$, λ_c – is the critical wavelength of the H_{10} wave.

If we consider expression (7.5) in expressions (7.3) and (7.4), we will find all the coefficients of the system of linear algebraic equations (7.2). We get the following expressions for the permeability of the slits and the magnetomotive force:

$$Y_{mm}^{Wg}(kL_m, kL_n) = \frac{2\pi}{ab} \sum_{\mathrm{m}=1,3\ldots}^{\infty} \sum_{\mathrm{n}=0}^{\infty} \frac{\varepsilon_{\mathrm{n}}(k^2 - k_x^2)}{kk_z} e^{-k_z \frac{d_m}{4}} I_{Wg}^2(kL_m), \tag{7.6}$$

$$\begin{aligned} Y_{mn}^{Wg}(kL_m, kL_n) = \frac{2\pi}{ab} \times \\ \times \sum_{\mathrm{m}=1,3\ldots}^{\infty} \sum_{\mathrm{n}=0}^{\infty} \frac{\varepsilon_{\mathrm{n}}(k^2 - k_x^2)}{kk_z} e^{-k_z|z_m - z_n|} I_{Wg}(kL_m) I_{Wg}(kL_n), \end{aligned} \tag{7.7}$$

$$\begin{aligned} Y^R_{\left\{\begin{smallmatrix} mm \\ mn \end{smallmatrix}\right\}}(kL_m) = \frac{2\pi}{L_m d_m} \sum_{\mathrm{m}=1,3\ldots}^{\infty} \sum_{\mathrm{n}=0}^{\infty} \frac{\varepsilon_{\mathrm{n}}(k^2 - k_{xR}^2)}{kk_{zR}} \left\{\begin{matrix} \coth(k_{zR}h) \\ 1/\mathrm{sh}(k_{zR}h) \end{matrix}\right\} \times \\ \times \cos\frac{\mathrm{n}\pi}{2} \cos\left(k_{yR}\frac{3d_m}{4}\right) I_R^2(kL_m), \end{aligned} \tag{7.8}$$

$$\begin{aligned} Y_{mn}^{Hs}(kL_m, kL_n) = \frac{1}{2k}\{(k\cos k_c L_m \sin kL_m - k_c \cos kL_m \sin k_c L_m) - \\ - \int_{-L_n}^{L_n} f_n(s_n') \left[G_{s_m}^{Hs}(L_m, s_n') + G_{s_m}^{Hs}(-L_m, s_n')\right] ds_n' - \\ - k_g \cos kL_m \int_{-L_m}^{L_m} \cos k_c s_m \left[\int_{-L_n}^{L_n} f_n(s_n') G_{s_m}^{Hs}(s_m, s_n') ds_n'\right] ds_m\Bigg\}, \end{aligned} \tag{7.9}$$

$$M_m(kL_m) = H_0 \frac{1}{k} e^{-\mathrm{i}k_g z_m} F(kL_m). \tag{7.10}$$

In expressions (7.6, 7.7, 7.8, 7.9 and 7.10), the following notations are accepted:

$$\begin{aligned} I_{Wg(R)}(kL_m) = \\ = 2\left\{\frac{k \sin kL_m \cos k_{x(xR)}L_m - k_{x(xR)} \cos kL_m \sin k_{x(xR)}L_m}{k^2 - k_{x(xR)}^2} \cos k_c L_m - \right. \\ \left. - \frac{k_c \sin k_c L_m \cos k_{x(xR)}L_m - k_{x(xR)} \cos k_c L_m \sin k_{x(xR)}L_m}{k_c^2 - k_{x(xR)}^2} \cos kL_m \right\}, \end{aligned}$$

$$\begin{aligned} F(kL_m) = 2\cos k_c L_m \frac{\sin kL_m \cos k_c L_m - (k_c/k)\cos kL_m \sin k_c L_m}{1 - (k_c/k)^2} - \\ - \cos kL_m \frac{\sin 2k_c L_m + 2k_c L_m}{2k_c/k}, \end{aligned}$$

$$k_x = \frac{\mathrm{m}\pi}{a}, \quad k_y = \frac{\mathrm{n}\pi}{b}, \quad k_z = \sqrt{k_x^2 + k_y^2 - k^2},$$

$\varepsilon_{\mathrm{n}} = (2 - \delta_{0\mathrm{n}})$ – Neumann score,

$$k_{xR} = \frac{m\pi}{2L_m}, \quad k_{yR} = \frac{n\pi}{d_m}, \quad k_{zR} = \sqrt{k_{xR}^2 + k_{yR}^2 - k^2},$$

m, n are integers, $z_m(z_n)$ is the coordinate of the axis of the m-th slot.

It should be noted that the proposed approximation function for the current distribution in the slits of the rectangular waveguide allows obtaining expressions in the form of quickly assembled series for the expressions Y_{mn}^{Wg}, Y_{mn}^{R}. In expression (7.8), the double integral can be converted into a single integral. This result allows to significantly reduce the time spent on calculations.

The reflection coefficient S_{11}, transmission coefficient and radiation power coefficient $|S_\Sigma|^2$ of the main type wave of the considered structure are as follows:

$$S_{11} = \frac{2\pi i k_g}{abk^3}\left\{\sum_{n=1}^{N} J_{0n} F(kL_n) e^{-ik_g z_n}\right\} e^{2ik_g z},$$
$$S_{12} = 1 + \frac{2\pi i k_g}{abk^3}\left\{\sum_{n=1}^{N} J_{0n} F(kL_n) e^{ik_g z_n}\right\}, \tag{7.11}$$

$$|S_\Sigma|^2 = 1 - |S_{11}|^2 - |S_{12}|^2. \tag{7.12}$$

The directional pattern of the antenna cage in the E plane is defined as follows:

$$F(\theta) = \sum_{n=1}^{N} J_{0n} \frac{\sin\left(\frac{kd_n}{2}\cos\theta\right)}{\frac{kd_n}{2}\cos\theta} e^{ikz_n\cos\theta}. \tag{7.13}$$

Figure 7.2 shows the theoretical and experimental ($h = 1.0$ mm) dependencies of the radiation power coefficient ($|S_\Sigma|^2$)of a rectangular waveguide with a cross-

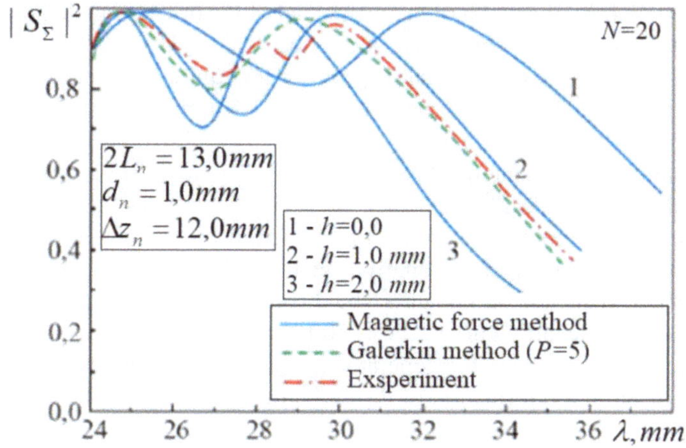

Fig. 7.2 Theoretical and experimental dependencies of radiation coefficient on power

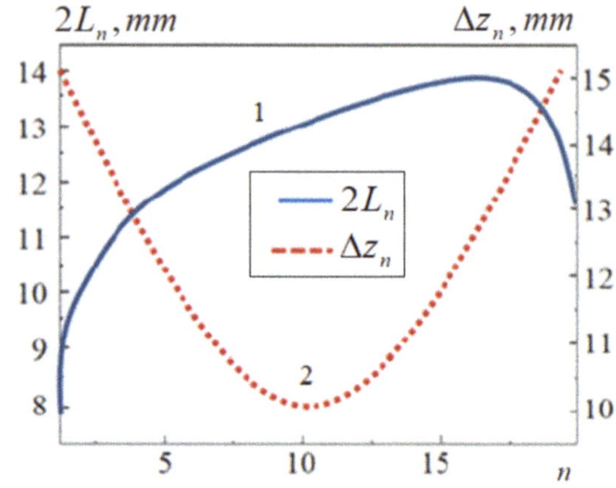

Fig. 7.3 Dependencies of variation of slit length (1 dependence) and distance between slits (2 dependence)

sectional area of $a \times b = 40 \times 20$ mm and slits on its large wall (with 20 slits) placed at a distance of ($2L_n/\lambda = 0.4$) from each other. Comparison of theoretical and experimental dependencies shows that the approximation function of the current in the slits varies according to the same law. It should be noted that increasing the thickness of the waveguide wall has a serious effect on the frequency characteristics of the considered structure.

At the same time, the dependencies of the length of the slit (dependency 1) and the distance between the slits (dependency 2) are shown in Fig. 7.3, and the effect of the change of individual parameters of the system on the orientation diagram is shown in Fig. 7.4.

Figure 7.4 also shows the graphs of the normalized distribution of the amplitude of the current along the length of the cage $\overline{J}_{0n} = |J_{0n}| / |J_{0n}|_{\max}$. It can be seen from Fig. 7.4 that when ($\Delta z_n = const = 12.0$ mm,Fig. 7.4, *a*) the combined length of the slits leads to a decrease in the level of both the near and far side lobes of the orientation diagram. At the same time, at different values of the distance between the slits (Fig. 7.4, *b*), the width of the main petal of the orientation diagram does not change, but the small side petal decreases to the value of $\theta \approx 90\,^\circ$.

Figure 7.5 shows the directional diagram of an equidistant antenna array. This cage consists of 100 slits with a length of $2L_n = 11.5$ mm. The distance between the axes is $\Delta z_n = 10.0$ mm (curve 1, $\lambda = 32.0$ mm; $d_n = 1.0$ mm; $h = 1.0$ mm). The power radiation coefficient of this system is equal to 0.9134, and the current amplitude distribution decreases slowly within the interval of $1.0 \div 0.3$ (Fig. 7.5). If we increase the number of slits in this structure by two times and reduce the distance between them accordingly, then $|S_\Sigma|^2$ increases significantly ($|S_\Sigma|^2 = 0.9972$). However, in this case, the directional effect coefficient of the cage will decrease significantly (curve 2, $N = 200$, $\Delta z_n = 5.0$ mm). This is explained by the fact that the slit closer to the generator is excited more effectively, while others are less involved in the formation of the radiation field of the system.

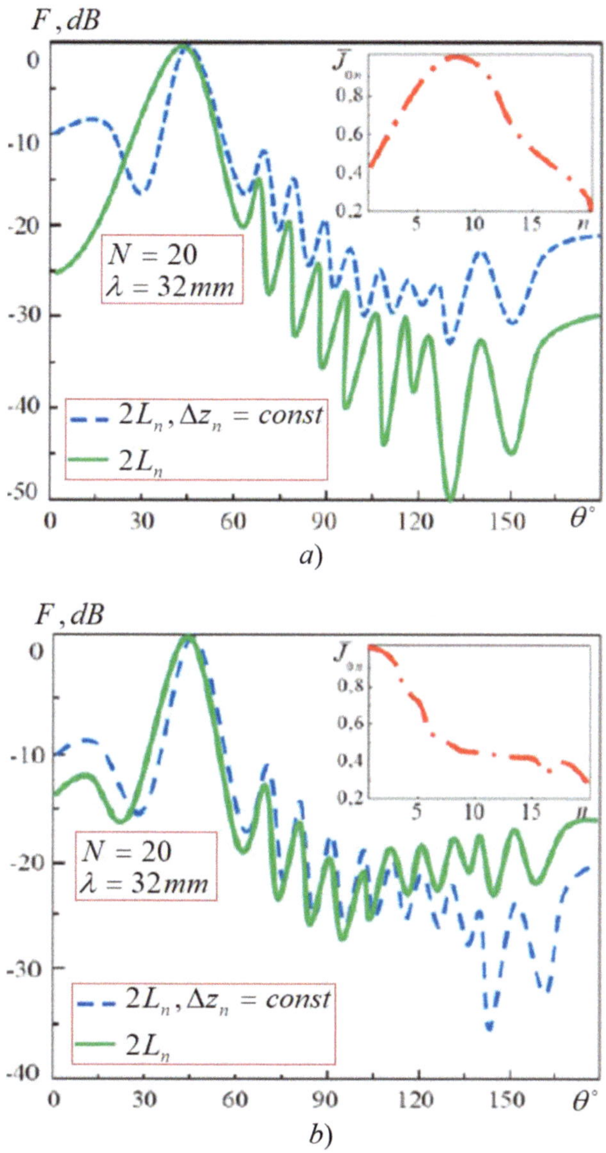

Fig. 7.4 The influence of the change of individual parameters of the system on the orientation diagram: (*a*) when it is $\Delta z_n = const = 12.0$ mm; (*b*) when it is $2L_n = const = 13.0$ mm

In the case of curve 3 shown in Fig. 7.5, $N = 200$, $\Delta z_n = 5.0$ mm, $2L_n = 9.98$ mm, the $|S_\Sigma|^2$ parameter value is equal to 0.9134. At this time, the level of the far side petals of the orientation diagram decreases.

In all of the above cases, the phase distribution of the currents in the antenna cage is essentially determined by the phase velocity in the excited transmission line and varies approximately according to the evanescent wave law.

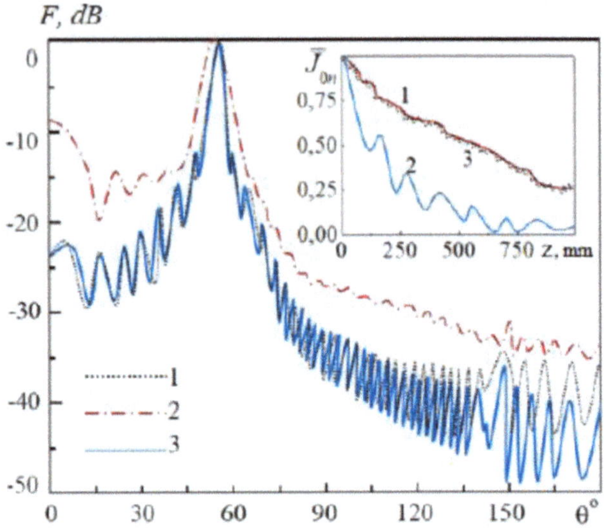

Fig. 7.5 Orientation diagram of equidistant antenna cage: curve 1, when $\lambda = 32.0\,\text{mm}$, $d_n = 1.0\,\text{mm}$, $h = 1.0\,\text{mm}$; curve 2, when $N = 200$, $\Delta z_n = 5.0\,\text{mm}$; when curve 3, $N = 200$, $\Delta z_n = 5.0\,\text{mm}$, $2L_n = 9.98\,\text{mm}$

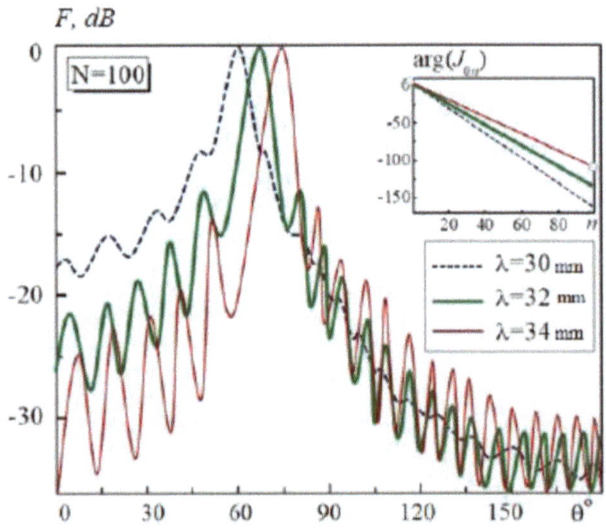

Fig. 7.6 Graphs of the change of the main petal of the orientation diagram as a result of the change of the value of the working wavelength (λ) : $2L_n = 11.5\,\text{mm}$; $d_n = 1.0\,\text{mm}$; $\Delta z_n = 10$, mm; $h = 1.0$ mm

It should be noted that as a result of changing the value of the working wavelength, the main petal of the orientation diagram will take the form as shown in Fig. 7.6 ($2L_n = 11.5$ mm; $d_n = 1.0$ mm; $\Delta z_n = 10.0\,\text{mm}$; $h = 1.0\,\text{mm}$). At this time, the direction and width of the main petal of the orientation diagram will change, as well as the level of the side petals.

7.2 Modeling of the Electromagnetic Field of a Rectangular Waveguide with Slotted Sidewalls

Let's look at a rectangular waveguide with a circular slot opening as the Z axis of its sidewalls increases (Fig. 7.7).

As can be seen from Fig. 7.7, the coordinate origin coincides with one vertex of the rectangle.

Let's accept the following notations: a – the size of the wide wall of the rectangular waveguide, b – the size of the narrow wall of the rectangular waveguide, R – the radius of the slits, h – the distance between the centers of the slits.

Let's assume that electromagnetic waves are propagated in the above waveguide and the surface impedance of the thin wall is known, and [12] is calculated by the expression shown in.

In such a waveguide, the electromagnetic field is determined by its longitudinal summation (H_z) and satisfies the Helmholtz equation [13].

$$\frac{\partial^2 \dot{H}_z}{\partial x^2}+\frac{\partial^2 \dot{H}_z}{\partial y^2}+\frac{\partial^2 \dot{H}_z}{\partial z^2}+k^2\dot{H}_z=0, \tag{7.14}$$

where $k=\frac{2\pi}{\lambda}$ – is the wave number.

Thus, the inhomogeneous impedance boundary condition under the condition $x=a$ is as follows:

$$\dot{E}_y=-Z_s(y,z) \mid \dot{H}_z. \tag{7.15}$$

Let the other walls be assumed to be ideal conductors. The periodic surface impedance can be expressed as a double Fourier series:

$$Z_s(y,z)=\sum_{p=0}^{\infty}\sum_{r=0}^{\infty}\left(A_{pr}\cos\frac{p\pi}{b}y\cdot\cos\frac{r\pi}{h}z+C_{pr}\cos\frac{p\pi}{h}y\cdot\sin\frac{r\pi}{h}z\right), \tag{7.16}$$

where A_{pr} and C_{pr} coefficients shown in [14] is determined by the expression.

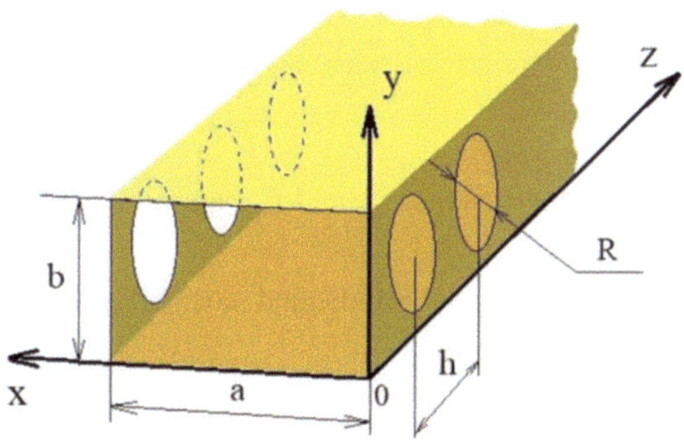

Fig. 7.7 A rectangular waveguide with a circular slot on the sidewall

So, since the impedance is a periodic function of the Z coordinate, then let's look for the solution of Eq. 7.14 in the form of Folke's series.

$$\dot{H}_z = \sum_{n=0}^{\infty}\sum_{s=0}^{\infty} A_n^{(s)}) \cos\gamma_{xn}^{(s)} x \cos\frac{n\pi}{b} y e^{-j\beta_s z}, \tag{7.17}$$

where $\beta_S = \beta_0 + p\frac{2\pi}{n}, \gamma_{xn}^{(s)2} = k^2 - \beta_s^2 - \left(\frac{n\pi}{b}\right)^2$,

β_0 – is a complex propagation constant.

Using Maxwell's equations for magnetic waves, we obtain the following expression relating the electric $\dot{E}_y$ and magnetic $\dot{H}_\tau$ summaries

$$\dot{E}_y = -j\omega\mu_0 \cdot \frac{1}{\gamma_{\perp n}^{(s)2}} \frac{\partial \dot{H}_z}{\partial x}, \tag{7.18}$$

where $\gamma_{\perp n}^{(s)2} = \gamma_{xn}^{(s)2} + \left(\frac{n\pi}{b}\right)^2$ – is the transverse wavenumber.

Considering expressions (7.17) and (7.18) for $\dot{E}_y$ components, we get

$$\dot{E}_y = -j\omega\mu_0 \sum_{n=0}^{\infty}\sum_{s=0}^{\infty} \frac{\gamma_{xn}^{(s)}}{\gamma_{\perp n}^{(s)2}} A_n^{(s)} \sin\gamma_{xn}^{(s)} x \cos\frac{n\pi}{b} y e^{-j\beta_s z}. \tag{7.19}$$

If we make simple transformations using the boundary condition (7.15) and the expression (7.19), we get

$$\begin{gathered}
\sum_{n=0}^{\infty}\sum_{s=-\infty}^{\infty} V_n^{(s)} A_n^{(s)} \cos\frac{p\pi}{b} y e^{-j\beta_s z} = \\
= \sum_{p=0}^{\infty}\sum_{r=0}^{\infty}\left(\overline{A}_{pr}\cos\frac{p\pi}{b}y \cdot \cos\frac{r\pi}{n}z + \overline{C}_{pr}\cos\frac{p\pi}{b}y \cdot \sin\frac{r\pi}{n}z\right) \times \\
\times \sum_{n=0}^{\infty}\sum_{s=0}^{\infty}\frac{\gamma_{xn}^{(s)}}{\gamma_{\perp n}^{(s)^2}} U_n^{(s)} A_n^{(s)} \cos\frac{n\pi}{b} y e^{-j\beta_s z},
\end{gathered} \tag{7.20}$$

where the coefficients $V_n^{(s)}$ and $U_n^{(s)}$ are determined using the following expressions:

$$U_n^{(s)} = \cos\gamma_{xn}^{(s)} d, \quad V_n^{(s)} = \frac{jkY_{xn}^{(s)}}{Y_{\perp n}^{(s)}} \sin Y_{xn}^{(s)} d. \tag{7.21}$$

The coefficients A_{pr} and C_{pr} are normalized to the $z_0 = \sqrt{\frac{\mu_0}{\varepsilon_0}}$-wave resistivity of free space.

If we replace the indicated $\cos\frac{r\pi}{n}z$, $\sin\frac{r\pi}{n}z$ trigonometric functions with exponential ones, we get

$$\sum_{n=0}^{\infty}\sum_{r=-\infty}^{\infty} V_n^{(s)} A_n^{(s)} \cos\frac{p\pi}{b} y e^{-j\beta_s z} =$$
$$= \frac{1}{2}\sum_{p=0}^{\infty}\sum_{r=0}^{\infty}\left(\overline{z}_{pr}^{+} e^{-j\frac{r\pi}{n}z} + \overline{z}_{pr}^{-} e^{-\frac{jr\pi}{n}z}\right)\cos\frac{n\pi}{b} y \times \tag{7.22}$$
$$\times \sum_{n=0}^{\infty}\sum_{r=-\infty}^{\infty} \frac{\gamma_{xn}^{(s)}}{\gamma_{\perp n}^{(s)^2}} U_n^{(s)} A_n^{(s)} \cos\frac{n\pi}{b} y e^{-j\beta_s z},$$

where $\overline{z}_{pr}^{+} = \overline{A}_{pr} + j\overline{C}_{pr}, \overline{z_{pr}^{-}} = \overline{A}_{pr} - j\overline{C}_{pr}$.

If we take into account the orthogonality of the transverse characteristic function of the waveguide and the function $e^{-j\beta_s z}$, we get a system of homogeneous linear algebraic equations with respect to the amplitude of the wave $A_k^{(s)}$.

$$\sum_{k=0}^{\infty}\left\{\left[\left(\widetilde{z}_{p0}^{+} + \widetilde{z}_{p0}^{-}\right) U_k^{(s)} - (1+\delta_{nk})\delta_{nk} \times \right.\right.$$
$$\left.\left. \times V_k^{(s)}\right] A_k^{(s)} + \frac{1}{2}\sum_{k=0}^{\infty}\left(\widetilde{z}_{pr}^{+} U_k^{(s-r)} A_k^{(s-r)} + \widetilde{z}_{pr}^{-} U_k^{(s+r)} A_k^{(s+r)}\right)\right\} = 0, \tag{7.23}$$

where $n = 0, 1, 2...; s = 0, \pm 1, \pm 2...; p = |n - k|$; δ_{nk} – is the Kronecker symbol.

If we open the definition in the system (7.23), we get the dispersion equation in the form of a matrix, and from here the diffusion coefficient β_0 is determined.

It should be noted that the cross-sectional area of the waveguide $a \times b = 40 \times 20$ mm; radius of auxiliary waveguides $R = 9$ mm; the distance between the centers of the holes $h = 19$ mm. The analysis of the dispersion Eq. 7.23 shows that the influence of high spatial harmonics on the propagation constant is very small for the selected parameters. The main contribution to this is given by the central elements of the system (7.23), whose diagonal is defined as follows:

$$\widetilde{z}_{00}\gamma_{\perp}^2 \cos\gamma_x a + jk\gamma_x \sin\gamma_x a = 0. \tag{7.24}$$

The dispersion Eq. 7.24 describes the volume (fast) wave in the guiding structure. If we substitute $\gamma_x \rightarrow j\gamma_x$ in expression (7.24) for surface waves propagating in an impedance waveguide, we get

$$\widetilde{z}_{00} Y_{\perp}^2 chY_s a - jkY_x shY_x a = 0, \tag{7.25}$$

where $\gamma_{\perp}^2 = -\gamma_x^2 + \left(\frac{n\pi}{b}\right)^2$.

Dispersion expressions (7.24) and (7.25) allow obtaining the spectrum of both volume and surface waves. As a result of the analysis, the frequency characteristics of the dependence of volume and propagation constant of surface waves on different

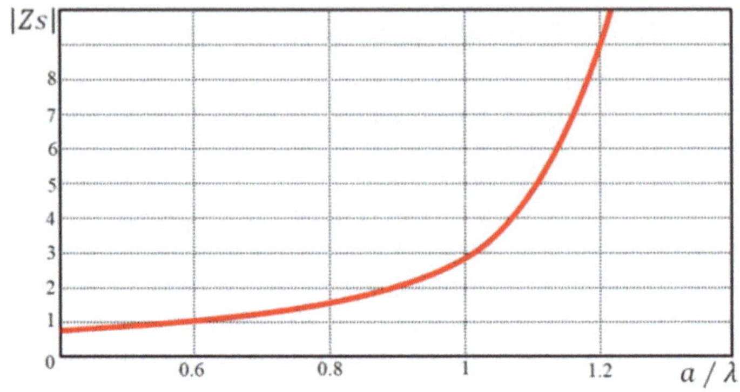

Fig. 7.8 $|Z_{seff}|$ of the modulus of effective impedance a/λ wavelength dependence

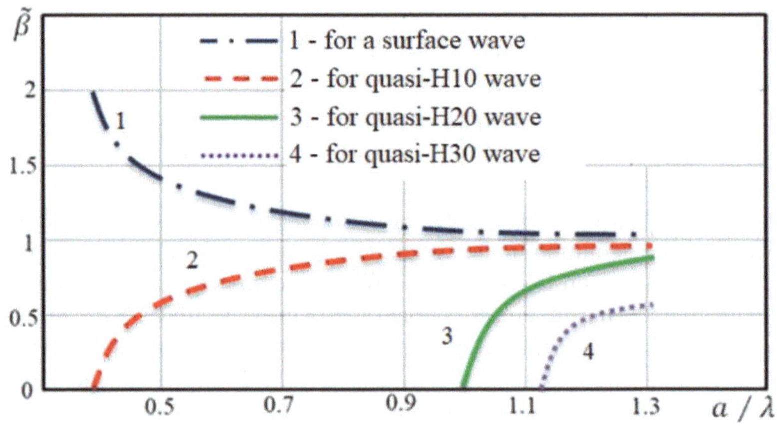

Fig. 7.9 Dependencies of $\widetilde{\beta}(a/\lambda)$ normalized propagation constants of surface and volume waves: 1 – for surface wave; 2 – for quasi H_{10} volume wave; 3 – for quasi H_{20} volume wave; 4 – for quasi H_{30} volume wave

values of impedance and wavelength were obtained. The results of numerical calculations are shown in Figs. 7.8, 7.9 and 7.10.

Figure 7.8 shows the modulus of the effective impedance as a function of length (for the radius shown above). Calculations show that the reactive part of the effective impedance is capacitive. Calculations [15] based on the following statement:

$$Z_{seff} = \sqrt{\frac{1}{b \cdot h} \cdot \int_0^b \int_0^h Z_s(y,z) \cdot Z_s^*(y,z) dy dz}, \tag{7.26}$$

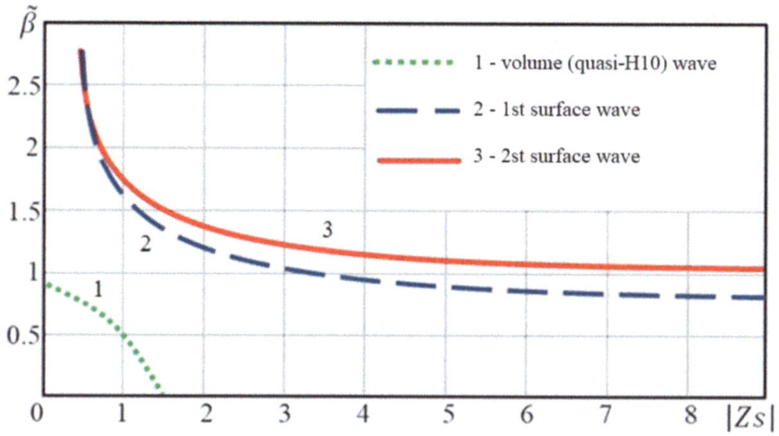

Fig. 7.10 $\widetilde{\beta}(|Z_{seff}|)$ dependencies for waveguide with two impedance walls in case $a/\lambda = 0.8$: 1 – volume (quasi H_{10}) wave; 2 – the first surface wave; 3 – the second surface wave

where $Z_s(y, z)$ – the surface impedance determined by the formula (7.16); $Z_s^*(y, z)$ is complex impedance.

Figure 7.9 shows the frequency dependencies of the normalized propagation constant of surface and volume waves for quasi H_{10}, H_{20} and H_{30} waves.

As can be seen from Fig. 7.9, the capacitance characteristic of the surface impedance of one wall of the waveguide tends to λ_{kr} towards the large value of the wavelength. As the value of the wavelength decreases, excitation of higher type waves occurs. Calculations show that an increase in the modulus of the effective impedance of $|Z_{seff}|$ leads to an increase in the propagation constant.

Numerical analysis was also carried out for the case of the impedance of the two walls of the waveguide. Calculations show that the propagation of two transverse surface waves is possible in this structure. Figure 7.10 shows the dependencies of $\widetilde{\beta}(|Z_{seff}|)$ for a waveguide with two impedance walls in case $a/\lambda = 0.8$.

It should be noted that in case $Z_{seff} \to 0$, the quasi H_{10} wave passes into the H_{10} wave. At this time surface waves are created. At the same time, it can be seen from the Fig. 7.10 that the volume wave (curve 1) is subject to interruption at a number of values of Z_{seff}. As a result of its increase, the value of the propagation constant β decreases. At this time, one of the surface waves (curve 3) tends to the unit, and the other wave passes from the surface wave to the volume wave (curve 2).

7.3 Modeling of Dielectric Type Circular Waveguide with Complex Structured Nonlinear Medium

Consider the three-dimensional space R^3 with Cartesian coordinates *OXYZ*. This space is filled with an isotropic medium with a dielectric constant of $\varepsilon_3 = const.$ $W = \{x : 0 \leq x^2 + y^2 \leq R_2^2\}$ waveguides with a circular cross-section are placed in this space.

$Or\phi z$ choose the cylindrical coordinate system so that the O3 axis coincides with the axis of the same name of the cylindrical coordinate system.

The cross-sectional area of the waveguide perpendicular to its axis is in the form of two concentric circles with radius R_1 and R_2, that is, the waveguide has two layers.

The electromagnetic field depends harmonically on time [16]:

$$\begin{aligned} \widetilde{E}(r,\phi,z,t) &= E_+(r,\phi,z)\cos\omega t + E_-(r,\phi,z)\sin\omega t,\\ \widetilde{H}(r,\phi,z,t) &= H_+(r,\phi,z)\cos\omega t + H_-(r,\phi,z)\sin\omega t, \end{aligned} \tag{7.27}$$

where ω– circular frequency; $\widetilde{E}, E_+, E_-$, are real functions. $\widetilde{H}, H_+, H_-$ $-\boldsymbol{E}, \boldsymbol{H}$ let's write the complex amplitudes of the fields

$$\begin{aligned} E &= E_+ + iE_-,\\ H &= H_+ + iH_-, \end{aligned} \tag{7.28}$$

where $E = (E_r, E_\phi, E_z)^T$, $H = (H_r, H_\phi, H_z)^T$ and $(\cdot)^T$ indicates a transponding operation. Each component of $\boldsymbol{E}, \boldsymbol{H}$ is a function of three spatial variables.

The electromagnetic field $\boldsymbol{E}$, $\boldsymbol{H}$ satisfies Maxwell's system of equations [17]

$$\begin{aligned} rot\vec{H} &= -i\omega\varepsilon\vec{E},\\ rot\vec{E} &= i\omega\mu\vec{H}, \end{aligned} \tag{7.29}$$

The electromagnetic field undergoes exponential decay in the case of $r \to \infty$. Let's use the following nonlinearity [18]:

$$\varepsilon = \left(1,8 + \chi E^2\right)\varepsilon_0, \tag{7.30}$$

where $\chi-$ is the nonlinearity coefficient.

Assume that the dielectric permittivity ε is a scalar function inside the waveguide and is defined inside and outside the waveguide as follows:

$$\varepsilon = \begin{cases} \varepsilon_1\varepsilon_0, 0 \leq r \leq R_1,\\ \left(1,8 + \chi E^2\right)\varepsilon_0, R_1 \leq r \leq R_2,\\ \varepsilon_3\varepsilon_0, r \geq R_2, \end{cases} \tag{7.31}$$

where ε_1, ε_2, ε_3- are positive constants.

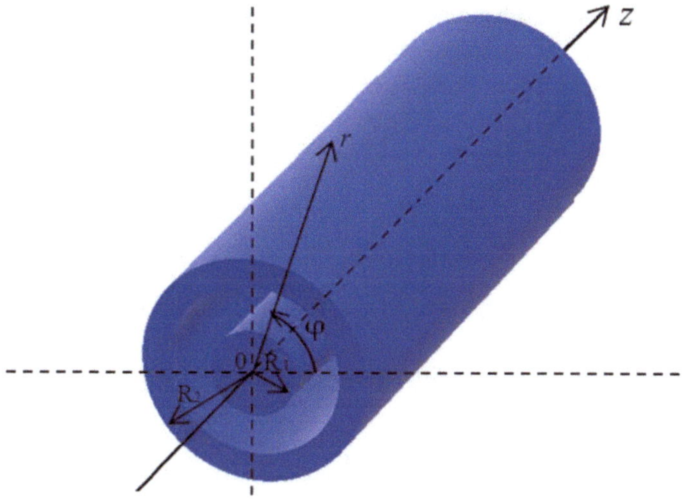

Fig. 7.11 Geometric representation of the problem

The medium is non-magnetic and isotropic. At the same time, let's assume that the whole space is $\mu = \mu_0$.

Where we define the surface wave propagating along the waveguide. The solution of Maxwell's equations is sought throughout space.

Figure 7.11 gives a geometric representation of the problem.

Let's look at the *TM* wave

$$E = (E_r, 0, E_z)^T, H = (0, H_\phi, 0)^T, \tag{7.32}$$

where $E_r = E_r(r, \phi, z),\ E_z = E_z(r, \phi, z),\ H_\phi = H_\phi(r, \phi, z)$.

It can be shown that the components of the field for the considered problem and selected nonlinearity can have the following form:

$$E_r = E_r(r)e^{i\gamma z},\ \ E_z = E_z(r)e^{i\gamma z},\ \ H_\phi = H_\phi(r)e^{i\gamma z}, \tag{7.33}$$

where $\gamma - k_0^2 = \omega^2 \mu_0 \varepsilon_0$ carry out the substitution. Considering (7.33) in (7.29), we get

$$\begin{aligned} \{\gamma^2 E_r + i\gamma \frac{\partial E_z}{\partial r} &= k_0^2 \widetilde{\varepsilon} E_r, \\ i\gamma \frac{1}{r}\frac{\partial}{\partial r}(rE_r) - \frac{1}{r}\frac{\partial}{\partial r}\left(r\frac{\partial E_z}{\partial r}\right) &= k_0^2 \widetilde{\varepsilon} E_z, \end{aligned} \tag{7.34}$$

where $k_0^2 = \omega^2 \mu_0 \varepsilon_0$ and $\varepsilon = \widetilde{\varepsilon}\varepsilon_0, \varepsilon_0, \mu_0 -$ are the dielectric and magnetic permeability of free space.

Make the following substitutions

$$\begin{aligned} E_r(r,\gamma) &= \Psi_1(r,\gamma), \\ iE_z(r,\gamma) &= \Psi_2(r,\gamma). \end{aligned} \tag{7.35}$$

Then we get from (7.34).

$$\begin{aligned} &\{\gamma\Psi_2' + (\gamma^2 - k_0^2\tilde{\varepsilon})\Psi_1 = 0, \\ &-\gamma\frac{1}{r}(r\Psi_1)' - \frac{1}{r}(r\Psi_2')' - k_0^2\tilde{\varepsilon}\Psi_2 = 0, \end{aligned} \tag{7.36}$$

where the derivative denotes the differentiation with respect to r, $\Psi_1(r,\gamma)$, $\Psi_2(r,\gamma)$– are material functions.

Let's look for the true value of the spectral parameter γ.

Assume that $\Psi_1(r,\gamma)$, $\Psi_2(r,\gamma)$ functions are differentiable in such a way that the following condition is satisfied for them:

$$\begin{aligned} \Psi_1 \in\ & C[0,R_1] \cap C[R_1,R_2] \cap C[R_2,+\infty) \cap \\ & \cap C_1[0,R_1] \cap C_1[R_1,R_2] \cap C_1[R_2,+\infty), \\ \Psi_2 \in\ & C[0,+\infty) \cap C_1[0,R_1] \cap C_1[R_1,R_2] \cap \\ & \cap C_1[R_2,+\infty] \cap C_2(0,R_1) \cap \\ & \cap C_2(R_1,R_2) \cap C_2(R_2,+\infty). \end{aligned} \tag{7.37}$$

Let's assume it is $\gamma^2 \geq \varepsilon_3$. When it is $r \leq R_1$, it becomes $\varepsilon = \varepsilon_1\varepsilon_0$, and the system (7.36) is as follows:

$$\begin{aligned} &\{\gamma\Psi_2' - k_1^2\Psi_1 = 0, \\ &-\gamma\frac{1}{r}(r\Psi_1)' - \frac{1}{r}(r\Psi_2')' - k_0^2\varepsilon_1\Psi_2 = 0, \end{aligned} \tag{7.38}$$

where $k_1^2 = \gamma^2 - k_0^2\varepsilon_1$.

If we write the first equation of (7.38) depending on the function Ψ_1 and substitute it in the second equation, we get the following equation for the function Ψ_2

$$-\frac{\gamma^2}{k_1^2}\frac{1}{r}(r\Psi_2')' - \frac{1}{r}(r\Psi_2')' - k_0^2\varepsilon_1\Psi_2 = 0. \tag{7.39}$$

This equation is called Bessel's equation and can easily be written as:

$$\frac{1}{r}(r\Psi_2')' + k_1^2\varepsilon\Psi_2 = 0. \tag{7.40}$$

Then the solution of the system of Eq. 7.38 is in the following form:

$$\begin{aligned}\Psi_1(r) &= \frac{\gamma}{k_1}\left(C_1 I_0'(k_1 r) + C_1 K_0'(k_1 r)\right),\\ \Psi_2(r) &= \left(C_1 I_0(k_1 r) + C_2 K_0(k_1 r)\right),\end{aligned} \tag{7.41}$$

where I_0 and K_0 are the modified Bessel and McDonald functions, respectively.

The McDonald function $K0(r)$ approaches infinity at $r \to \infty$, while the Bessel function $I_0(r)$ is bounded at $r \to \infty$. If we accept the area restriction condition and $I_0'(r) = I_1(r)$ along the entire computational domain, we get from (7.38)

$$\begin{aligned}\Psi_1(r) &= -\frac{\gamma}{k_1} C_1 K_1(k_1 r),\\ \Psi_2(r) &= C_1 I_0(k_1 r).\end{aligned} \tag{7.42}$$

When it is $R_1 \leq r \leq R_2$, we get $\varepsilon = (1,8 + \chi E^2)\varepsilon_0$.

Then the system (7.36) will take the following form:

$$\{\gamma\Psi_2' + \left(\left(\gamma^2 - k_0^2\left(1,8 + \chi\left(\Psi_1^2 + \Psi_2^2\right)\right)\right)\Psi_1 = 0,$$

$$-\gamma\frac{1}{r}(r\Psi_1)' - \frac{1}{r}\left(r\Psi_2'\right)' - k_0^2\left(1,8 + \chi\left(\Psi_1^2 + \Psi_2^2\right)\right)\Psi_2 = 0,$$

When it is $r \geq R_2$, we get $\varepsilon = \varepsilon_3\varepsilon_0$. Then the system (7.36) will take the following form:

$$\begin{aligned}&\{\gamma\Psi_2' + k_3^2\Psi_1 = 0,\\ &-\gamma\frac{1}{r}(r\Psi_1)' - \frac{1}{r}\left(r\Psi_2'\right)' - k_0^2\varepsilon_3\Psi_2 = 0,\end{aligned} \tag{7.44}$$

where $k_3^2 = \gamma^2 - k_0^2\varepsilon_3$.

The solution of the system of equations (7.44) is as follows:

$$\begin{aligned}\Psi_1(r) &= -\frac{\gamma}{k_3^2}\left(C_3 I_0'(k_3 r) + C_4 K_0'(k_3 r)\right),\\ \Psi_2(r) &= C_3 I_0(k_3 r) + C_4 K_0(k_3 r).\end{aligned} \tag{7.45}$$

It is known that the function $I_0(r)$ converges to infinity when it is $r \to \infty$, and the function $K_0(r)$ converges to zero when it is $r \to \infty$. Considering that $K_0(r) = -K_1(1)$, we get from (7.45)

$$\begin{aligned}\Psi_1(r) &= \frac{\gamma}{k_3^2} C_4 K_1(k_3 r),\\ \Psi_2(r) &= C_4 K_0(k_3 r).\end{aligned} \tag{7.46}$$

At the same time, it is known that the tangential summaries of the electromagnetic field are continuous at the separation boundary of the medium. The components Ez

and Hv are the components that touch the case being viewed. We will buy for these conditions

$$E_z(R_1+0)=E_z(R_1-0), H_\phi(R_1+0)=H_\phi(R_1-0),$$
$$E_z(R_2+0)=E_z(R_2-0), H_\phi(R_2+0)=H_\phi(R_3-0). \quad (7.47)$$

The Er component is called the normal component and is continuous at the separation boundary of the medium.

From the above we get for functions Ψ_1 and Ψ_2:

$$[\widetilde{\varepsilon}\Psi_1]|_{r=R_1}=0, [\Psi_2]|_{r=R_1}=0, [\widetilde{\varepsilon}\Psi_1]|_{r=R_2}=0, [\Psi_2]|_{r=R_2}, \quad (7.48)$$

where $[f]_{x=x_0}=\lim\limits_{x\to x_0-0} f(x)-\lim\limits_{x\to x_0+0} f(x)$.

Then we get from (7.48).

$$\varepsilon_1\Psi_1(R_1-0)=(1,8+\chi\Psi)\Psi_1(R_1+0),$$
$$\Psi_2(R_1-0)=\Psi_2(R_1+0), \quad (7.49)$$

$$(1,8+\chi\Psi_1)\Psi_1(R_2-0)=\varepsilon_3\Psi_1(R_2+0),$$
$$\Psi_2(R_2-0)=\Psi_2(R_2+0), \quad (7.50)$$

where $\Psi=\Psi_1+\Psi_2$, $\Psi=(\Psi_1,\Psi_2)^T$.

It is easy to see that multiplying the fields $\boldsymbol{E}$, $\boldsymbol{H}$ in (7.28) by $C_0\neq 0$ arbitrary constants and C_0^{-2} nonlinearities will not change the system of Maxwell's equations. However, this case does not allow the selection of an additional normalization condition. So, during the calculations, it becomes clear that the coefficient χ is normalized according to the unknown constant C_0^{-2}. This indicates that the considered nonlinear problem is strongly dependent on the initial condition (amplitude of the incident wave). This, in turn, is the biggest aspect that distinguishes the considered nonlinear problem from the analogous linear problem.

If we consider the constant $C1$ as given and take it equal to unity, then we get the following dispersion equation from expressions (7.42), (7.46), (7.49), (7.50):

$$\Delta(\gamma)\equiv\left(\varepsilon_2+\chi\Psi|_{r=R_2-0}\right)\Psi_1(R_2-0)-$$
$$-\gamma\frac{\varepsilon_3}{k_3}\frac{K_1(k_3R_2)}{K_0(k_3R_2)}\Psi_2(R_2-0). \quad (7.51)$$

To obtain the numerical results, the system of equations (7.43) was solved for step h_0 in case

$$\gamma_j=\gamma_0+jh_0, j=\overline{0,N-1}.$$

where $\gamma\in(\varepsilon_3,\gamma^*)$, $\gamma^*\leq\max(\varepsilon_1,\varepsilon_2)$.

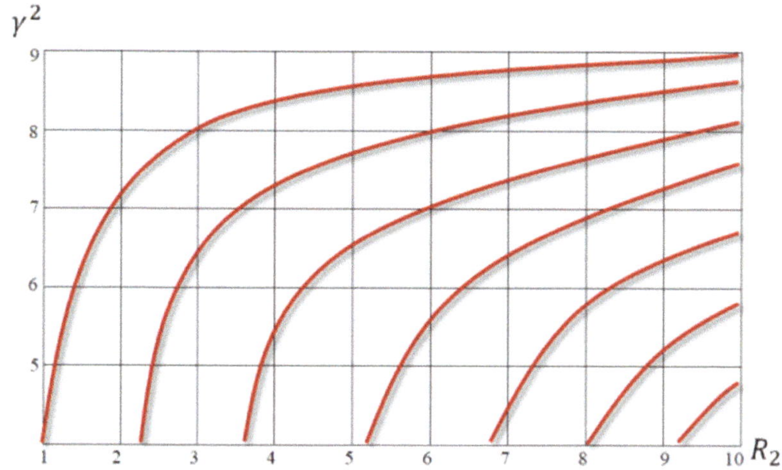

Fig. 7.12 γ^2 dependence of propagation constant on R_2 radius

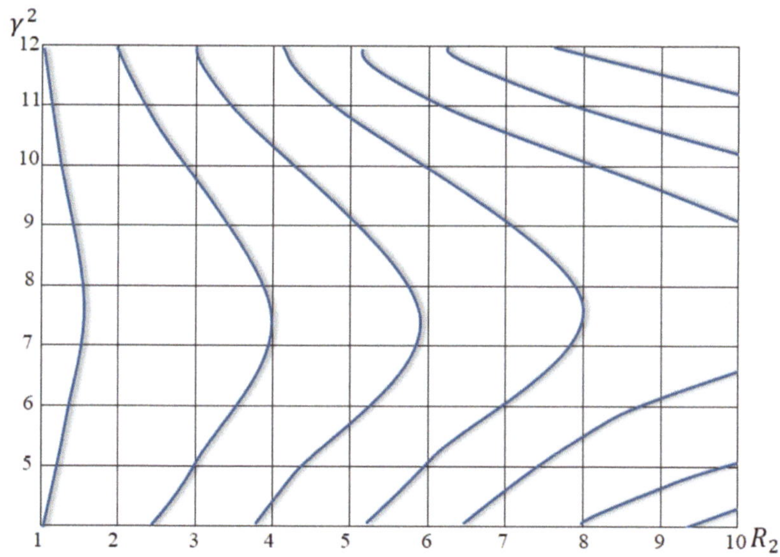

Fig. 7.13 γ^2 dependence of propagation constant on R_2 radius

After that, the value of parameter $\Delta(\gamma_j)$was calculated.
The results of calculations are shown in Figs. 7.12, 7.13, 7.14, 7.15 and 7.16.
The values of the following parameters were used during the calculations:

$$\varepsilon_1 = 4, \varepsilon_2 = 9, \varepsilon_3 = 1, R_1 = 2, 2 \leq R_2 \leq 12, \chi = 0, k_0 = 1$$

The values of the following parameters were used during the calculations:

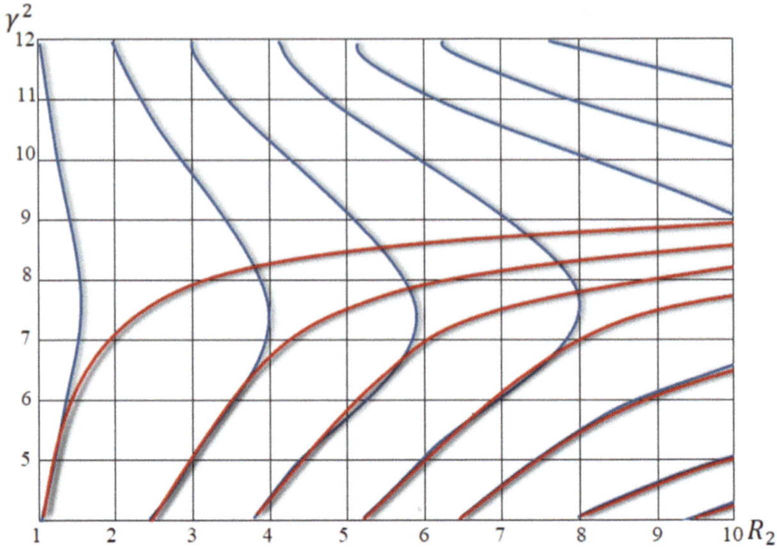

Fig. 7.14 Combined illustration of the graphs shown in Figs. 7.12 and 7.13

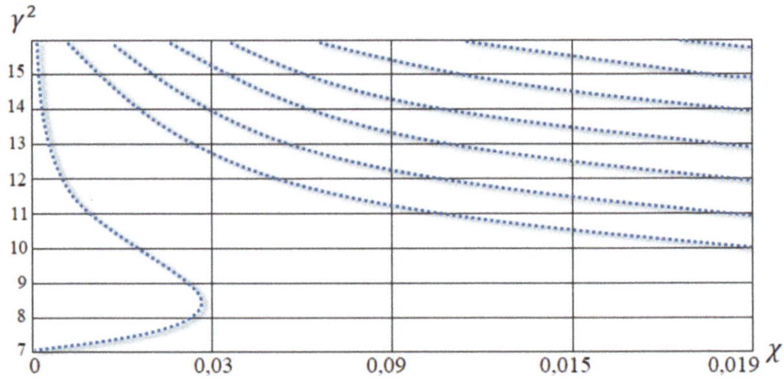

Fig. 7.15 γ^2 dependence of the diffusion constant on the χ nonlinearity coefficient

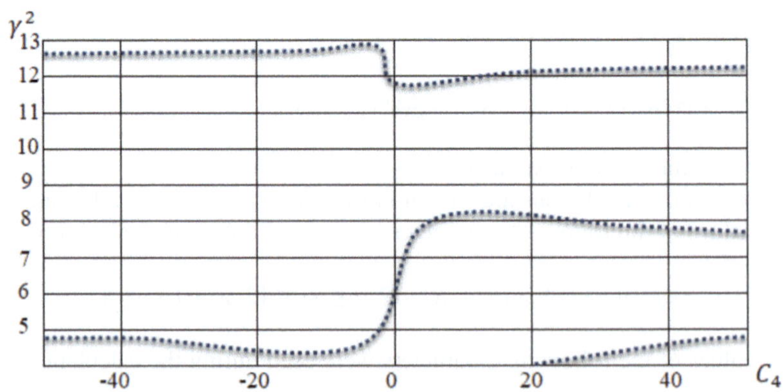

Fig. 7.16 γ^2 dependence of the propagation constant on the C_4 constant

$$\varepsilon_1 = 4, \varepsilon_2 = 9, \varepsilon_3 = 1, R_1 = 2.2 \le R_2 \le 12, \chi = 0.01, k_0 = 1$$

The values of the following parameters were used during the calculations:

$$\varepsilon_1 = 4, \varepsilon_2 = 9, \varepsilon_3 = 1, R_1 = 2, R_2 = 4, k_0 = 1$$

The values of the following parameters were used during the calculations:

$$\varepsilon_1 = 4, \varepsilon_2 = 9, \varepsilon_3 = 1, R_1 = 2, R_2 = 4, \chi = 0.01, k_0 = 1$$

7.4 Calculation of the Electromagnetic Field of a Rectangular Waveguide with an Anisotropic Medium

A biisotropic medium is a general type of linear isotropic medium. Biisotropic medium differs from ordinary dielectric or magnetic medium in presence of magnetoelectric coupling. Therefore, additional terms appear in the material equations [19]:

$$\mathbf{D} = \varepsilon\mathbf{E} + \xi\mathbf{H}, \ \ \mathbf{B} = \varsigma\mathbf{E} + \mu\mathbf{H}, \tag{7.52}$$

where $\xi = (k - i\chi)$, $\varsigma = (k + i\chi)$.

In these statements, ε is the dielectric permeability of the medium, μ is the magnetic permeability, k is the Tellegen parameter, and χ is the chiral parameter of the medium. The material equations are written for theexp($-i\omega t$)-form harmonic time dependence in the frequency domain [20].

The electromagnetic field $\{\mathbf{E}, \mathbf{H}\}$ propagated in a chiral medium is divided into two independent $\{\mathbf{E}_\pm, \mathbf{H}_\pm\}$ components. Statements $\varepsilon_\pm = \varepsilon(1 \pm \chi_r)$, $\mu_\pm = \mu(1 \pm \chi_r)$ are true for this environment, where $\chi_r = \chi/\sqrt{\varepsilon\mu}$. In artificial chiral environments, the parameter χ is not a small adjustment, and at certain frequencies $|\chi_r| > 1$ is paid. In this case, one of the $\{\mathbf{E}_\pm, \mathbf{H}_\pm\}$ waves is the opposite wave.

A waveguide with infinite ideal walls spaced apart. Let's denote the distance between the walls of this waveguide by a and assume that this waveguide is filled with a chiral medium. At the same time, the environment is located in the $z \in [0, z_0]$ area (Fig. 7.17).

Assume that the internal material of the waveguide is characterized by parameters ε, μ and χ. The outer material of the transducer is characterized by parameters ε_0 and μ_0.

Now let's study the issue of diffraction of the electromagnetic wave falling from the $z = -\infty$ region. Assume that the incident wave is one of the normal waves of the waveguide or a linear combination of them. Conditions $z = 0$ and $z = z_0$ should be included to formulate the border issue in the $D = \{x \in [0, a], z \in [0, z_0]\}$ region.

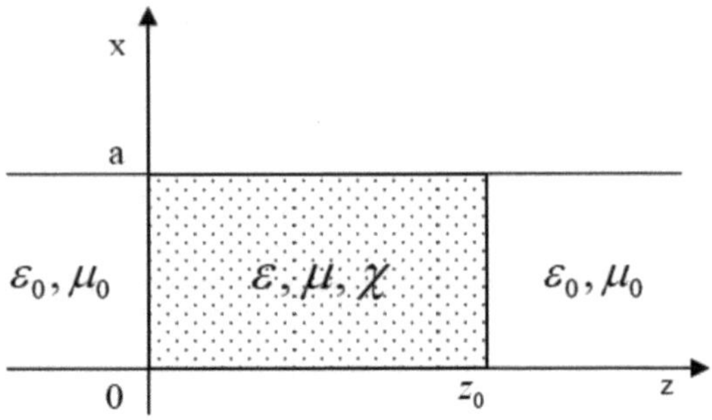

Fig. 7.17 Cross section of a chiral medium waveguide

For normal waves of the considered waveguide *TE*, we can write [47, pp. 3–40].

$$\mathbf{E}^m_{n,\pm} = (i\omega\mu_0/c)\sqrt{2/a}\exp(\pm i\widehat{\gamma}_n z)(\pi n/a)\sin(\pi nx/a)\mathbf{e}_y, \tag{7.53}$$

$$\begin{aligned}\mathbf{H}^m_{n,\pm} &= \mp i\widehat{\gamma}_n\sqrt{2/a}\exp(\pm i\widehat{\gamma}_n z)(\pi n/a)\sin(\pi nx/a)\mathbf{e}_x + \\ &+(\pi n/a)^2\sqrt{2/a}\exp(\pm i\widehat{\gamma}_n z)\cos(\pi nx/a)\mathbf{e}_z,\end{aligned} \tag{7.54}$$

where $\widehat{\gamma}_{\rm n} = \sqrt{k_0^2 - (\pi n/a)^2}$, $\mathrm{k}_0^2 = \varepsilon_0\mu_0\omega^2/c^2, \mathrm{n} = 1, 2, \ldots,$.

For *TM*-type normal waves of a waveguide, we can write:

$$\begin{aligned}\mathbf{E}^e_{n,\pm} &= \mp i\gamma_n\exp(\pm i\gamma_n z)\sqrt{2/a}(\pi n/a)\cos(\pi nx/a)\mathbf{e}_x + \\ &+(\pi n/a)^2\sqrt{2/a}\exp(\pm i\gamma_n z)\cos(\pi nx/a)\mathbf{e}_z,\end{aligned} \tag{7.55}$$

$$\mathbf{H}^e_{n,\pm} = (i\omega\varepsilon_0/c)\sqrt{2/a}\exp(\pm i\gamma_n z)(\pi n/a)\cos(\pi nx/a)\mathbf{e}_y, \tag{7.56}$$

where $\gamma_n = \sqrt{k_0^2(\pi n/a)^2}, n = 1, 2, \ldots.$

In region $z \in (0, z_0)$, the electromagnetic field satisfies Maxwell's system of equations and material equations:

$$\begin{cases} rot\mathbf{E} = (i\omega/c)\mathbf{B}, & div\mathbf{D} = 0, \\ rot\mathbf{H} = -(i\omega/c)\mathbf{D}, & div\mathbf{B} = 0 \end{cases} \tag{7.57}$$

$$\mathbf{D} = \varepsilon\mathbf{E} - i\chi\mathbf{H}, \quad \mathbf{B} = \mu\mathbf{H} + i\chi\mathbf{E} \tag{7.58}$$

Thus, the plane surrounding the waveguide is an ideal conductor, and the boundary conditions in the case of $x = 0$, $x = a$ are as follows:

$$\mathrm{E_y}\big|_{x=0} = E_y\big|_{x=a} = 0, \quad \mathrm{E_z}|_{x=0} = E_z|_{x=a} = 0. \tag{7.59}$$

$z = 0$ və $z = z_0$ olanda aşağıdakı şərtlərdən istifadə edək

$$[\mathrm{e_z}, \mathbf{E}]|_{z=0} = [e_z, \mathbf{E}^{ext}]|_{z=0}, [\mathrm{e_z}, \mathbf{E}]|_{z=z_0} = [e_z, \mathbf{E}^{ext}]|_{z=z_0}, \tag{7.60}$$

$$[\mathrm{e_z}, H]|_{z=0} = [e_z, H^{ext}]|_{z=0}, \quad [\mathrm{e_z}, H]|_{z=z_0} = [e_z, H^{ext}]|_{z=z_0}, \tag{7.61}$$

where $\{\mathbf{E}^{\mathrm{ext}}, \mathbf{H}^{ext}\}$– is the area outside the transducer.

It should be noted that in the considered issue, wave $\{\mathbf{E}_+^{ins}, \mathbf{H}_+^{ins}\}$ falls into the chiral environment from region $\mathrm{z} = -\infty$. At this time, when $z < 0$,the $\mathbf{E}^{ext}$ field is in the form of the sum of the incident and reflected waves:

$$\begin{aligned}
\mathbf{E}^{ext} &= \sum_{n=1}^{\infty} \{R_n^m \mathbf{E}_{n,-}^m + R_n^e \mathbf{E}_{n,-}^e\} + \mathbf{E}_+{}^{ins}, \\
\mathbf{H}^{ext} &= \sum_{n=1}^{\infty} \{R_n^m \mathbf{H}_{n,-}^m + R_n^e \mathbf{H}_{n,-}^e\} + \mathbf{H}_+^{ins}
\end{aligned} \tag{7.62}$$

When $\mathrm{z} > \mathrm{z_0}$, the $\mathbf{E}^{ext}$ field is in the form of the sum of the passing waves:

$$\mathbf{E}^{ext} = \sum_{n=1}^{\infty} \{T_n^m \mathbf{E}_{n,+}^m + T_n^e \mathbf{E}_{n,+}^e\}, \mathbf{H}^{ext} = \sum_{n=1}^{\infty} \{T_n^m \mathbf{H}_{n,+}^m + T_n^e \mathbf{H}_{n,+}^e\}. \tag{7.63}$$

Let's express the reflection and transmission coefficients as a dependence on the value of the sought field $\boldsymbol{E}$. For this, let's use condition (7.60) when $z = 0$ and $z = z_0$:

$$\begin{aligned}
&\mathbf{e}_y E_x\big|_{z=0} - \mathbf{e}_x E_y\big|_{z=0} = \\
&= \sum_{n=1}^{\infty} \Big\{ - \mathbf{e}_x R_n^m E_{y,n,-}^m \big|_{z=0} + \mathbf{e}_y R_n^e E_{x,n,-}^e \big|_{z=0} \Big\} + \mathbf{e}_y E_{x,+}^{ins}\big|_{z=0} - \mathbf{e}_x E_{y,-}^{ins}\Big|_{z=0}.
\end{aligned} \tag{7.64}$$

If we compare the right and left sides of Eq. 7.64, we get:

$$\begin{aligned}
&\left(E_{x,+}^{ins} + \sum_{n=1}^{\infty} R_n^e E_{x,n,-}^e \right)\Bigg|_{z=0} = E_x|_{z=0}, \\
&\left(E_{y,+}^{ins} + \sum_{n=1}^{\infty} R_n^m E_{y,n,-}^m \right)\Bigg|_{z=0} = E_x\big|_{z=0},
\end{aligned} \tag{7.65}$$

Using these equations, we get for the normal components of the electric field

$$E^{ins}_{x,+}(x,0) - \sum_{n=1}^{\infty} R^e_n i\gamma_n \sqrt{2/a}(\pi n/a)\cos(\pi nx/a) = E_x(x,0), \tag{7.66}$$

$$E^{ins}_{y,+}(x,0) + \sum_{n=1}^{\infty} R^m_n (i\omega\mu_0/c)\sqrt{2/a}(\pi n/a)\sin(\pi nx/a) = E_y(x,0). \tag{7.67}$$

We determine from (7.67).

$$R^e_n = (ia/\gamma_n \pi n)\sqrt{2/a}\int_0^a \left[E_x(x,0) - E^{ins}_{x,+}(x,0)\right] \times \\ \times \cos(\pi nx/a)dx, \tag{7.68}$$

$$R^m_n = -(ica/\omega\mu_0 \pi n)\sqrt{2/a}\int_0^a \left[E_y(x,0) - E^{ins}_{y,+}(x,0)\right] \times \\ \times \sin(\pi nx/a)dx. \tag{7.69}$$

Analogously, for the dependence of the transition coefficients T^e_n and T^m_n on the field $\boldsymbol{E}$ when $z = z_0$, we get:

$$\sum_{n=1}^{\infty} T^e_n E^e_{x,n,+}|_{z=z_0} = E_x|_{z=z_0}, \\ \sum_{n=1}^{\infty} T^m_n E^m_{y,n,+}|_{z=z_0} = E_y|_{z=z_0}. \tag{7.70}$$

Taking into account these equations, we get for the normal components of the electric field.

$$\sum_{n=1}^{\infty} T^e_n i\gamma_n \exp(i\gamma_n z_0)\sqrt{2/a}(\pi n/a)\cos(\pi nx/a) = E_x(x,z_0), \tag{7.71}$$

$$\sum_{n=1}^{\infty} T^m_n (i\omega\mu_0/c)\exp(i\hat{\gamma}_n z_0)\sqrt{2/a}(\pi n/a)\sin(\pi nx/a) = E_y(x,z_0), \tag{7.72}$$

From here, we get the following expressions for the transition coefficients:

$$T^e_n = -(ia/\gamma_n \pi n)\exp(-i\gamma_n z_0)\sqrt{2/a}\int_0^a E_x(x,z_0) \times \\ \times \cos(\pi nx/a)dx, \tag{7.73}$$

$$T_n^m = -(ica/\omega\mu_0\pi n)\exp(-i\widehat{\gamma}_n z_0)\sqrt{2/a}\int_0^a E_y(x, z_0)\times \\ \times\sin(\pi nx/a)dx. \tag{7.74}$$

To solve the problem under consideration by the finite element method, let's write Maxwell's equations for a chiral medium using the material equations as follows:

$$rot\mathbf{H} = -(i\omega/c)(\varepsilon\mathbf{E} - i\chi\mathbf{H}), \\ rot\mathbf{E} = (i\omega/c)(\mu\mathbf{H} + i\chi\mathbf{E}). \tag{7.75}$$

If we eliminate the vector ***H*** from the Eq. 7.75, we get:

$$rot\frac{1}{\mu}rot\mathbf{E} + \frac{\omega}{c}rot\frac{\chi}{\mu}\mathbf{E} + \frac{\omega}{c}\frac{\chi}{\mu}rot\mathbf{E} - \\ -\frac{\omega^2}{c^2}\left(\varepsilon - \frac{\chi^2}{\mu}\right)\mathbf{E} = 0. \tag{7.76}$$

Let's take $\mathbf{E}^*$ smooth test functions that satisfy (7.59). Multiply Eq. 7.76 by the vector $\mathbf{E}^*$ and integrate the result over the domain D:

$$\iint_D \mathbf{E}^* rot\frac{1}{\mu}rot\mathbf{E}dS + \frac{\omega}{c}\iint_D \mathbf{E}^* rot\frac{\chi}{\mu}\mathbf{E}dS + \\ +\frac{\omega}{c}\iint_D \mathbf{E}^*\frac{\chi}{\mu}rot\mathbf{E}dS - \frac{\omega^2}{c^2}\iint_D\left(\varepsilon - \frac{\chi^2}{\mu}\right)\mathbf{E}^*\mathbf{E}dS = 0. \tag{7.77}$$

Let's convert the first integral on the left side of Eq. 7.77:

$$\int_D \mathbf{E}^* rot\frac{1}{\mu}rot\mathbf{E}dxdz = \int_D \frac{1}{\mu}rot\mathbf{E}^* rot\mathbf{E}dxdz - \\ -\int_{\partial D}\frac{1}{\mu}(\mathbf{n}, [\mathbf{E}^*, rot\mathbf{E}])dl = \int_0^{z_0}\int_0^a \frac{1}{\mu}rot\mathbf{E}^* rot\mathbf{E}dxdz - \\ -\int_0^a \frac{1}{\mu}[\mathbf{e}_z, \mathbf{E}^*]rot\mathbf{E}|_{z=z_0}dx + \int_0^a \frac{1}{\mu}[\mathbf{e}_z, \mathbf{E}^*]rot\mathbf{E}|_{z=0}dx. \tag{7.78}$$

Analogously, let's perform the following transformation:

$$\int_D \mathbf{E}^* rot\frac{\chi}{\mu} rot\mathbf{E}dS = \int_D \frac{\chi}{\mu}\mathbf{E}rot\mathbf{E}^* -$$
$$- \int_{\partial D} \frac{\chi}{\mu}(\mathbf{n}, [\mathbf{E}^*, \mathbf{E}])dl = \int_0^{z_0}\int_0^a \frac{\chi}{\mu}\mathbf{E}rot\mathbf{E}^* dxdz - \tag{7.79}$$
$$- \int_0^a \frac{\chi}{\mu}[\mathbf{e}_z, \mathbf{E}^*]\cdot \mathbf{E}|_{z=z_0}dx + \int_0^a \frac{\chi}{\mu}[\mathbf{e}_z, \mathbf{E}^*]\cdot \mathbf{E}|_{z=0}dx.$$

Thus, Eq. 7.77 has the following form:

$$\int_0^{z_0}\int_0^a \frac{1}{\mu} rot\mathbf{E}^* rot\mathbf{E}dxdz + \frac{\omega}{c}\int_c^{z_0}\int_0^a \frac{\chi}{\mu}\mathbf{E}rot\mathbf{E}^* dxdz +$$
$$\frac{\omega}{c}\int_0^{z_0}\int_0^a \frac{\chi}{\mu}\mathbf{E}^* rot\mathbf{E}dxdz - \frac{\omega^2}{c^2}\int_0^{z_0}\int_0^a \left(\varepsilon - \frac{\chi^2}{\mu}\right)\mathbf{E}^*\mathbf{E}dxdz - \tag{7.80}$$
$$- \int_0^a \frac{1}{\mu}[e_z, \mathbf{E}^*]\left\{rot\mathbf{E}\Big|_{z=0}^{z=z_0} + \frac{\omega}{c}\chi\mathbf{E}\Big|_{z=0}^{z=z_0}\right\} = 0.$$

To transform the last integral in Eq. 7.80, let's use conditions (7.60) and (7.61) and Maxwell's equations when $z = z_0$ and $z = 0$.

$$rot\mathbf{E} = (i\omega/c)(\mu\mathbf{H} + i\chi\mathbf{E}),\ \ z \in (0, z_0), \tag{7.81}$$

$$rot\mathbf{E}^{ext} = (i\omega/c)\mu_0\mathbf{H}^{ext},\ \ z \in (-\infty, 0) \cup (z_0, +\infty). \tag{7.82}$$

Multiply these equations by the $\boldsymbol{e}_z$ vector and write them in the following form:

$$\begin{aligned}&[\mathbf{e}_z, (rot\mathbf{E} + (\omega\chi/c)\mathbf{E})]/\mu = (i\omega/c)[\mathbf{e}_z, \mathbf{H}],\ \ z \in (0, z_0),\\ &[\mathbf{e}_z, rot\mathbf{E}^{ext}]/\mu_0 = (i\omega/c)[\mathbf{e}_z, \mathbf{H}^{ext}],\ \ z \in (-\infty, 0) \cup (z_0, +\infty).\end{aligned} \tag{7.83}$$

Using the condition (7.61), we obtain the expression (7.83).

$$\mu_0[\mathbf{e}_z, (rot\mathbf{E} + (\omega\chi/c)\mathbf{E})]|_{z=0,z_0} = \mu[\mathbf{e}_z, rot\mathbf{E}^{ext}]|_{z=0,z_0}. \tag{7.84}$$

Accordingly, Eq. 7.80 takes the following form:

$$\int_0^{z_0}\int_0^{a} \frac{1}{\mu} rot\mathbf{E}^* rot\mathbf{E} dxdz + \frac{\omega}{c}\int_0^{z_0}\int_0^{a} \frac{\chi}{\mu}\mathbf{E} rot\mathbf{E}^* dxdz +$$
$$+\frac{\omega}{c}\int_0^{z_0}\int_0^{a} \frac{\chi}{\mu}\mathbf{E}^* rot\mathbf{E} dxdz -$$
$$-\frac{\omega^2}{c^2}\int_0^{z_0}\int_0^{a}\left(\varepsilon - \frac{\chi^2}{\mu}\right)\mathbf{E}^*\mathbf{E} dxdz + \tag{7.85}$$
$$+\int_0^{a}\frac{1}{\mu_0}\mathbf{E}^*\left[\mathbf{e}_z, rot\mathbf{E}^{ext}\right]\Big|_{z=0}^{z=z_0} dx = 0.$$

Let's use the following expression of the $\mathbf{E}^{ext}$ field:

$$rot\mathbf{E}^{ext} = \begin{cases} rotE_+^{ins} + \sum\limits_{n=1}^{\infty}\left\{R_n^e rot\mathbf{E}_{n,-}^e + R_n^m rot\mathbf{E}_{n,-}^m\right\}, \\ \sum\limits_{n=1}^{\infty}\left\{T_n^e rot\mathbf{E}_{n,+}^e + T_n^m rot\mathbf{E}_{n,+}^m\right\}, \end{cases} \tag{7.86}$$

$$\begin{cases} z<0, \\ z>z_0, \end{cases}$$

Then we will buy

$$\left[\mathbf{e}_z, rot\mathbf{E}^{ext}\right]\Big|_{z=0} = \mathbf{e}_y\left\{-\frac{\partial E_{y,+}^{ins}}{\partial z} - \sum_{n=1}^{\infty} R_n^m\frac{\partial E_{y,n,-}^m}{\partial z}\right\}\Big|_{z=0} -$$
$$-\mathbf{e}_x\left\{\frac{\partial E_{x,+}^{ins}}{\partial z} - \frac{\partial E_{z,+}^{ins}}{\partial x} + \sum_{n=1}^{\infty} R_n^e\left(\frac{\partial E_{x,n,-}^e}{\partial z} - \frac{\partial E_{z,n,-}^e}{\partial x}\right)\right\}\Big|_{z=0} =$$
$$=\mathbf{e}_y\left\{-\frac{\partial E_{y,+}^{ins}}{\partial z}\Big|_{z=0} + \frac{2i}{a}\sum_{n=1}^{\infty}\widehat{\gamma}_n\sin\frac{\pi n x}{a}\int_0^{a}\left(E_y(x',0) - E_{y,+}^{ins}(x',0)\right)\sin\frac{\pi n x'}{a}dx\right\} -$$
$$-e_x\left\{\begin{aligned} &\left(\frac{\partial E_{x,+}^{ins}}{\partial z} - \frac{\partial E_{z,+}^{ins}}{\partial x}\right)\Big|_{z=0} - \\ &-k_0^2\frac{2i}{a}\sum_{n=1}^{\infty}\frac{1}{\gamma_n}\cos\frac{\pi n x}{a}\int_0^{a}\left(E_x(x',0) - E_{x,+}^{ins}(x',0)\right)\cos\frac{\pi n x'}{a}dx' \end{aligned}\right\} \tag{7.87}$$

and

$$
\begin{aligned}
&[\mathbf{e}_z, rot\mathbf{E}^{ext}]\big|_{z=z_0} = -\mathbf{e}_y \sum_{n=1}^{\infty} T_n^m \frac{\partial E^m_{y,n,+}}{\partial z}\Big|_{z=z_0} - \\
&- \mathbf{e}_x \sum_{n=1}^{\infty} T_n^e \left\{ \frac{\partial E^e_{x,n,+}}{\partial z} - \frac{\partial E^e_{z,n,+}}{\partial x} \right\} = \\
&= -\mathbf{e}_y \frac{2i}{a} \sum_{n=1}^{\infty} \widehat{\gamma}_n \sin\frac{\pi n x}{a} \int_0^a E_y(x', z_0) \sin\frac{\pi n x'}{a} dx' - \\
&= \mathbf{e}_x k_0^2 \frac{2i}{a} \sum_{n=1}^{\infty} \frac{1}{\gamma_n} \cos\frac{\pi n x}{a} \int_0^a E_x(x', 0) \cos\frac{\pi n x'}{a} dx'.
\end{aligned}
\tag{7.88}
$$

Using the Eqs. 7.87 and 7.88 and considering the vectors $\boldsymbol{E}$ and $\boldsymbol{E*}$ in (7.85), we get the general statement of the problem: within the condition $E^* \in H(rot, D)$, it is necessary to determine such vector $E \in H(rotD)$ that satisfies the condition (7.59) so that the following equation is satisfied:

$$
\begin{aligned}
&\int_0^a\int_0^{z_0} \frac{1}{\mu} \left\{ \begin{array}{l} \frac{\partial E_y^*}{\partial z}\frac{\partial E_y}{\partial z} + \frac{\partial E_y^*}{\partial x}\frac{\partial E_y}{\partial x} + \frac{\partial E_x^*}{\partial z}\frac{\partial E}{\partial z} + \frac{\partial E_z^*}{\partial x}\frac{\partial E_z}{\partial x} - \\ - \frac{\partial E_x^*}{\partial z}\frac{\partial E_z}{\partial x} - \frac{\partial E_z^*}{\partial x}\frac{\partial E_x}{\partial z} \end{array} \right\} dxdz + \\
&+ \frac{\omega}{c} \int_0^a\int_0^{z_0} \frac{\chi}{\mu} \left\{ -\frac{\partial E_y^*}{\partial z} E_x + \left(\frac{\partial E_x^*}{\partial z} - \frac{\partial E_z^*}{\partial x} \right) E_y + \right. \\
&\left. + \frac{\partial E_y^*}{\partial x} E_z - \frac{\partial E_y}{\partial z} E_x^* + E_y^* \left(\frac{\partial E_x}{\partial z} - \frac{\partial E_z}{\partial x} \right) + E_z^* \frac{\partial E_y}{\partial x} \right\} dxdz - \\
&- \frac{\omega^2}{c^2} \int_0^a\int_0^{z_0} \left(\varepsilon - \frac{\chi^2}{\mu} \right) \mathbf{E}^*\mathbf{E} dxdz - \\
&- \frac{2ik_0^2}{\mu_0 a} \sum_{n=1}^{\infty} \frac{1}{\gamma_n} \int_0^a E_x^*(x, z_0) \cos\frac{\pi n x}{a} dx \int_0^a E_x(x, z_0) \cos\frac{\pi n x}{a} dx - \\
&- \frac{2ik_0^2}{\mu_0 a} \sum_{n=1}^{\infty} \frac{1}{\gamma_n} \int_0^a E_x^*(x, 0) \cos\frac{\pi n x}{a} dx \int_0^a E_x(x, 0) \cos\frac{\pi n x}{a} dx - \\
&- \frac{2i}{\mu_0 a} \sum_{n=1}^{\infty} \gamma_n \int_0^a E_y^*(x, z_0) \sin\frac{\pi n x}{a} dx \int_0^a E_y(x, z_0) \sin\frac{\pi n x}{a} dx - \\
&- \frac{2i}{\mu_0 a} \sum_{n=1}^{\infty} \gamma_n \int_0^a E_y^*(x, 0) \sin\frac{\pi n x}{a} dx \int_0^a E_y(x, 0) \sin\frac{\pi n x}{a} dx =
\end{aligned}
$$

$$= -\frac{1}{\mu_0}\left\{\int_0^a E_x^*(x,0)\cdot\left(\frac{\partial E_x^{ins}}{\partial z}-\frac{\partial E_z^{ins}}{\partial x}\right)|_{z=0}dx+\int_0^a E_y^*(x,0)\frac{\partial E_y^{ins}}{\partial z}|_{z=0}dx\right\}-$$

$$-\frac{2ik_0^2}{\mu_0 a}\sum_{n=1}^{\infty}\frac{1}{\gamma_n}\int_0^a E_x^*(x,0)\cos\frac{\pi nx}{a}dx\int_0^a E_x^{ins}(x,0)\cos\frac{\pi nx}{a}dx-$$

$$-\frac{2i}{\mu_0 a}\sum_{n=1}^{\infty}\gamma_n\int_0^a E_y^*(x,0)\sin\frac{\pi nx}{a}dx\int_0^a E_y^{ins}(x,0)\sin\frac{\pi nx}{a}dx. \tag{7.89}$$

Let's choose the finite element method to solve the problem. Let's draw a rectangular grid in the Oxz plane of the transducer: $z^j = j\cdot\Delta z, j=\overline{0,J},\ \ x^i = i\cdot\Delta x,\ \ i=\overline{0,I}$. Choose a basis function as follows:

$$N_i(\xi)=\begin{cases}(\xi_{i+1}-\xi)/(\xi_{i+1}-\xi_i),\\ (\xi-\xi_{i-1})/(\xi_i-\xi_{i-1}),\\ 0,\xi\in(-\infty,\xi_{i-1})\cup(\xi_{i+1},+\infty),\end{cases} \tag{7.90}$$
$$P_{i,i+1}(\xi)=\begin{cases}1,\xi\in[\xi_i,\xi_{i+1}],\\ 0,\xi\in(-\infty,\xi_i)\cup(\xi_{i+1}),+\infty.\end{cases}$$

Let's look for the approximate solution of the problem (7.89) in the form of separation of $N_i(\xi)$ and $P_{i,\,i\,+\,1}(\xi)$ basis functions:

$$\begin{cases}E_x(x,z)=\sum\limits_{i=0}^{I-1}\sum\limits_{j=0}^{J}E_x^{i,j}N_j(z)P_{i,i+1}(x),\\ E_y(x,z)=\sum\limits_{i=1}^{I-1}\sum\limits_{j=0}^{J}E_y^{i,j}N_j(z)N_i(x),\\ E_z(x,z)=\sum\limits_{i=1}^{I-1}\sum\limits_{j=0}^{J-1}E_z^{i,j}P_{j,j+1}(z)N_i(x).\end{cases} \tag{7.91}$$

Let's look for the coefficients of the unknown $E_x^{i,j}, E_y^{i,j}, E_z^{i,j}$ in the form of the solution of the system of algebraic equations

$$1)\mathbf{E}^*(x,z)=\{N_j(z)P_{i,i+1}(x)\}\mathbf{e}_x,\quad i=\overline{0,(I-1)},\ j=\overline{0,J}; \tag{7.92}$$

$$2)\mathbf{E}^*(x,z)=\{N_j(z)N_i(x)\}\mathbf{e}_y,\ \ i=\overline{1,(I-1)},\ j=\overline{0,J}; \tag{7.93}$$

$$3)\mathbf{E}^*(x,z)=\{P_{i,i+1}(z)N_j(x)\}\mathbf{e}_z,\quad i=\overline{1,(I-1)},\quad j=\overline{0,(J-1)}. \tag{7.94}$$

Let's write the required vector of the field at the nodes of the network in the following form:

$$\psi=\left\{\begin{array}{l}\underbrace{\left[E_x^{i,0}\right]}_{i=\overline{0,(I-1)}},\left[E_x^{i,1}\right],\ldots,\left[E_x^{i,J}\right];\left[E_y^{i,0}\right],\underbrace{\left[E_y^{i,1}\right]}_{i=\overline{1,(I-1)}},\ldots,\\ \left[E_y^{i,J}\right];\underbrace{\left[E_z^{i,0}\right]}_{i=\overline{1,(I-1)}},\left[E_z^{i,1}\right],\ldots,\left[E_z^{i,J-1}\right]\end{array}\right\}^T. \tag{7.95}$$

Then the system of equations for searching the coefficients can be written in the following matrix form:

$$A\psi=F, \tag{7.96}$$

where F is the column of the first part and is determined from the parameters of the electric component $\mathbf{E}^{ins}(x,z)$ falling into the chiral environment.

If the incident wave is a fundamental *TE*-type wave ($n_0=1$).

$$\begin{aligned}E_+^{ins}&=\left\{0,E_{y,+}^{ins},0\right\},\\ E_{y,+}^{ins}(x,z)&=A_0\exp\left(i\widehat{\gamma}_{n_0}z\right)\sin(\pi n_0x/a).\end{aligned} \tag{7.97}$$

The results obtained for the above values are shown in Figs. 7.18, 7.19, 7.20, 7.21 and 7.22.

It is $\omega/c=5$ at this time. At the same time, except for the chirality parameter χ,other parameters of the waveguide remain unchanged, i.e., $a=1,\ z_0=3,\ \varepsilon_0=\mu_0=1$.

If the incident wave is a fundamental *TM*-type wave

$$\mathbf{E}_+^{ins}=\{E_{x,+}^{ins},0,E_{z,+}^{ins}\}, \tag{7.98}$$

$$E_{x,+}^{ins}(x,z)=\widetilde{A}_0i\gamma_{n_0}(\pi n_0/a)\exp\left(i\gamma_{n_0}z\right), \tag{7.99}$$

$$E_{z,+}^{ins}(x,z)=\widetilde{A}_0(\pi n_0/a)^2\exp\left(i\gamma_{n_0}z\right)\sin(\pi n_0x/a). \tag{7.100}$$

The results obtained for the above values are shown in Figs. 7.23, 7.24, 7.25, 7.26 and 7.27.

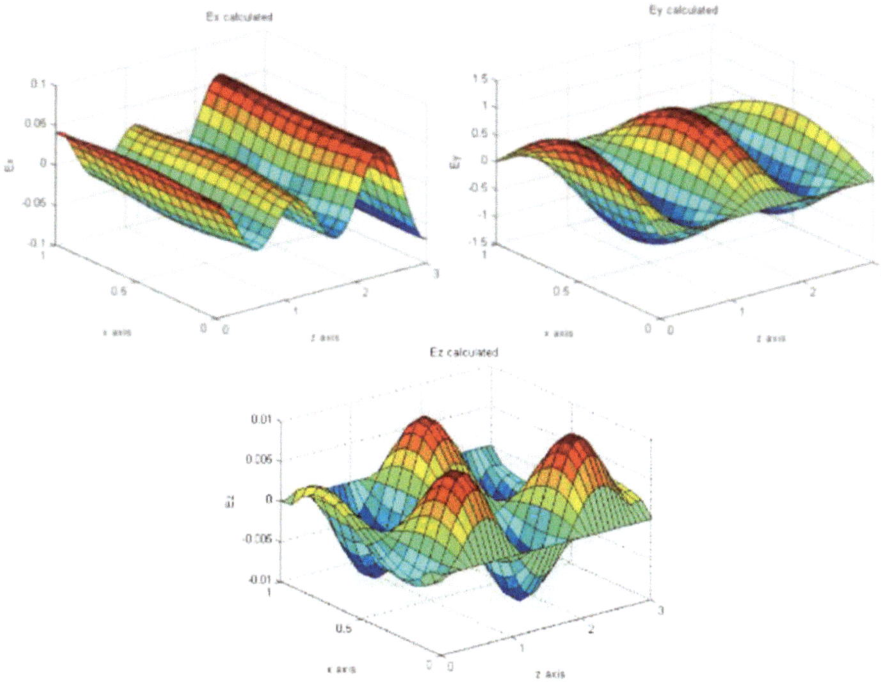

Fig. 7.18 Components of the field inside the waveguide in case $\chi = 0.01$

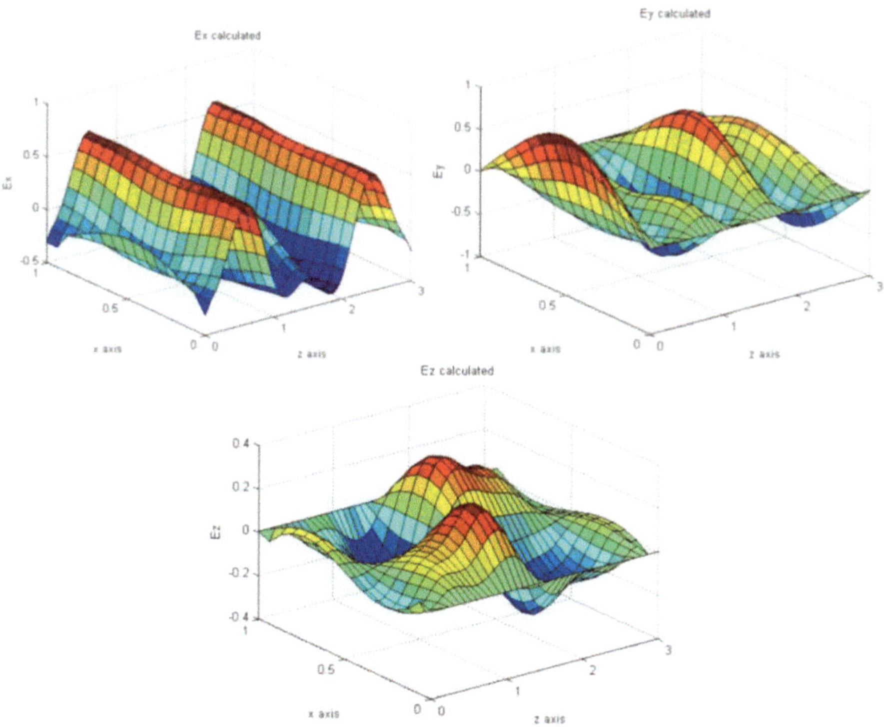

Fig. 7.19 Components of the field inside the waveguide in case $\chi = 0.5$

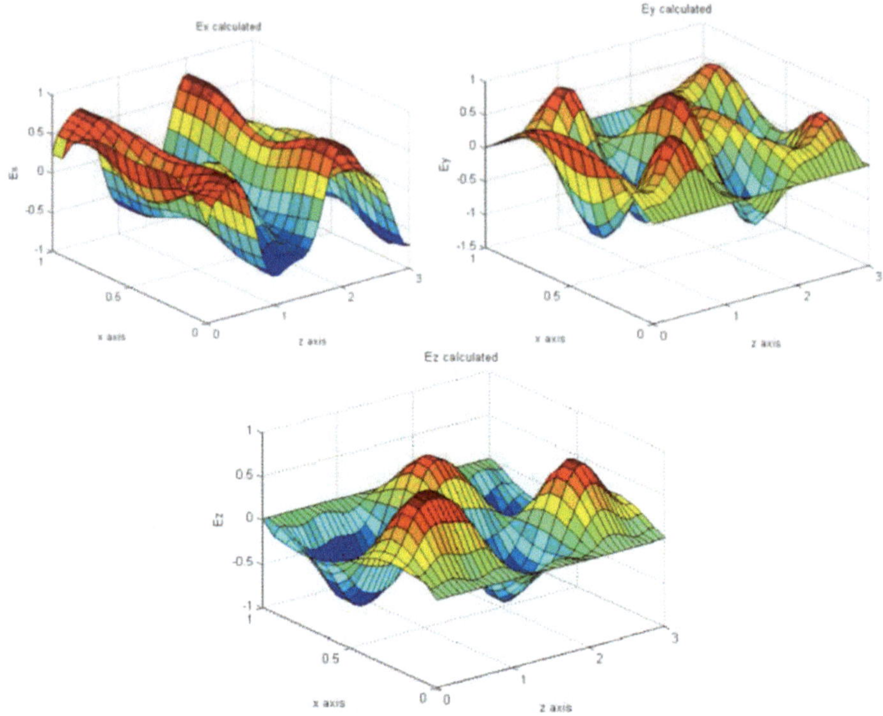

Fig. 7.20 Components of the field inside the waveguide in case $\chi = 0.8$

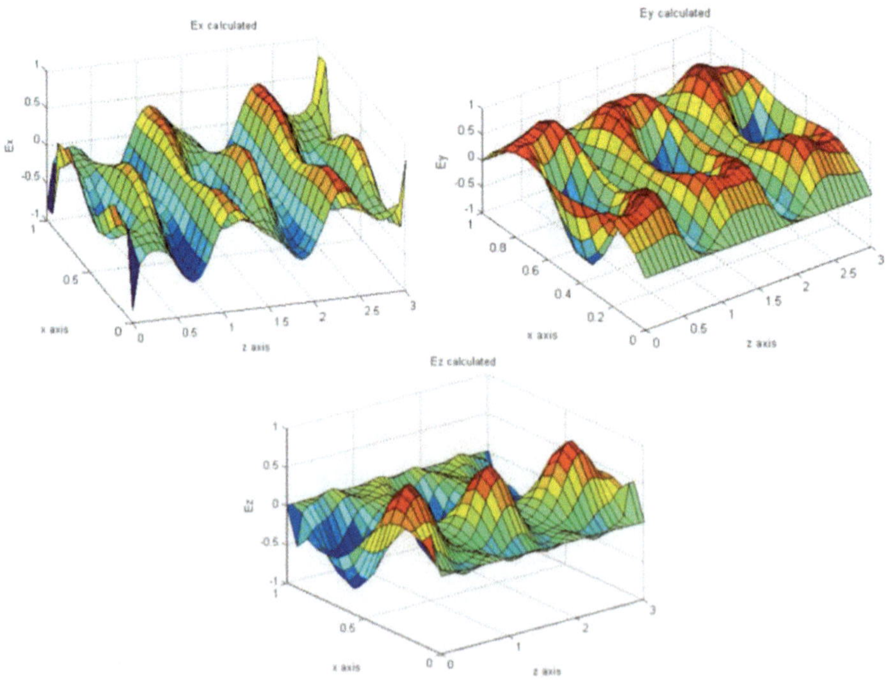

Fig. 7.21 Components of the field inside the waveguide in case $\chi = 1.01$

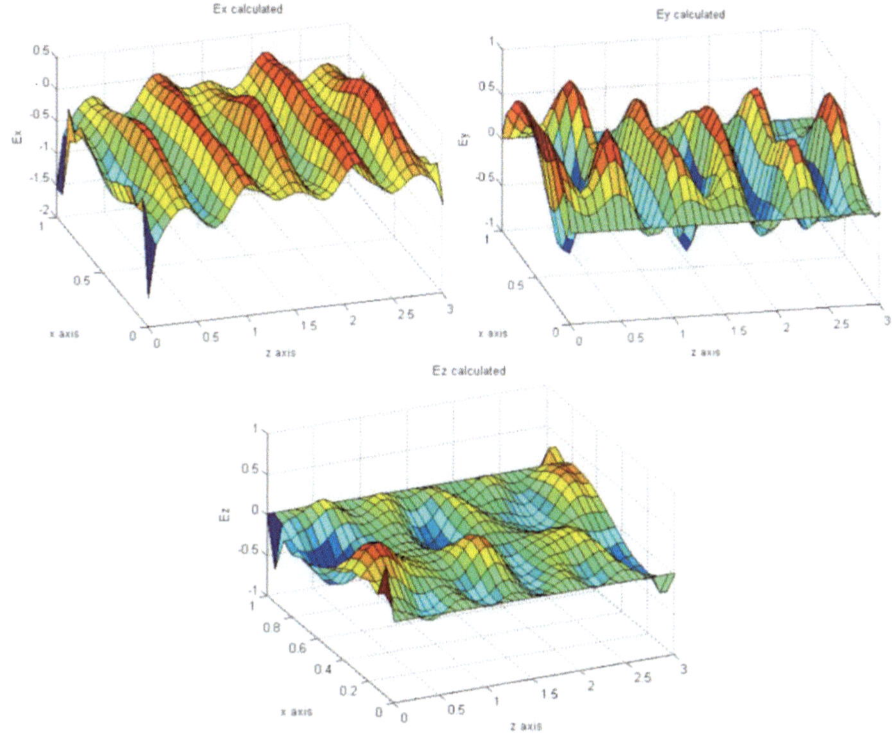

Fig. 7.22 Components of the field inside the waveguide in case $\chi = 1.1$

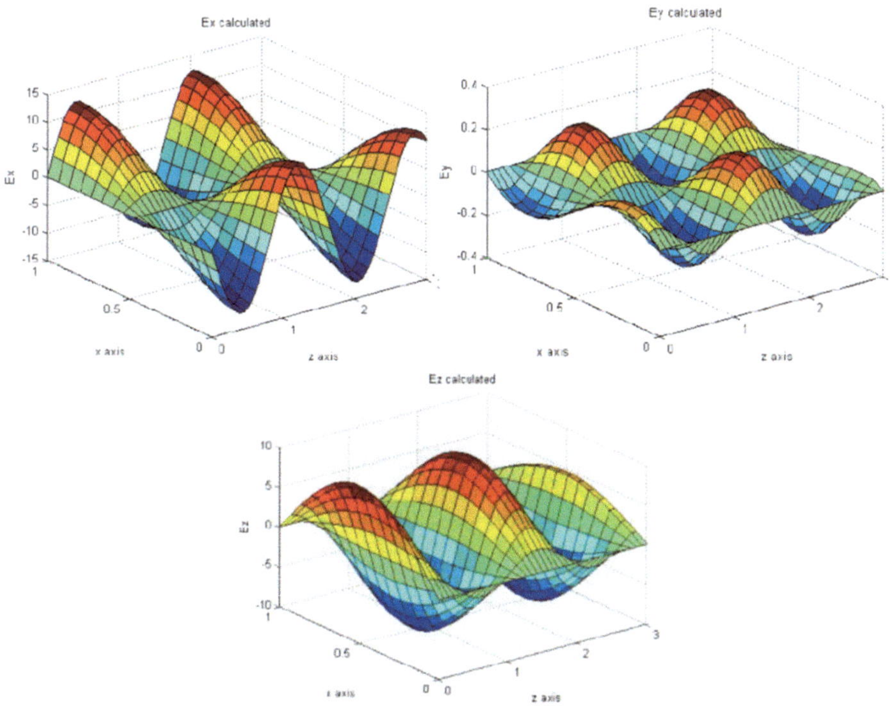

Fig. 7.23 Components of the field inside the waveguide in case $\chi = 0.01$

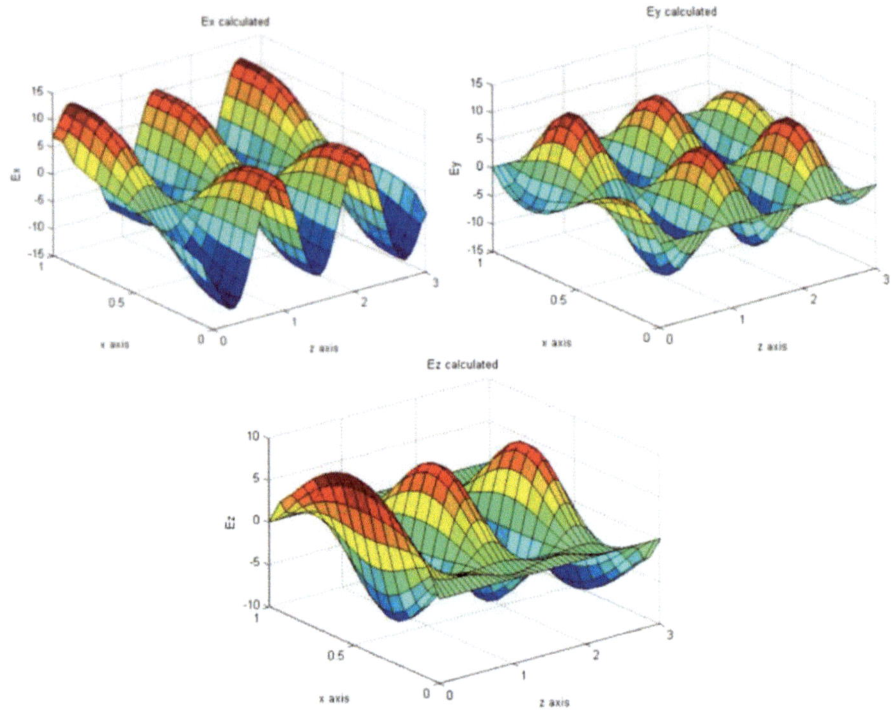

Fig. 7.24 Components of the field inside the waveguide in case $\chi = \mathbf{0.5}$

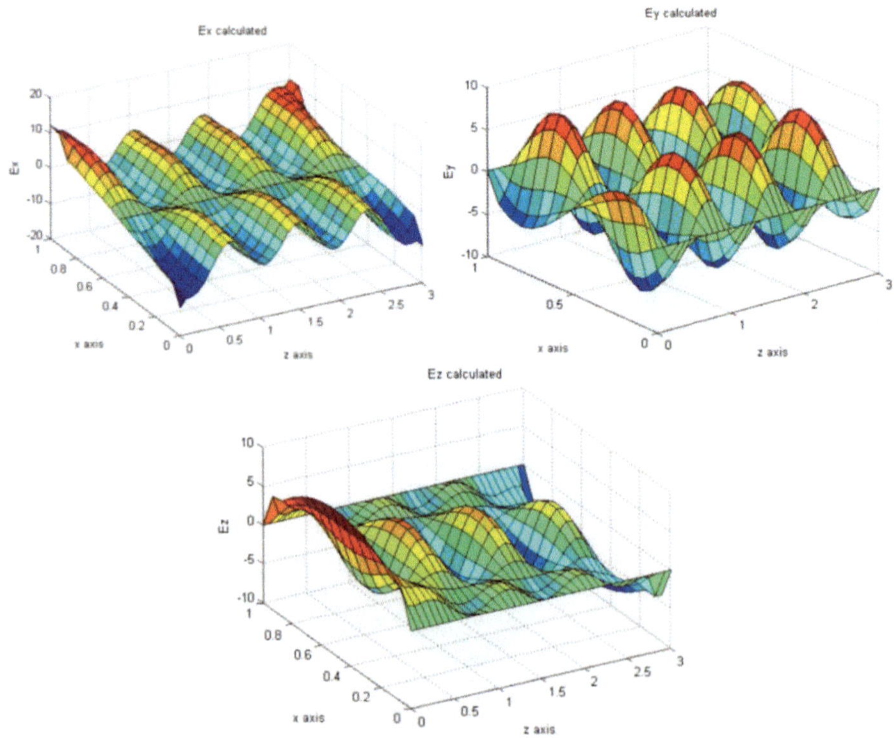

Fig. 7.25 Components of the field inside the waveguide in case $\chi = 0.8$

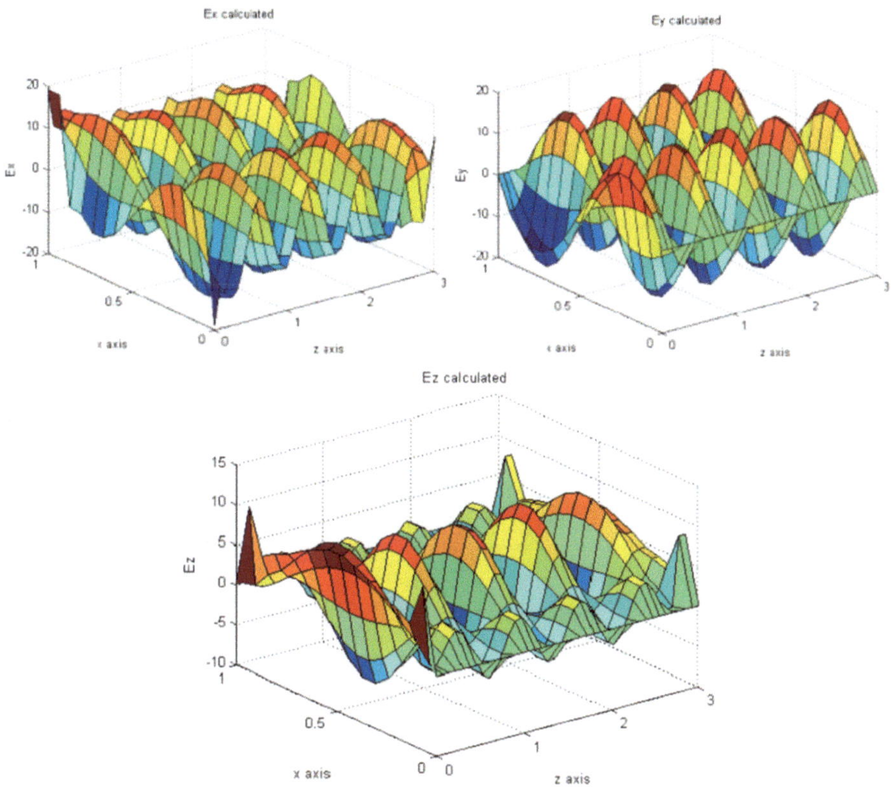

Fig. 7.26 Components of the field inside the waveguide in case $\chi = 1.01$

7.5 Numerical Simulation of Characteristics of Propagation of Symmetric Waves in Microwave Circular Shielded Waveguide with a Radially Inhomogeneous Dielectric Filling

Circular non-uniformly filled waveguides, possessing a number of unique features (anomalous dispersion, complex waves, complex resonance [21]), are widely used [22] in the construction of microwave devices such as attenuators, delay lines, bandpass filters, resonators for radio spectroscopes, etc. Calculation and optimization of the parameters of such devices require the development of numerical and analytical methods for studying waveguides with arbitrary dielectric filling. The possibility of calculating the characteristics of waveguides with filling described by arbitrary analytical functions makes it possible to pose problems of parametric synthesis aimed at the implementation of devices with given characteristics. In

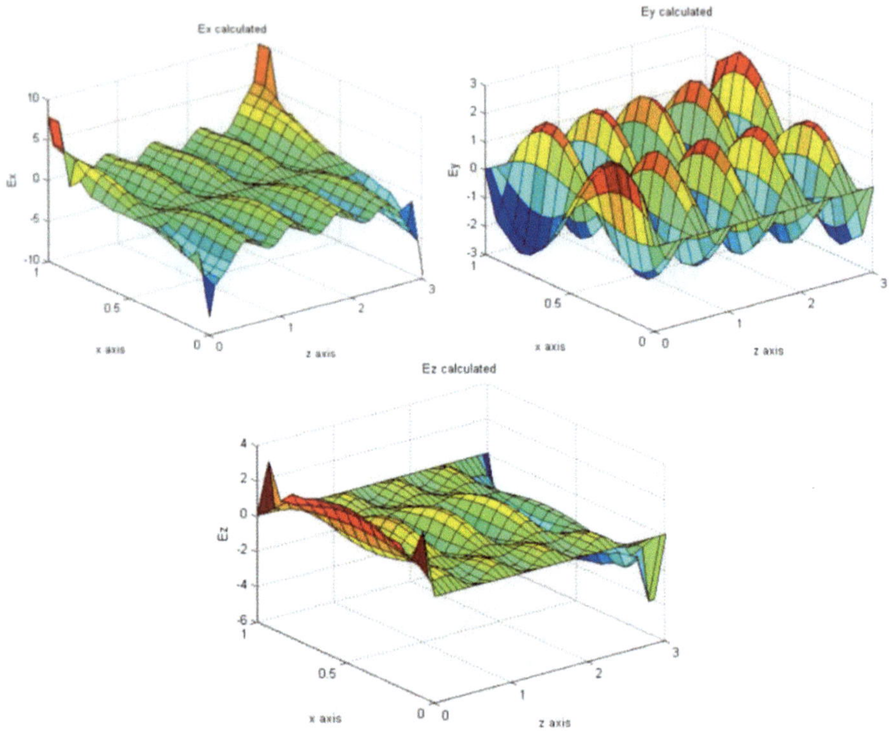

Fig. 7.27 Components of the field inside the waveguide in case $\chi = \mathbf{1.1}$

addition, algorithms for calculating inhomogeneously filled circular waveguides can be used to study gradient optical fibers [23]. This paper proposes a method for calculating the characteristics of wave propagation of a circular shielded waveguide with a radially inhomogeneous dielectric filling, based on a modified Galerkin method as a variant of the spectral method.

To calculate an inhomogeneously filled circular shielded waveguide, it is proposed to use a modified Galerkin method [24], which is a variant of the general spectral method.

Let's consider the problem of the propagation of symmetric E and H-waves in a circular shielded waveguide with partial dielectric filling, the value of the dielectric constant of which arbitrarily depends on the radial coordinates $\varepsilon(r,\ z,\ \varphi) = \varepsilon(r)$ (Fig. 7.28). The value of the magnetic permeability is assumed to be constant.

From Maxwell's equations we get:

$$rotrotE = k_0^2 \varepsilon(r, \phi)E. \tag{7.101}$$

Using the following expressions.

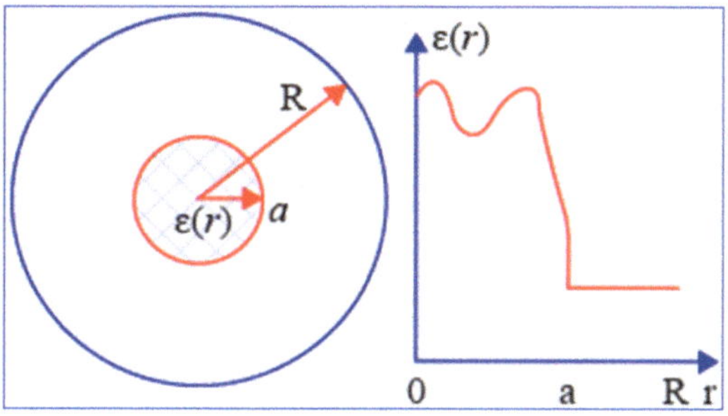

Fig. 7.28 The distribution function of the dielectric constant in the cross section of the waveguide

$$grad(\psi) = r\frac{d\psi}{dr} + \varphi\frac{1}{r}\frac{d\psi}{d\varphi} + z\frac{d\psi}{dz},$$

$$div(E) = \frac{\partial E_r}{\partial r} + \frac{1}{r}\frac{\partial E_\varphi}{\partial \varphi} + \frac{\partial E_z}{\partial z},$$

$$rot(E) = r\left(\frac{1}{r}\frac{\partial E_z}{\partial \varphi} - \frac{\partial E_\varphi}{\partial z}\right) + \varphi\left(\frac{\partial E_r}{\partial z} - \frac{\partial E_z}{\partial r}\right) + z\left(\frac{\partial E_\varphi}{\partial r} + \frac{1}{r}E_\varphi - \frac{1}{r}\frac{\partial E_r}{\partial \varphi}\right),$$

we write Eq. 7.101 for the field components in a cylindrical coordinate system:

$$\begin{aligned} rotrot(E)|_r = & \frac{1}{r}\frac{\partial^2 E_\phi}{\partial r \partial \phi} + \frac{\partial^2 E_z}{\partial r \partial z} - \frac{1}{r^2}\frac{\partial^2 E_r}{\partial \phi^2} + \frac{1}{r^2}\frac{\partial E_\phi}{\partial \phi} - \frac{\partial^2 E_r}{\partial z^2} = \\ & = k_0^2 \varepsilon(r,\phi) E_r, \end{aligned} \tag{7.102}$$

$$\begin{aligned} rotrot(E)|_\phi = & \frac{1}{r^2}E_\phi + \frac{1}{r}\frac{\partial^2 E_r}{\partial r \partial \phi} - \frac{1}{r^2}\frac{\partial E_r}{\partial \phi} + \frac{1}{r}\frac{\partial^2 E_z}{\partial \phi \partial z} - \frac{\partial^2 E_\phi}{\partial r^2} - \\ & - \frac{1}{r}\frac{\partial E_\phi}{\partial r} - \frac{\partial^2 E_\phi}{\partial z^2} = k_0^2 \varepsilon(r,\phi) E_\phi, \end{aligned} \tag{7.103}$$

$$\begin{aligned} rotrot(E)|_z = & \frac{\partial^2 E_r}{\partial r \partial z} + \frac{1}{r}\frac{\partial E_r}{\partial z} + \frac{1}{r}\frac{\partial^2 E_\phi}{\partial \phi \partial z} - \frac{\partial^2 E_z}{\partial r^2} - \\ & - \frac{1}{r}\frac{\partial E_z}{\partial r} - \frac{1}{r^2}\frac{\partial^2 E_z}{\partial \phi^2} = k_0^2 \varepsilon(r,\phi) E_z. \end{aligned} \tag{7.104}$$

We represent the wave fields of the guiding structure in the form of expansions in terms of eigenfunctions of the Dirichlet and Neumann boundary value problems for a uniformly filled circular waveguide. The connection between the components of the electric field, in accordance with the spectral method, is established through the coefficients of the series of expansions substituted in (7.101).

In the absence of the angular dependence of the field, we assume $\frac{\partial}{\partial \varphi} = 0, E_r = 0, \ \ E_z = 0$. In this case, Eq. 7.101 will be reduced to a single equation for the $\varphi-$ component of the electric field.

$$\frac{1}{r^2}E_\varphi - \frac{\partial^2 E_\varphi}{\partial r^2} - \frac{1}{r}\frac{\partial E_\varphi}{\partial r} - \frac{\partial^2 E_\varphi}{\partial z^2} = k_0^2\varepsilon(r)E_\varphi.$$

Writing $E_\varphi(r,\varphi,z) = E_\varphi(r, \ \varphi)e^{-i\beta z}$, we obtain an equation for the transverse coordinate function

$$\frac{\partial^2 E_\varphi}{\partial r^2} + \frac{1}{r}\frac{\partial E_\varphi}{\partial r} - \frac{1}{r^2}E_\varphi + \left(k_0^2\varepsilon(r) - \beta^2\right)E_\varphi = 0. \tag{7.105}$$

where $\varepsilon(r) = \begin{cases} \varepsilon_1 - \dfrac{\varepsilon_1 - \varepsilon_2}{a^2}r^2, & r \le a \\ 1, a \le r \le R. \end{cases}$.

$\varepsilon-$ relative dielectric constant.

Assuming the dependence of the field on the longitudinal coordinate and time, we obtain equations for the components of the electric field:

$$\begin{cases} \dfrac{\partial^2 E_\varphi}{\partial a^2} + \left(k_0^2\varepsilon(r,a) - \beta^2\right)E_\varphi - \dfrac{\partial^2 E_a}{\partial r \partial a} + i\beta\dfrac{\partial E_\theta}{\partial r} = 0, \\ \dfrac{\partial^2 E_a}{\partial r^2} + \left(k_0^2\varepsilon(r,a) - \beta^2\right)E_a - \dfrac{\partial^2 E_r}{\partial r \partial a} + i\beta\dfrac{\partial E_\theta}{\partial a} = 0, \\ \dfrac{\partial^2 E_\theta}{\partial r^2} + \dfrac{\partial^2 E_\theta}{\partial a^2} + k_0^2\varepsilon(r,a)E_\theta + i\beta\dfrac{\partial E_r}{\partial r} + i\beta\dfrac{\partial E_a}{\partial a} = 0. \end{cases} \tag{7.106}$$

The solution to Eq. 7.106 will be sought [25] in the form:

$$E_\varphi(r) = \sum_{n=0}^{N} b_n J_1(\alpha_n r), \tag{7.107}$$

where $J_1(\alpha_n r)-$ is the Bessel function of the first order, the coefficients α_n are determined taking into account the boundary condition $E_\varphi(r = R)$ from equation $J_1(\alpha_n R) = 0$.

Substituting (7.107) into (7.105), we obtain.

$$-\sum_{n=0}^{N} b_n\left[\frac{\partial^2 J_1(\alpha_n r)}{\partial r^2} + \frac{1}{r}\frac{\partial J_1(\alpha_n r)}{\partial r} - \frac{1}{r^2}J_1(\alpha_n r)\right] =$$
$$= \sum_{n=0}^{N} b_n k_0^2\varepsilon(r)J_1(\alpha_n r) - \sum_{n=0}^{N} b_n\beta^2 J_1(\alpha_n r).$$

Considering that

$$\frac{\partial^2 J_1(\alpha_n r)}{\partial r^2} + \frac{1}{r}\frac{\partial J_1(\alpha_n r)}{\partial r} - \frac{1}{r^2} J_1(\alpha_n r) = -\alpha_n^2 J_1(\alpha_n r),$$

we get

$$\sum_{n=0}^{N} b_n \left(\alpha_n^2 + \beta^2\right) J_1(\alpha_n r) = \sum_{n=0}^{N} b_n k_0^2 \varepsilon(r) J_1(\alpha_n r). \tag{7.108}$$

Multiplying both sides of Eq. (7.108) by $rJ_1(\alpha_q r)$ and integrating within $r \in [0; R]$, we obtain the equation

$$\left(\alpha_q^2 + \beta^2\right) Q_q b_q = \sum_{n=0}^{N} b_n k_0^2 \int_0^R \varepsilon(r) r J_1(\alpha_n r) J_1(\alpha_n r) dr. \tag{7.109}$$

Here we used the orthogonality condition for the Bessel functions:

$$\int_0^R r J_1(\alpha_n r) J_1\left(\alpha_q r\right) dr = \begin{cases} Q_q, q = n \\ 0, q \neq n \end{cases},$$

where $Q_q = 0.5R^2 J_0^2(\alpha_n R)$, which takes place, since in this case the Bessel functions are a solution to the homogeneous boundary value problem on the Bessel equation.

Equation 7.109 can be represented in matrix form:

$$M \cdot b = T \cdot b, \tag{7.110}$$

where

$$M_{q,n} = \begin{cases} \left(\alpha_q^2 + \beta^2\right) Q_q, q = n, \\ 0, q \neq n, \end{cases}$$

$$T_{q,n} = k_0^2 \int_0^R \varepsilon(r) r J_1(\alpha_n r) J_1\left(\alpha_q r\right) dr.$$

Writing Eq. 7.110 in the form $(M - T) \cdot b = 0$ and equating the determinant of matrix $(M - T)$ to zero, we obtain the dispersion equation for symmetric H-waves propagating in a circular waveguide with an arbitrary dependence of ε on r : .

$$Det(\beta) = |M - T| = 0. \tag{7.111}$$

Note that the matrix $\boldsymbol{T}$ does not depend on β, therefore, when solving the dispersion Eq. 7.111, it is calculated only once, which significantly reduces the search time for the roots of the dispersion equation. Note that, when deriving Eqs. 7.110 and 7.111, no restrictions were imposed on the form of dependence $\varepsilon(r)$, i.e. This method allows one to calculate symmetric H-waves with a completely arbitrary nature of the change in the dielectric constant along the transverse coordinate, while ε can also be a complex quantity, which allows, for example, calculating waveguides with a complex absorption distribution in the cross section, that is, to solve non-self-adjoint boundary value problems, in which the identity of the differential operators of the direct and adjoint boundary value problems is not satisfied.

For symmetric E-waves, we put

$$\frac{\partial}{\partial \varphi}=0, E_{\varphi}=0, H_{r}=H_{z}=0.$$

In this case, Eq. (7.101) transform into a system of two equations:

$$i\beta\frac{\partial E_z}{\partial r}+\left(k_0^2\varepsilon(r)-\beta^2\right)E_r=0,$$
$$i\beta\frac{\partial E_r}{\partial r}+i\beta\frac{1}{r}E_r+\frac{\partial^2 E_{\varphi}}{\partial r^2}+\frac{1}{r}\frac{\partial E_z}{\partial r}+k_0^2\varepsilon(r)E_z=0.$$

Introducing variable $\widetilde{E}_z=i\beta\cdot E_z$, we arrive at the equations:

$$\begin{gathered}\frac{\partial \widetilde{E}_z}{\partial r}+\left(k_0^2\varepsilon(r)-\beta^2\right)E_r=0,\\ \frac{\partial^2 \widetilde{E}_z}{\partial r^2}+\frac{1}{r}\frac{\partial \widetilde{E}_z}{\partial r}+k_0^2\varepsilon(r)\widetilde{E}_z-\beta^2\left(\frac{\partial E_r}{\partial r}+\frac{1}{r}E_r\right)=0.\end{gathered} \tag{7.112}$$

The boundary conditions on an ideally conducting surface for the tangential and normal components of the electric field $E_\tau|_s=0$, $\left.\frac{\partial E_n}{\partial n}\right|_n=0$ [26], in this case lead to the equation

$$E_z|_{r=R}=0. \tag{7.113}$$

The components of the electric field in accordance with the spectral method will be sought in the form of autonomous expansions:

$$\widetilde{E}_z=\sum_{n=0}^{N}A_n J_0(\alpha_n r), E_r=\sum_{m=0}^{N}B_m J_1(\alpha_m r). \tag{7.114}$$

Taking into account the first boundary condition (7.113), the wave numbers α_n are determined from equation $J_0(\alpha_n R)=0$.

Substituting (7.114) into (7.112), we obtain a system of two functional equations:

$$
\begin{gathered}
-\sum_{n=0}^{N} A_n \alpha_n J_1(\alpha_n r) + \sum_{m=0}^{N} B_m \left(k_0^2 \varepsilon(r) - \beta^2\right) J_1(\alpha_m r) = 0, \\
\sum_{n=0}^{N} A_n \left(\frac{\partial^2 J_0(\alpha_n r)}{\partial r^2} + \frac{1}{r} \frac{\partial J_0(\alpha_n r)}{\partial r} + k_0^2 \varepsilon(r) J_0(\alpha_n r) \right) - \\
-\beta^2 \sum_{m=0}^{N} B_m \left(\frac{\partial J_1(\alpha_m r)}{\partial r} + \frac{1}{r} J_1(\alpha_m r) \right) = 0.
\end{gathered} \tag{7.115}
$$

Taking into account the equalities

$$
\begin{gathered}
\frac{\partial^2 J_0(\alpha_n r)}{\partial r^2} + \frac{1}{r} \frac{\partial J_0(\alpha_n r)}{\partial r} = -\alpha_0^2 J(r), \\
\frac{\partial J_1(\alpha_m r)}{\partial r} = \alpha_m J_0(\alpha_m r) - \frac{1}{r} J_1(\alpha_m r),
\end{gathered}
$$

system (7.115) can be rewritten as

$$
-\sum_{n=0}^{N} A_n \alpha_n J_1(\alpha_n r) + \sum_{m=0}^{N} B_m \left(k_0^2 \varepsilon(r) - \beta^2\right) J_1(\alpha_m r) = 0, \tag{7.116}
$$

$$
\sum_{n=0}^{N} A_n \left(k_0^2 \varepsilon(r) - \alpha^2\right) J_0(\alpha_n r) - \beta^2 \sum_{m=0}^{N} B_m \alpha_m J_0(\alpha_m r) = 0. \tag{7.117}
$$

Multiplying Eq. 7.116 by $r J_1(\alpha_q r) = 0$, Eq. 7.117 by $r J_0(\alpha_q r) = 0$ and integrating within $r \in [0; R]$, we obtain the system of equations:

$$
\begin{gathered}
-A_q \alpha_q Q_q + k_0^2 \sum_{m=0}^{N} B_m \int_0^R r \varepsilon(r) J_1(\alpha_m r) J_1(\alpha_q r) dr - \\
-B_q \beta^2 Q_q = 0, \\
k_0^2 \sum_{n=0}^{N} A_n \int_0^R r \varepsilon(r) J_0(\alpha_n r) J_0(\alpha_q r) dr - \\
-A_q \alpha_q^2 Q_q - B_q \beta^2 \alpha_q Q_q = 0.
\end{gathered} \tag{7.118}
$$

Where we used the orthogonality conditions for the Bessel functions

$$\int_0^R rJ_0(\alpha_n r)J_0(\alpha_q r)dr = \int_0^R rJ_1(\alpha_n r)J_1(\alpha_q r)dr =$$
$$= \begin{cases} Q_q, q = n \\ 0, q \neq n \end{cases},$$

where $Q_q = \frac{R^2}{2} J_1^2(\alpha_q R)$, since in this case the Bessel functions are a solution to a homogeneous boundary value problem.

The system of equations (7.118) can be written in matrix form:

$$\begin{bmatrix} T^{(0,0)} & T^{(0,1)} \\ T^{(1,0)} & T^{(1,1)} \end{bmatrix} \cdot \begin{bmatrix} A \\ B \end{bmatrix} = 0, \tag{7.119}$$

where

$$\begin{aligned} T_{q,m}^{(0,0)} &= -\alpha_q Q_q \delta_{q,m}, \\ T_{q,m}^{(0,1)} &= k_0^2 \int_0^R r\varepsilon(r)J_1(\alpha_m r)J_1(\alpha_q r)dr - \beta^2 Q_q \delta_{q,m}, \\ T_{q,m}^{(1,0)} &= k_0^2 \int_0^R r\varepsilon(r)J_0(\alpha_m r)J_0(\alpha_q r)dr - \alpha_q^2 Q_q \delta_{q,m}, \\ T_{q,m}^{(1,1)} &= -\beta^2 \alpha_q Q_q \delta_{q,m}, \end{aligned} \tag{7.120}$$

where $\delta_{q,\ n}$– Kronecker symbol.

Equating the determinant of matrix Eq. 7.119 to zero, we obtain a dispersion equation describing the symmetric *E*-waves of a circular waveguide with an arbitrary radial dielectric filling.

Two-layer shielded waveguide. As an example, we use Eqs. 7.111 and 7.119 to calculate the simplest test structure – a circular waveguide with a homogeneous dielectric rod (i.e., $\varepsilon(r) = \varepsilon = const$, Fig. 7.29) and compare the results with the exact ones obtained by the classical method of partial regions.

The calculations were carried out for a waveguide with parameters: $R = 20$ mm, $a = 10$ mm, $\varepsilon = 3$, at a frequency of $f = 10$ GHz.

The classical calculation method gives the following results: for symmetric *H*-waves $\beta_H = 237,6891\ \text{m}^{-1}$, for symmetric *E*-waves $\beta_E = 227,55000\ \text{m}^{-1}$.

The calculation of test structures using the proposed technique was carried out by substituting the function $\varepsilon(r) = \begin{cases} 3, & r \leq a \\ 1, & a \leq r \leq R. \end{cases}$ into Eqs. 7.110 and 7.119.

The convergence of solutions obtained by the modified Galerkin method for symmetric *E* and *H*-waves is shown in Table 7.1 and in Fig. 7.30.

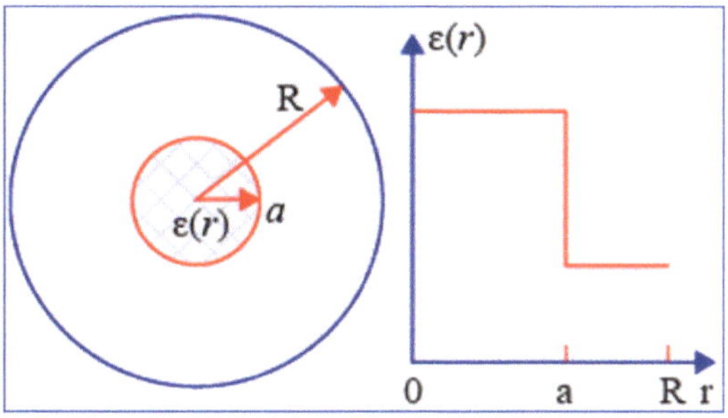

Fig. 7.29 Circular waveguide with a dielectric rod

Table 7.1 Calculation by the modified Galerkin method

№	Symmetric E-waves ($\beta_E = 227,55000\,\mathrm{m}^{-1}$)	Symmetric H-waves ($\beta_H = 237,6891\,\mathrm{m}^{-1}$)
1	233,1366	234,7509
2	234,3096	236,5369
3	232,3624	237,2738
4	231,8978	237,4027
5	230,9273	237,5506
6	230,7095	237,5777
7	230,1335	237,6259
8	230,0207	237,6347
9	229,6397	237,655
10	229,5748	237,6586
11	229,304	237,6686
12	229,2637	237,6704
13	229,0613	237,6758
14	229,0347	237,6768
15	228,8776	2,376,800

From Table 7.1 and Fig. 7.30, it follows that the convergence of the modified Galerkin method is monotonic and occurs rather quickly (already at $N = 5$ the difference between longitudinal wave numbers does not exceed 1.5%).

Figure 7.30 also shows that in the case of symmetric H-waves, convergence occurs faster, which, apparently, is associated with the difference in the number of equations to be solved (one equation (7.105) for symmetric H-waves and two equations (7.112) for symmetric E-waves).

In Fig. 7.31, the dotted line shows the dependencies of the field components H_z and E_φ on the coordinate r, calculated for the symmetric H-wave at $N = 5$.

From the graphs shown in Fig. 7.31, the field distributions calculated by two different methods practically coincide.

Thus, using the example of a test problem with an exact solution, a high accuracy, efficiency of the method and fast convergence of the solution obtained using the modified Galerkin method are shown.

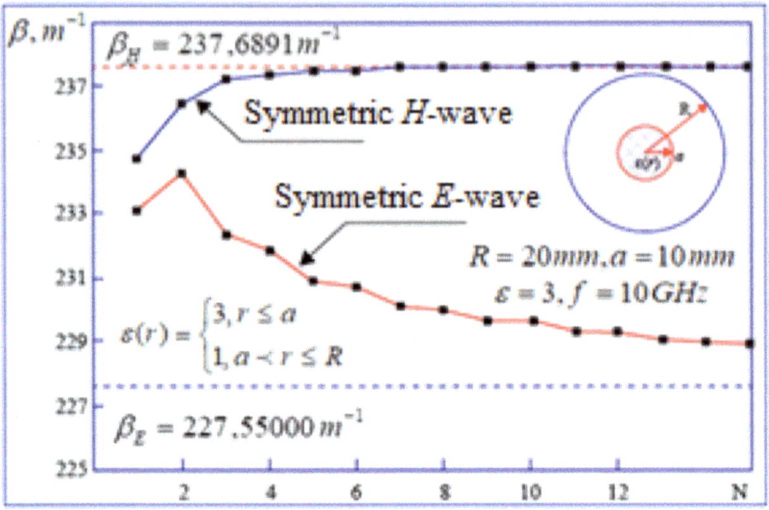

Fig. 7.30 Convergence in integral characteristic

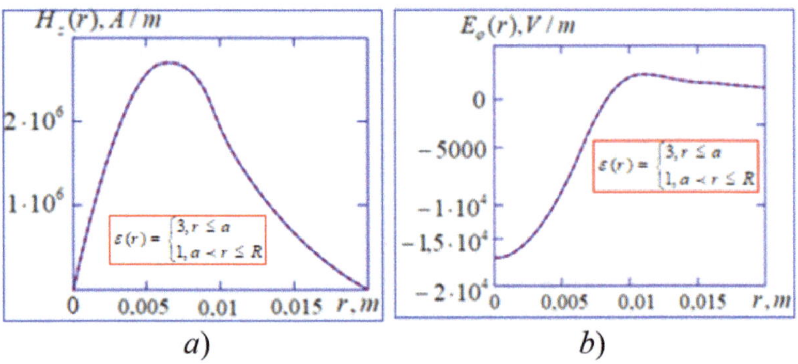

Fig. 7.31 Field distribution of the first symmetric *H*-wave: dotted line - partial domain method, solid line - modified Galerkin method: *a*) for component H_z; *b*) for component E_φ

Based on Eq. 7.120, the dispersion characteristics of symmetric *E*-waves propagating in a circular waveguide with partial dielectric filling, the permeability of which changes according to the parabolic law, is described by the equation:

$$\varepsilon(r) = \begin{cases} \varepsilon_1 - \dfrac{\varepsilon_1 - \varepsilon_2}{a^2} r^2, r \leq a \\ 1, a \leq r \leq R. \end{cases}$$

Substituting this expression in (7.120) and calculating the integrals (numerically or analytically), we obtain a solution to the dispersion problem. Note that for any

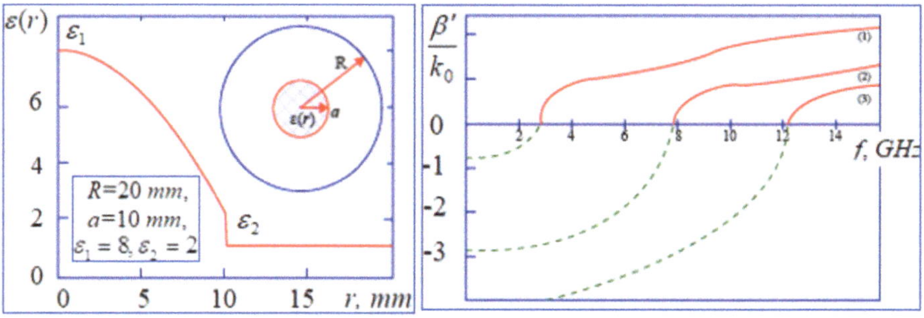

Fig. 7.32 Dispersion characteristics of symmetric E-waves of a circular waveguide with a parabolic profile of dielectric filling

Fig. 7.33 Distribution of power flux density symmetric E-waves at frequency $f = 14$ GHz

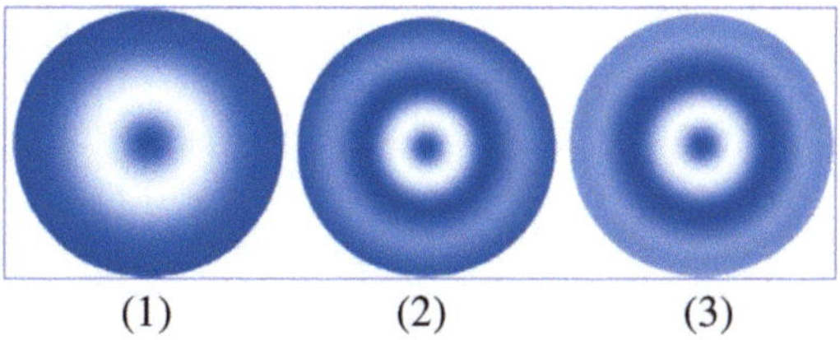

calculation of the integrals from (7.120) is carried out only once, since they do not depend on either the frequency or the longitudinal wavenumber, and are determined only by the filling parameters. This is an unconditional advantage of this method, which makes it possible to significantly reduce the time for calculating the characteristics of the structure.

The results of calculating the dispersion characteristics of symmetric E-waves of a circular waveguide with a parabolic profile of the dielectric filling are shown in Fig. 7.32.

Fig. 7.33 shows the distribution of the Umov-Poynting vector over the cross section of the waveguide, calculated for three modes at frequency $f = 14$ GHz (points 1, 2, 3 in Fig. 7.32).

Based on Eq. 7.111, the structure is calculated in the form of a circular waveguide with partial dielectric filling, the permeability of which varies linearly (Fig. 7.34) within $r \in [0 \div a]$. The calculations were carried out for a waveguide with parameters:

$$R = 20\,\text{mm}, a = 10\,\text{mm},$$

$$\varepsilon(r) = \varepsilon_1 - \frac{\varepsilon_1 - \varepsilon_2}{a} r, \varepsilon_1 = 6, \varepsilon_2 = 2,$$

$$f = 10\,\text{GHz}.$$

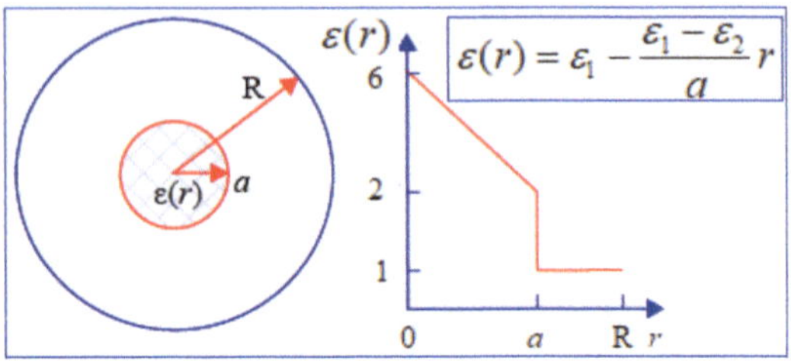

Fig. 7.34 Dielectric constant function

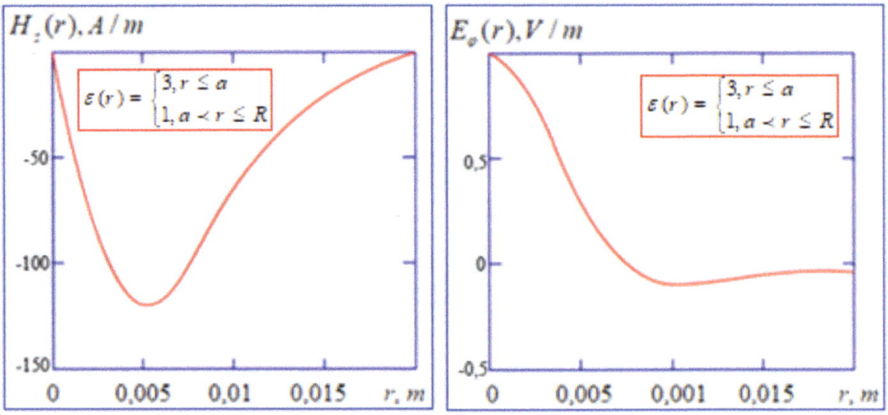

Fig. 7.35 Results of calculating the distribution of the wave field H_{01} of a waveguide with a linear distribution of permeability

For comparison, the calculation of the same structure was performed with the representation of the linear profile of the dielectric constant in the form of *a* step approximation (Fig. 7.34) with the number of steps equal to 20. The results of the calculation of the field distribution obtained by solving the dispersion Eq. 7.111 are shown in Fig. 7.35. The results of calculating the field distribution, performed according to the proposed technique and using the partial domain method, coincide with the graphic accuracy.

On the basis of the method developed in this work, algorithms have been developed for calculating the characteristics of symmetric waves of a cylindrical waveguide with an axisymmetric dielectric filling, which has a radial dependence of the dielectric constant. The procedure for composing algorithms is a modified Galerkin method, in which a variational procedure is applied to functional relations following directly from Maxwell's equations, and corresponds to the canons of the spectral method.

On the example of three boundary value problems, the correctness and efficiency of the modified Galerkin method as a variant of the spectral method are confirmed.

The method is an alternative partial domain method in cases where the latter requires a multilayer approximation of the dielectric filling function, and can be extended to all waveguides with coordinate screening surfaces that provide complete sets of eigen functions of boundary value problems for comparison waveguides.

7.6 Simulation of Antenna Arrays

To the radio technical characteristics of antennas operating in GPS navigation systems are subject to a number of specific requirements. They must ensure the reception/transmission of waves with right-hand circular polarization, have close to uniform radiation patterns in the upper half-space, and ensure the systems operability in the required frequency ranges. In this area of antenna technology, microstrip antennas [27], which is mainly due to their planar structure, relative cheapness and ease of manufacture, high repeatability, and the possibility of minimizing the size of the antenna due to the use of a substrate with a high dielectric constant.

Today, much attention is also paid to the development of antenna arrays for signal receivers of satellite navigation systems. In particular, a number of specific applications [28] require antenna systems with a relatively high gain with a fixed amplitude-phase distribution in the aperture. But the most widely used antenna arrays for receiving navigation signals are found in adaptive noise protection systems with digital beamforming of the required shape [29]. This direction is relevant for both military and civilian facilities. Note that the maximum number of interference that can be suppressed by a digital array is one less than the number of array antenna elements [30].

The simplest antenna arrays designed for adaptive noise protection systems include a small number of elements $N \leq 10$ and consist of microstrip antenna elements. In this case, the characteristic interelement distances are half the wavelength or less, which, together with the weak directivity of the radiation patterns of the antenna elements themselves, does not allow us to neglect the mutual influence of the antennas on each other. As a result, in order to obtain an adequate theoretical estimate of the radio technical characteristics, it is necessary to carry out an electrodynamic calculation of the entire model of antenna arrays as a whole. The close location of the boundaries of antenna arrays to peripheral elements and the possibility of making antenna elements in the form of three-dimensional planar structures that do not have a common dielectric substrate, in turn, limits the possibility of using numerical two-dimensional methods for analyzing flat layered structures to solve such problems.

A priori, it may seem that modern software packages for three-dimensional electrodynamic modeling make it possible to analyze and synthesize (using parametric optimization) microstrip antenna arrays with a small number of elements even on non-modern personal computers, because the characteristic dimensions of such problems do not exceed several wavelengths. However, in practice, to analyze such structures, rather significant hardware resources are required, as well as significant computer time due to the need to divide each of the array elements into a sufficiently large number of grid cells, especially if the antenna elements used are small in size, for example, due to the use of radiators of complex indented shape and/or the use of materials with high dielectric constant.

In the proposed work, we consider the simulation of a seven-element planar antenna array in the $L1$ range (1574–1610 MHz), which consists of commercially available small-sized ceramic antenna elements and can be used in receivers with adaptive noise reduction algorithms. In the first part of the article, the characteristics of the antenna elements used are given, a solitary element is calculated, and the layout of the antenna elements in the array is described. In the second part, the results of modeling the entire antenna arrays are described, the material of the common dielectric base and the relative position of the antenna elements are selected. There is a difference between the characteristics of the radiation field in the far zone and the reflection coefficients for antennas in the composition of antenna arrays and one solitary antenna element. For numerical calculations, the method of finite differences in the time domain was used. Special attention is paid to recommendations on the use of manual settings for partitioning the lattice model into cells. In the third part of the work, the results of measuring the elements of the scattering matrix of the manufactured sample of antenna arrays are presented.

Modeling of antenna arrays was carried out using microstrip antenna elements. They consist of a silver emitting element, a ceramic substrate, and an excitation pin. An integral part of such a microstrip antenna is a metal "screen" with a standard size of 70 × 70 mm, which is not supplied by the manufacturer along with these antennas.

Note that circular polarization of the radiation field of a microstrip antenna can be achieved in the case when two orthogonal modes with a phase shift of 90° are excited in the formed antenna resonator. In our case, to obtain a field polarization close to circular, asymmetry is used, namely, a cut of two edges of a square element. This also makes it possible to separate the resonant frequencies of the two excited modes. The single-point excitation system used is quite simple to implement, but it is inferior to the two-point excitation system in terms of the operating frequency band [31] and, often, the value of the maximum achievable coefficient of ellipticity.

Neighboring elements of the developed antenna arrays should be at a distance of half a wavelength from each other, which for the center frequency of the $L1$ range is $d = 93.2$ mm. In the center of the antenna arrays is its central element, the remaining six antenna elements are located at the vertices of a regular hexagon. Figure 7.36 schematically shows the arrangement of the elements of such a lattice. Due to the need to locate the centers of six peripheral antenna elements on a circle of radius d relative to the center of the central element, the dimensions of the sides of square metal screens are chosen to be 65 mm (instead of the standard 70 mm according to the documentation, at which the edges of adjacent screens will intersect), as a result of which, as computer simulation showed that the resonant frequencies of one solitary antenna element are shifted by several megahertz towards higher frequencies.

To check the adequacy of numerical calculations and determine of the required "quality" of the mesh of the model partition, we simulated one solitary microstrip antenna in the CST Microwave Studio environment [32] using the finite integration in the time domain [33], which is nothing more than the more widely known method of finite difference time domain [34].

The constructed model is shown in Fig. 7.37. Dimensions of the metal screen are 65 × 65 × 5 mm, ceramic substrate – 25 × 25 × 4 mm. For excitation, an air 50-Ohm coaxial line was used, located inside the screen. As the material of the radiating

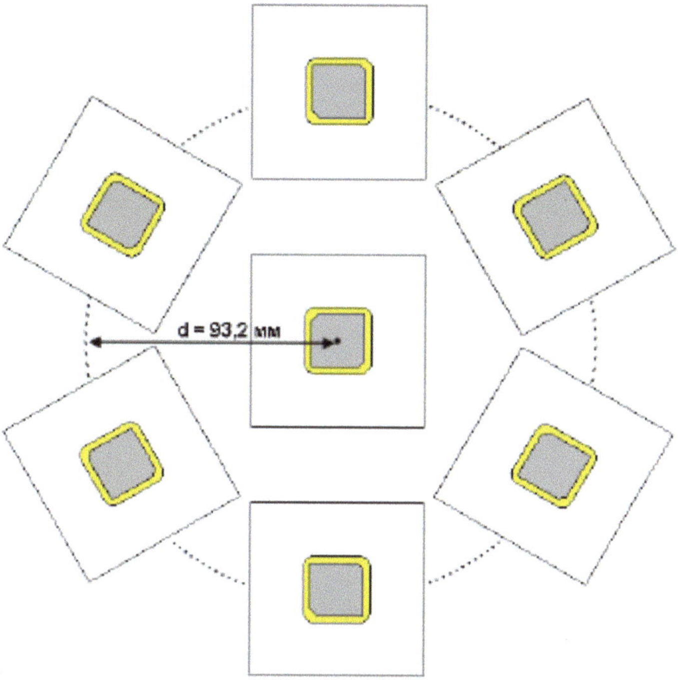

Fig. 7.36 Layout of elements in the developed antenna arrays

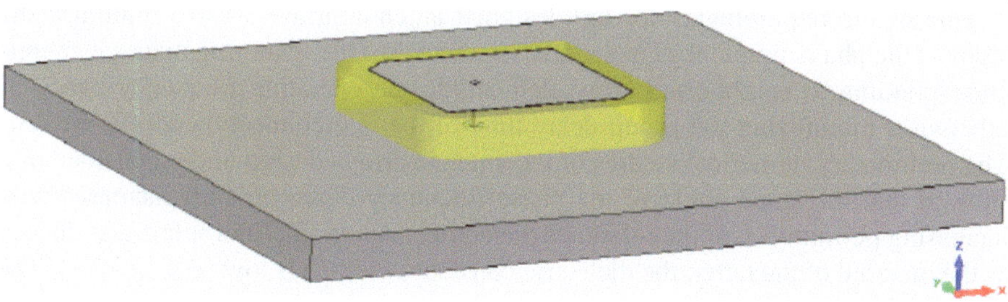

Fig. 7.37 Microstrip antenna model

element and the excitation pin were set to silver, the screen material was aluminum, the parameters of the ceramic substrate were set as follows: dielectric constant $\varepsilon = 20.5$ (average value according to the specification for the antenna) and dielectric loss tangent $\tan\delta = 0.002$.

Figure 7.38 shows the calculated right-hand circular polarization and cross-polarization (left-hand circular polarization) radiation patterns of the antenna elements. The antenna gain is about 4 *dB*. The calculated frequency characteristics of the reflection coefficient have two characteristic minima and, considering the possible spread of the parameters of the antenna elements, are in good agreement with the

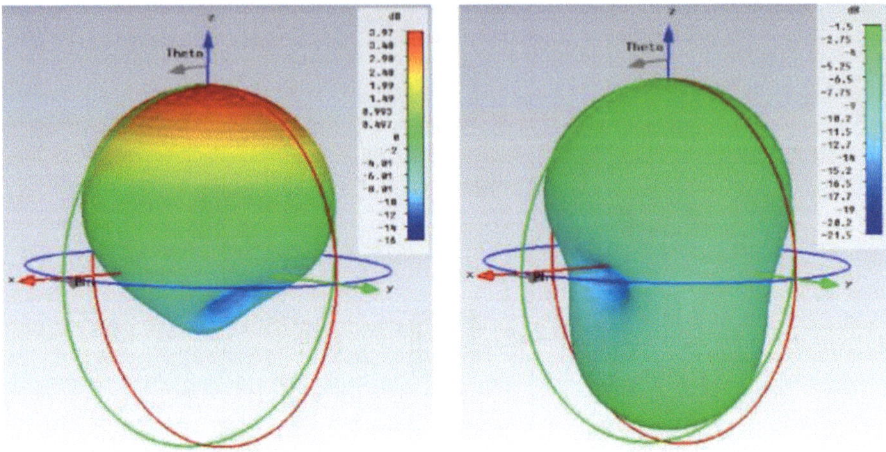

a) *b*)

Fig. 7.38 Spatial radiation patterns of antenna elements at a frequency of 1.6 GHz: right circular polarization (*a*) and cross-polarization (*b*) components

measurement data (for a model with a metal screen with dimensions 70 × 70) given in the documentation for the antenna elements.

In order to reduce the possible effect of multipath (interference from metal surfaces), it was decided to place the elements of the antenna system on a common supporting dielectric substrate with no metallization on any of its sides.

One of the requirements for the designed antenna arrays was to minimize the slope of the phase-frequency characteristics $arg(S_{ij})$ of the scattering matrix elements corresponding to neighboring array elements in the operating frequency band. Or otherwise, minimizing the group delay time of the interconnection coefficients S_{ij}. The preliminary numerical studies of the antenna arrays design under consideration showed that the slope angle of the phase-frequency characteristics increases with increasing permittivity of the reference dielectric. As a result, fluoroplast was chosen as the material of the latter, the dielectric constant of which is low ($\varepsilon = 2...2{,}1$). The sheet thickness was 5 mm.

To obtain the results of calculating the matrix of *S*-parameters, which are closer to the real antenna arrays, the excitation circuit of the antenna elements was slightly modified relative to that used above for one solitary element. Thus, the pin that excites the antenna elements, which is the central residential supply coaxial line in the cross section of the screen, changes its stepwise radius at the border of the aluminum screen and the ceramic substrate. Figure 7.39 shows the power supply circuit of the antenna elements used in the manufacture of antenna arrays, as well as a section of the corresponding area of the electrodynamic model. The air-filled coaxial line still has a characteristic impedance of 50 Ohm. Note that a more significant complication of the excitation circuit was not carried out due to the lack of information about the internal structure of the used TNC connectors. As shown below,

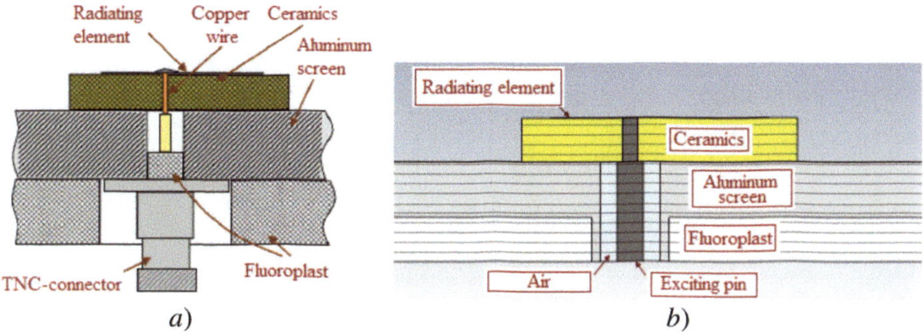

Fig. 7.39 The power supply circuit of antenna elements during manufacture (*a*) and the power supply model used for modeling (*b*)

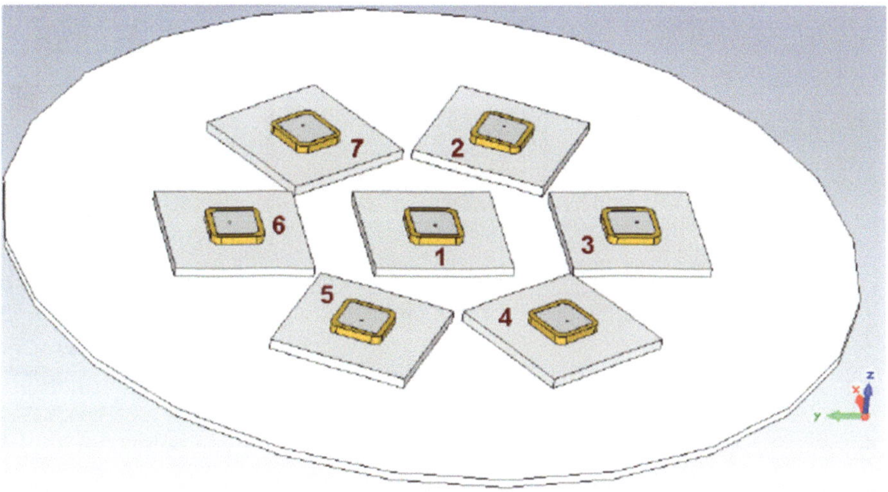

Fig. 7.40 Model a in Microwave Studio with antenna element numbering used

this, however, did not prevent us from obtaining a fairly good agreement between the results of calculation and measurements of the matrix of *S*-parameters.

The mutual arrangement of elements in the composition of antenna arrays also has a certain effect on the frequency characteristics $arg(S_{ij})$ of interest to us. As a result of modeling, from several possible options for the location of peripheral antenna elements, the structure shown in Fig. 7.40. This option is characterized by the maximum remoteness of the power points of the peripheral elements from the central antenna element and allows, among other things, to minimize the mutual connections $|S_{ij}|$ between antennas.

Let's dwell on some issues related to the modeling process itself. When using the finite difference method in the time domain to calculate a multiport device (2 *N*-terminal), it is necessary to carry out *N* calculations of the transient propagation of

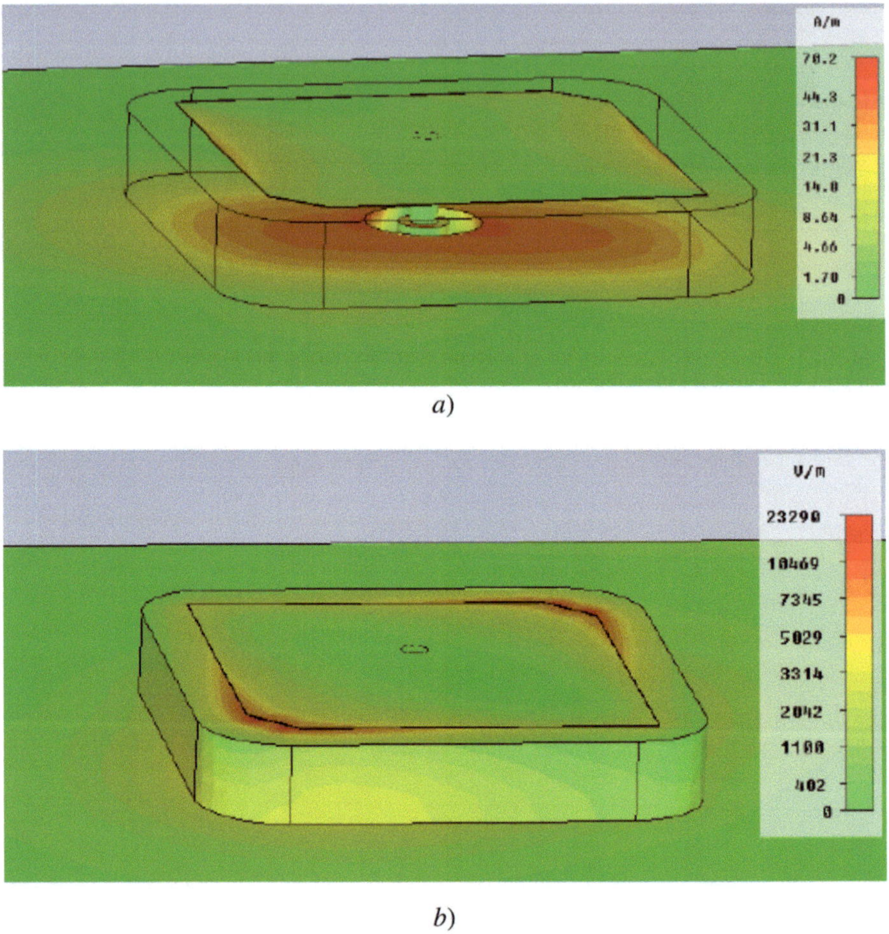

Fig. 7.41 Calculated distributions of surface currents (*a*) and electric field vector (*b*) averaged over a period for one solitary antenna element when a power of 1 *Wt* is applied to its input.

the input signal, one for each port. In our case, we have $N = 7$. As a result, when using an adaptive algorithm for partitioning an electrodynamic model into cells, which involves several steps of the iterative procedure for compacting the mesh, the time spent on calculating the considered antenna array can be significant. To reduce the time required to calculate the lattice, the possibility of manually setting the model partitioning parameters was used.

The highest energy density of the electromagnetic field of the antenna elements is observed in the space between the radiating elements and the screen, as well as near the edges of the radiating element (Fig. 7.41). Therefore, it is precisely near these areas that it is first necessary to condense the mesh of the model partition with respect to some rough initial partition. In addition, to obtain a more accurate result of calculating the matrix of S-parameters, it is also necessary to densify the grid inside

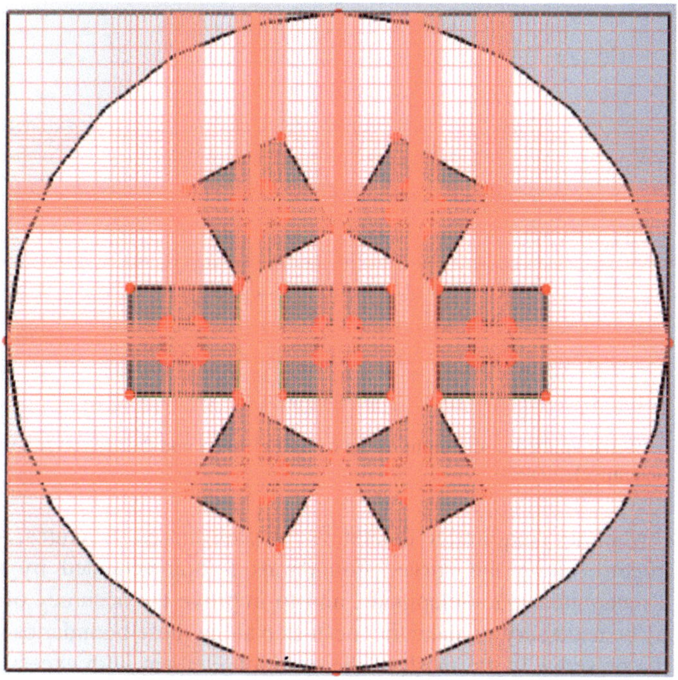

Fig. 7.42 Cell division of the antenna array model

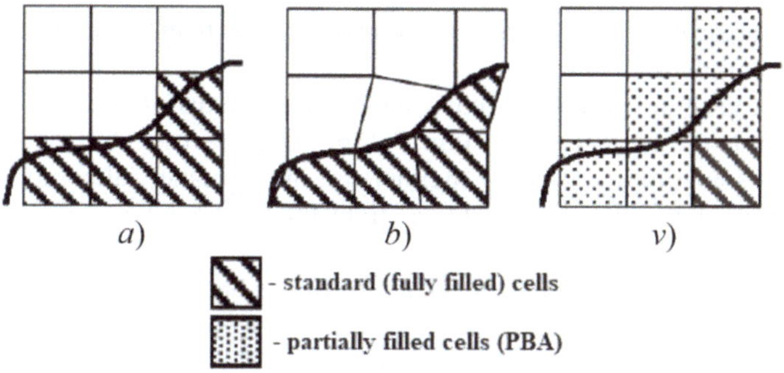

Fig. 7.43 Grid approximation of curved boundaries in the finite difference method in the time domain: standard (*a*), using a non-orthogonal grid (*b*) and PBA (*v*)

the coaxial power line. Figure 7.42 shows the partition mesh used for the entire model in the lattice plane. The total number of partition elements – rectangular parallelepipeds – was more than 1.2 million. A preliminary check of the "quality" of the resulting grid can be carried out using the example of calculating one solitary antenna element.

We also note the following feature of the numerical calculation method used. The disadvantage of the classical finite difference method in the time domain in three-dimensional space using Cartesian orthogonal grids (Fig. 7.43*a*) is the lack of

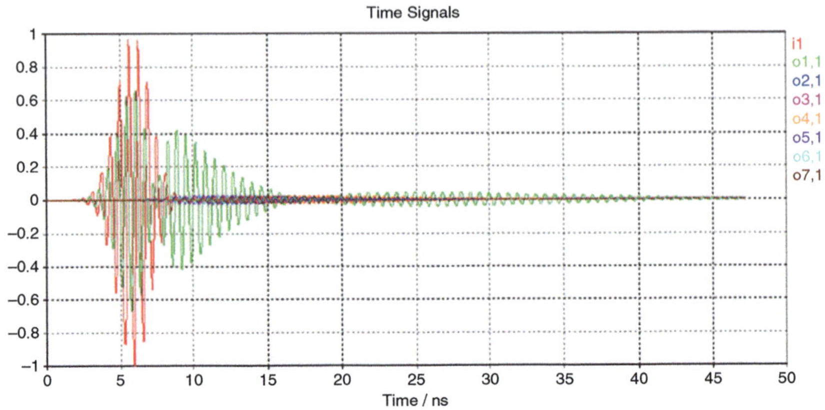

Fig. 7.44 The signal applied to the input of the central element and the signals reflected from the inputs of all antenna elements

"flexibility" when it is required to discretize complex curved structures. As a result, the real structure along the curved boundaries is approximated by orthogonal elements of rectangular parallelepipeds, and to improve the accuracy of the calculation, it may be necessary to make the mesh step along these boundaries very small. However, the presence in the mesh of the partition element with excessively small dimensions compared to the wavelength will inevitably lead to an increase in the calculation time, since to maintain the stability of calculations, it will be necessary to reduce the calculation step of the time process [35]. Moreover, the numerical complexity of the problem also increases due to the growth in the number of partition elements themselves.

The most general approach to consider curved boundaries is to use generalized conformal non-orthogonal grids (Fig. 7.43*b*). But the use of a non-orthogonal algorithm is sometimes limited by the increase in the numerical complexity of the problem and the need to create a tight-fitting boundary of a structured non-orthogonal mesh.

As a more efficient approach, the technique of ideal boundary approximation was proposed [36]. With this approach, there is no need to match the orthogonal computational grid with rounded boundaries (Fig. 7.44*v*). Instead, additional information about the contents of the cells in the space is considered, resulting in a second-order accurate algorithm for free-form boundaries. When calculating the considered antenna array in the Microwave Studio program, the option of using the technique of ideal approximation of the PBA boundary was forcibly set, which made it possible to fairly accurately consider the presence in the array model of geometry elements whose faces are curved or do not coincide with the directions of the axes of the Cartesian coordinate system, and also consider the finite thickness of the metal layer of the radiating elements.

Thus, when setting adequate partitioning parameters, taking into account the features of the structure of the electromagnetic field in the device and the features

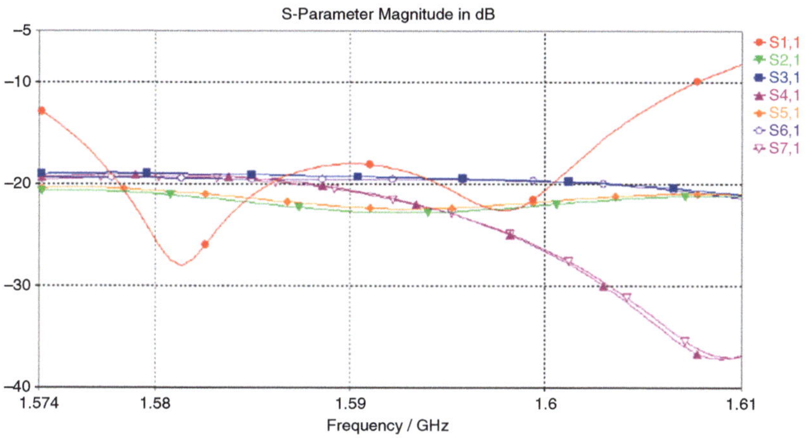

Fig. 7.45 Calculated modules of the elements of the first row of the matrix of *S*-parameters

of the numerical finite difference method itself in the time domain, it is possible to reduce the time for analyzing the model of a particular device compared to the case of using adaptive mesh compaction while maintaining a comparable level of calculation accuracy. In the Microwave Studio environment, you can use a similar design approach when solving problems using the finite integration method in the frequency domain, which also uses an orthogonal mesh. This approach seems to be especially effective at the stage of optimizing the device model.

The experience of modeling microstrip antennas with circular polarization shows that such a calculated antenna characteristic as the ellipticity factor is often more sensitive to the accuracy of the partition grid compared to the matrix of *S*-parameters and radiation patterns. Therefore, in order to be completely confident in the accuracy of the results obtained, the final (optimized) device model should still be calculated using a finer mesh, the parameters of which are set manually, or an automatic adaptive meshing algorithm.

Consider the results of the electrodynamic modeling. Figure 7.44 shows the calculated time dependence of the transient when the first port of the antenna array is excited. Figures 7.45 and 7.46 show the calculated frequency dependencies of the modules and arguments of the elements of the first row of the lattice *S*-parameter matrix in the operating frequency range. Note that, in order to reduce the time of analysis of the array model by the finite difference method in the time domain, the calculation was carried out in a frequency band several times wider than the width of the operating band *L*1 (1574–1610 MHz).

Let's compare the calculated characteristics of the reflection coefficients and radiation fields of one solitary antenna element and two antenna elements in the antenna array. Figure 7.47 shows the frequency dependencies of the reflection coefficients. A comparison of the graphs shows that the bandwidth of the central antenna element in terms of the modulus of the reflection coefficient – 10 *dB* is narrower compared to the bandwidth of the solitary antenna element. The frequency

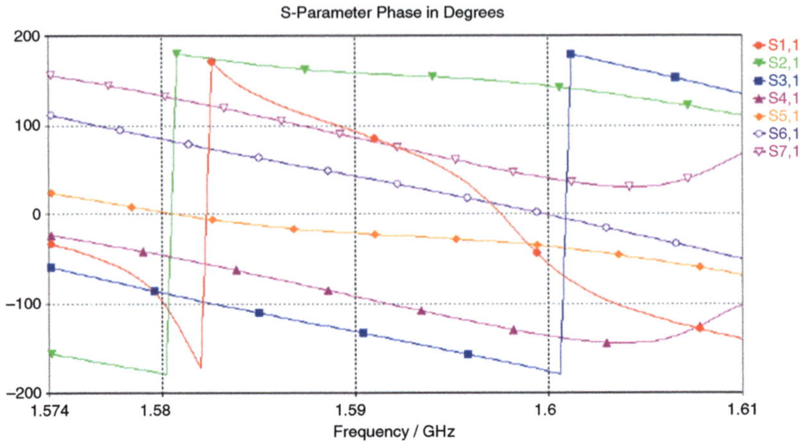

Fig. 7.46 Calculated arguments of the elements of the first row of the matrix of *S*-parameters

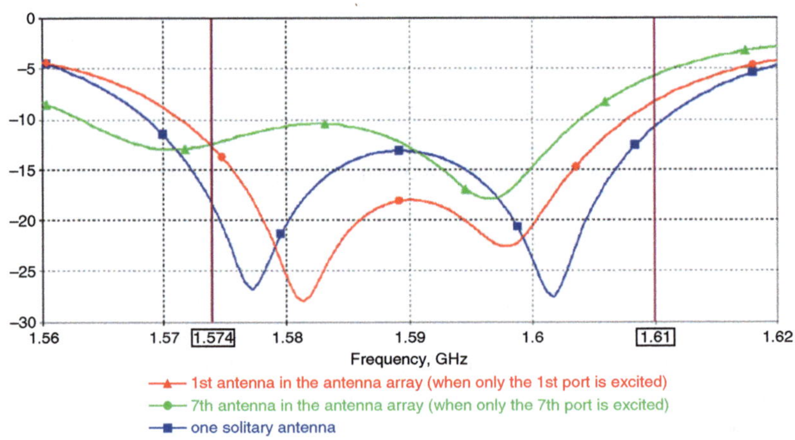

Fig. 7.47 The reflection coefficients of one solitary antenna element and two separate antenna elements in the antenna array (in *dB*)

band of the peripheral element is comparable to that of the solitary element, but it is shifted to lower frequencies and is characterized by "floating" of the first minimum of the reflection coefficient.

The graphs shown in Figs. 7.48 and 7.49 show the modification of the radiation patterns and the final element when moving from a single antenna element to elements in the array. It can be seen that in the two presented planes, the radiation patterns of the central element differ from the radiation patterns of a solitary antenna element insignificantly, at the same time, the radiation patterns of the peripheral element of the lattice are asymmetric and "jagged." Finite element graphs differ to a greater extent. We note a significant difference in the finite elements in two orthogonal planes for the central antenna element: in directions close to the sliding angles

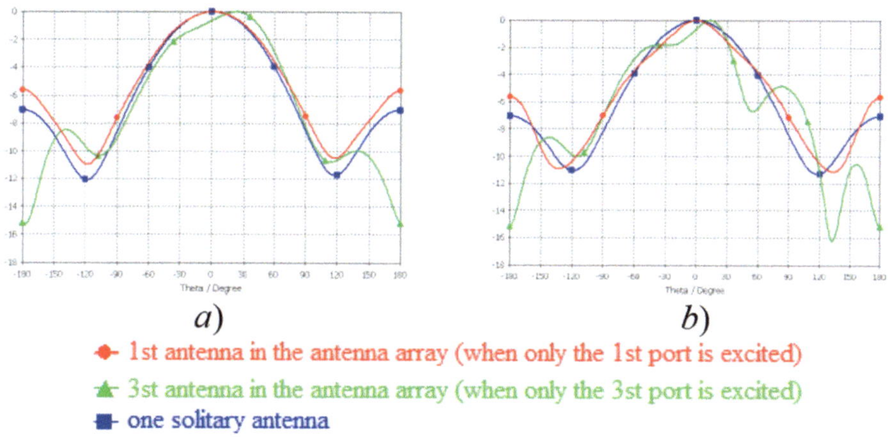

Fig. 7.48 Normalized radiation patterns of one solitary antenna element and two separate antenna elements as part of an antenna array in the plane of the feed point (*a*) and orthogonal to it plane (*b*) on a logarithmic scale

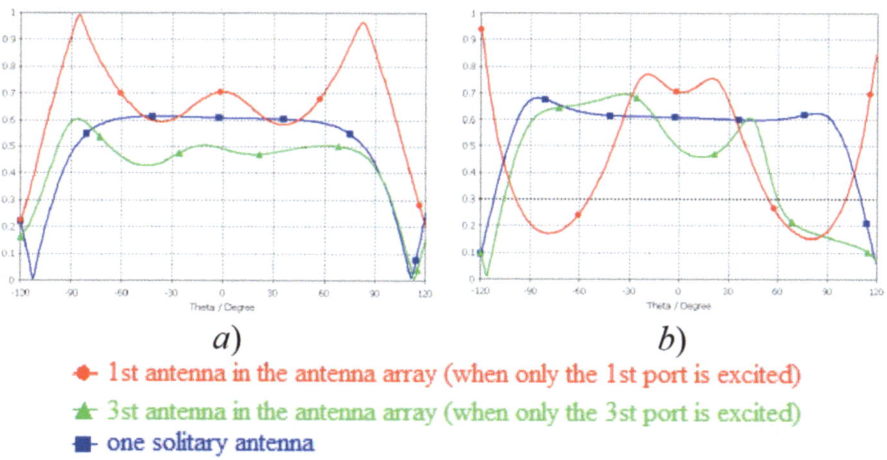

Fig. 7.49 The finite element of one solitary antenna element and two separate antenna elements as part of an antenna array in the plane of the feed point location (*a*) and the plane orthogonal to it (*b*)

$\theta = \pm 90°$, in the plane passing through the feed point, the finite element is close to 1, while in the perpendicular plane of the finite element decreases to the level of 0.2.

Thus, the radiation patterns of the elements in the considered antenna array differ from each other. Knowledge of the characteristics of the radiation field of all elements can be taken as the basis for calculating the complex vector of weight coefficients for the formation of the required radiation patterns of the entire array, for example, for the synthesis dip in the direction of arrival of interference signals in the presence of a priori information about the angular coordinates of targets and interference.

Fig. 7.50 Photo of the manufactured antenna array

In cases where the direction of interference arrival is unknown, approaches to constructing adaptive algorithms are widely used, which are based on the element-by-element principle of searching for the optimal vector of weight coefficients according to the criteria for minimizing the signal-to-noise ratio or other target functional. The disadvantages of such methods include the minimal use of a priori information about the structure of the antenna array, as well as a sharp increase in the computational complexity of the algorithms with an increase in the number of antenna elements.

The so-called "group" adaptation method [37], the essence of which consists in successive scanning of dips in predetermined directions of directional diagrams (for example, in the direction of side lobes) using the method of aperture orthogonal polynomials [38–45], by means of which it is possible to synthesize the required distribution the vector of weight coefficients at once on the entire opening of the antenna array. In this case, it is important to know the radiation patterns of each array element. However, the relevance of such an approach for the case of an antenna array with a small number of elements N (in our case $N = 7$) is beyond dispute.

Figure 7.50 shows a photograph of an antenna array made in accordance with the design features noted above. Due to lack at the time of the measurements of the technical base necessary to determine the characteristics of the radiation field of the antenna array in the far zone, the tests were limited to measuring the elements of the matrix of S-parameters. The corresponding measurements were carried out using a vector network analyzer, while the antenna array itself was placed in an anechoic chamber.

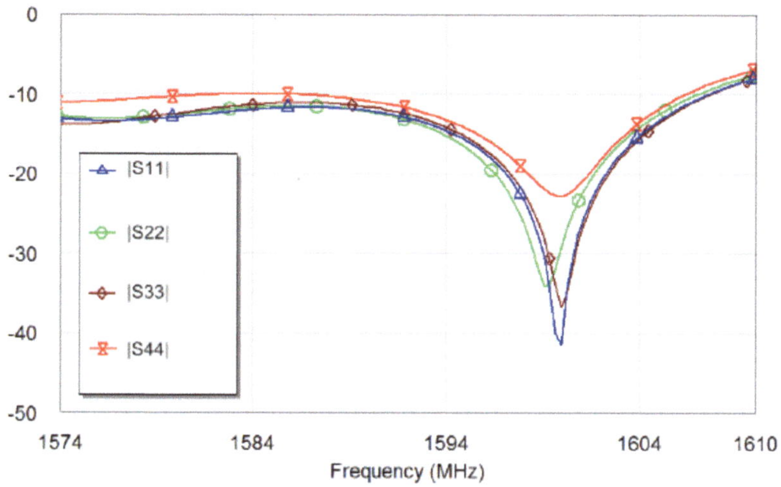

Fig. 7.51 Measured moduli of the diagonal elements of the matrix of *S*-parameters

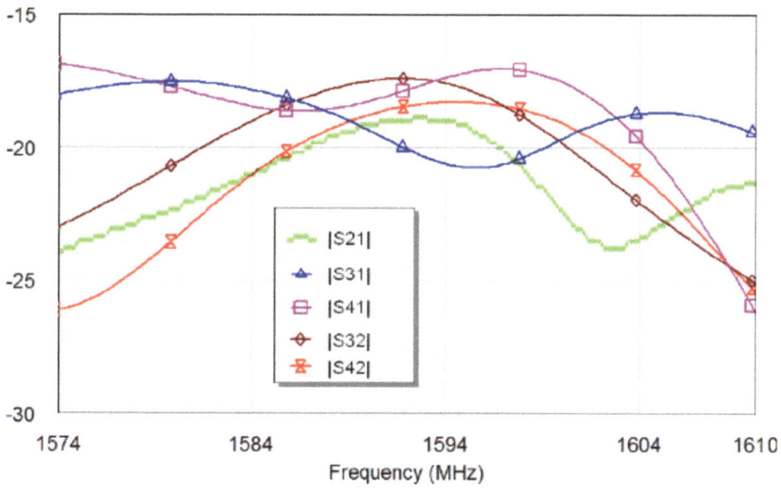

Fig. 7.52 Measured modules of coupling coefficients between antenna elements

Below are the results of measurements of *S*-parameters: in Fig. 7.51 – frequency dependencies of the modules of several diagonal elements of the scattering matrix (in *dB*), in Figs. 7.52 and 7.53 – frequency responses of modules (in *dB*) and arguments of *S*-parameters (in degrees) characterizing mutual relations between antenna array elements. These dependencies allow us to conclude that in the range of operating frequencies, the levels of mutual connections (expressed by the modules of *S*-parameters) are less than minus 17 *dB*. The corresponding differences in the levels of the phase-frequency characteristics at the boundaries of the range *L*1 are on

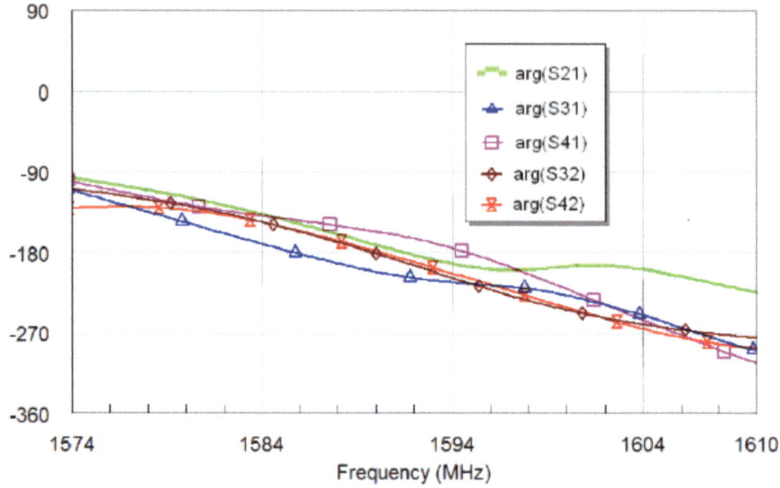

Fig. 7.53 Measured arguments of the coupling coefficients between antenna elements

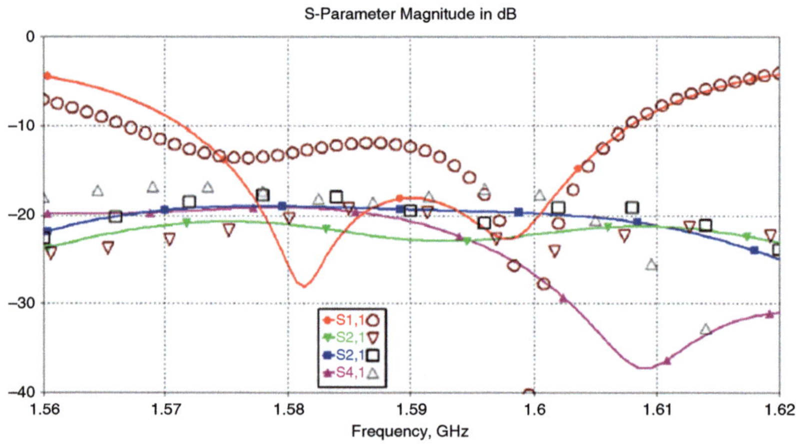

Fig. 7.54 Comparison of *S*-parameter modules: solid curves are numerical calculations, hollow symbols are measurements

average about 180° (Fig. 7.53), which corresponds to a linear relationship with a slope of 5 degrees per megahertz.

Let's compare the results obtained with modeling, and measurements. Figures 7.54 and 7.55 show the corresponding dependencies of the modules and arguments of the S-parameters; in this case, the simulation results are displayed as solid curves, and the measurement results are displayed as hollow symbols. Note that in order to adequately measure/calculate the levels of the arguments of the *S*-parameters, it is necessary to count the phases relative to the position points of the

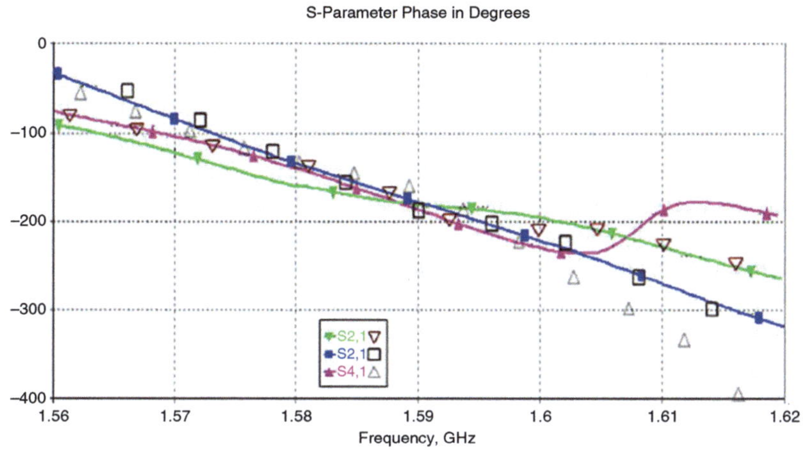

Fig. 7.55 Matching S-parameter arguments: solid curves are numerical calculations, hollow symbols are measurements

phase centers (radiation centers) of the antenna elements (which is not considered in the simulation results, shown in Fig. 7.46).

Due to the spacing of neighboring antenna elements by half the wavelength corresponding to the center frequency f_0 = 1592 MHz of the $L1$ band, at the frequency f_0 one should expect a phase difference of the S-parameters corresponding to neighboring elements of about 180°. Note that this was considered when processing the measurement results shown in Fig. 7.53, where for a visual comparison of the slope of the graphs, the curve corresponding to the argument S_{42} was similarly shifted. On the graphs of Fig. 7.55, the noted fact was also considered for the simulation results, which allows an adequate comparison of the calculations and measurements of the arguments of the S-parameters.

Sufficiently significant differences are observed between the curves corresponding to the modules of the coefficient S_{11}, as well as the modules and arguments of the coefficient S_{41}. However, the reasons for the occurrence of these differences include errors in the manufacture of the antenna array, a possible spread in the parameters of the used ceramic microstrip antennas. All this should have a greater effect on the diagonal elements of the S-matrix, which characterize the degree of matching of the antenna elements with the power line. Thus, the presented results of calculations and measurements corresponding to the parameters S_{21} and S_{31} are in good agreement with each other. Not very good correspondence in the dependencies of the modulus and argument S_{41}, apparently, is due to insufficiently high accuracy manufacturing of the fourth antenna element, which also affected the deterioration of the parameter $|S_{44}|$ (Fig. 7.51).

The process of modeling a seven-element antenna array of the range 1574–1610 MHz, which can be used in receivers of GPS navigation systems with adaptive noise suppression algorithms, is considered. The design features of the manufactured sample of the antenna array are noted. Selected the material of the common dielectric base and the mutual position of the antenna elements, which make it possible to minimize the slope of the frequency dependencies of the arguments $arg(S_{ij})$ of the scattering matrix elements characterizing the mutual relations between neighboring elements. The difference between the characteristics of the radiation field in the far zone and the coefficients of the scattering matrix for antennas in the composition of an antenna array and one solitary antenna element is noted. Numerical calculations were carried out by the finite difference method in the time domain. Special attention is paid to the process of manually partitioning the antenna array model into cells, and the characteristic features of the numerical finite difference method itself in the time domain are noted. The results of numerical calculations of the elements of the matrix of S-parameters are in good agreement with the measurement results.

7.7 Conclusions

1. New mathematical models for the conductivity of the slits of the rectangular waveguide-slot system were developed, which made it possible to obtain the proposed expressions for the current distribution in the slits of the rectangular waveguide in the form of quickly assembled series, to convert the double integral into a single integral, and to reduce the time spent on calculations.
2. The theoretical and experimental dependencies of the coefficient of the radiation of the electromagnetic field of a rectangular waveguide with a cross-sectional area of $a \times b = 40 \times 20$ mm and slits on its large wall (with 20 slits) placed at a distance of $(2L_n/\lambda = 0.4)$ from each other were established. The comparison of theoretical and experimental dependencies showed that the approximation function of the current in the cracks varies according to the same law.
3. The dispersion equation in the form of a matrix was developed for a rectangular waveguide with an infinite number of slits on its small wall. The effective surface impedance was calculated for this structure and the dispersion characteristics were obtained depending on the value of the effective surface impedance and the wavelength. It is shown that both volume and surface waves can be propagated in such an extremely high-frequency structure, and two surface waves can be propagated in a waveguide with two impedance walls. It was found that the increase in the modulus of the effective impedance leads to an increase in the propagation constant.
4. The dispersion equation for the propagation coefficient of electromagnetic waves of a dielectric waveguide with a circular cross-section was obtained. The dependence of the diffusion coefficient on the radius was established.
5. New mathematical models were developed for *TE* and *TM* normal waves of the electromagnetic field of a chiral medium rectangular waveguide with infinite ideal walls located at a certain distance from each other. The mathematical models obtained using the finite element method were solved and the

distributions of the field components inside the waveguide were established for different values of the chirality parameter of the medium.

6. It is shown that the field is hybrid in the chiral region of the waveguide. As a result, such a system can perform the function of a converter to convert one type of wave into another type of wave.
7. Presents a numerical simulation of the propagation characteristics of symmetric *E*-type and *H*-type waves in microwave circular shielded waveguide with radially inhomogeneous dielectric filling. Using the modified Galerkin method, the calculation of a circular two-layer shielded waveguide was carried out, as a result of which the distribution of the electromagnetic field of the waveguide with linear and parabolic distribution of permeability was determined. The results obtained using the modified Galerkin method were compared with the results obtained using the classical partial domain method, which agree well enough.

References

1. Wang, Z., et al. (2018). A short-range range-angle dependent beampattern synthesis by frequency diverse array. *IEEE Access, 6*, 22664–22669.
2. Wong, H., et al. (2015). Substrate integrated meandering probe-fed patch antennas for wideband wireless devices. *IEEE Transactions on Components, Packaging and Manufacturing Technology, 5*(3), 381–388.
3. Wong, K. L., et al. (2016). 8-antenna and 16-antenna arrays using the quad-antenna linear array as a building block for the 3.5-GHz LTE MIMO operation in the smartphone. *Microwave and Optical Technology Letters, 58*(1), 174–181.
4. Wu, C.-Y., et al. (2020). Distributed antenna system using sigma-delta intermediate-frequency-over-fiber for frequency bands above 24 GHz. *Journal of Lightwave Technology, 38*(10 & 15), 2765–2773.
5. Wu, K., et al. (2021). Substrate integrated transmission lines: Review and applications. *IEEE Journal of Microwaves, 1*, 345–363.
6. Wu, S., et al. (2021). Low-dimensional organic metal halide hybrids with excitation-dependent optical waveguides from visible to near-infrared emission. *ACS Applied Materials & Interfaces, 13*(22), 26451–26460.
7. Wu, Y. D. (2020). A general method for analyzing arbitrary planar negative-refractive-index multilayer slab optical waveguide structures. *Scientific Reports, 10*, 14964.
8. Wu, Y.-D., et al. (2022). Analytical and numerical analyses of multilayer photonic metamaterial slab optical waveguide structures with Kerr-type nonlinear cladding and substrate. *Crystals, 12*, 628.
9. Wu, Z., et al. (2013). A TE 01 mode generator for testing high power transmission devices. *The Review of Scientific Instruments, 84*(11), 112–127.
10. Wu, Z., et al. (2020). A circular TE 02 mode generator with high purity for gyro-TWT study. *IEEE Microwave and Wireless Components Letters, 30*(2), 137–140.
11. Xu, Y., et al. (2019). Design and test of broadband rectangular waveguide TE 10 to circular waveguide TE 21 and TE 01 mode converters. *IEEE Transactions on Electron Devices, 66*(8), 3573–3579.
12. Xu, Y., et al. (2020). Proof-of-principle experiment of a 20-kW-average-power Ka-band gyro-traveling wave tube with a cut-off waveguide section. *IEEE Electron Device Letters, 41*(5), 769–772.

13. Yakubu, A., et al. (2015). Effect of ZnO-PCL nanocomposite thickness on attenuation in a rectangular waveguide at microwave frequency using FEM. *Journal of Microwave Power and Electromagnetic Energy, 49*(2), 112–118.
14. Yang, G., et al. (2021). Wide-band and wide-angle scanning phased array antenna for mobile communication system. *IEEE Open Journal of Antennas and Propagation, 2*, 203–212.
15. Yeap, K. H. (2017). Attenuation in circular and rectangular waveguides. *Electromagnetics, 37*(3), 171–184.
16. Yeap, K. H., et al. (2016). Propagation in dielectric rectangular waveguides. *Optica Applicata, 46*(2), 317–330.
17. Yin, J.-Y., et al. (2021). Wideband single-layer substrate integrated waveguide filtering antenna with U-shaped slots. *IEEE Antennas and Wireless Propagation Letters, 20*, 1726–1730.
18. Yin, Y. H., & Zhu, P. (2019). The streamline-diffusion finite element method on graded meshes for a convection-diffusion problem. *Applied Numerical Mathematics, 138*, 19–29.
19. You, K. Y. (2017). *Materials characterization using microwave waveguide system in microwave systems and applications* (pp. 341–358). InTech.
20. Yousefian, M., et al. (2019). Compact broadband coaxial to rectangular waveguide transition. *Journal of Electromagnetic Waves, 33*(9), 1239–1247.
21. Yu, C., et al. (2020). A super-wideband and high isolation MIMO antenna system using a windmill-shaped decoupling structure. *IEEE Access, 8*, 115767–115777.
22. Yuan, W., et al. (2019). Rectangular grating waveguide slot array antenna for SATCOM applications. *IEEE Transactions on Antennas and Propagation, 67*(6), 3869–3880.
23. Yuan, X., et al. (2022). Ultra-compact multichannel optical waveguide crossings designed by a particle swarm optimized method. *Optics Communications, 503*, 127458.
24. Yuan, X.-T., et al. (2020). Ultra-wideband MIMO antenna system with high element-isolation for 5G smartphone application. *IEEE Access, 8*, 56281–56289.
25. Zada, M., & Yoo, H. (2018). A miniaturized triple-band implantable antenna system for bio-telemetry applications. *IEEE Transactions on Antennas and Propagation, 66*(12), 7378–7382.
26. Zada, M., et al. (2021). Ultra-compact implantable antenna with enhanced performance for leadless cardiac pacemaker system. *IEEE Transactions on Antennas and Propagation, 69*(2), 1152–1157.
27. Zakharchenko, O. S., et al. (2018). Generalized mathematical model of thin asymmetric inductive diaphragm in rectangular waveguide. *Visnyk NTUU KPI, 72*, 13–22.
28. Zayed, E. M. E., et al. (2020). Solitons and conservation laws in magneto-optic waveguides with triple-power law nonlinearity. *Journal of Optics, 49*, 584–590.
29. Zeeshan, A., et al. (2022). Comments on frequency diverse array Beampattern synthesis with Taylor windowed frequency offsets. *IEEE Antennas and Wireless Propagation Letters, 21*(8), 1713–1714.
30. Zemlyakov, V., et al. (2019). A design of waveguide elliptic filter based on resonant diaphragms with a complex aperture. *International Journal of Circuit Theory and Applications, 47*, 55–64.
31. Zemlyakov, V., et al. (2018). Complex geometry apertures for resonant diaphragms in rectangular waveguides. *Journal of Electromagnetic Waves and Applications, 32*(18), 2470–2480.
32. Zhang, Y., & Cui, M. (2019). Refractive index sensor based on the symmetric MIM waveguide structure. *Journal of Electronic Materials, 48*(2), 1005–1010.
33. Zhang, C., et al. (2017). *Empirical formulas for propagation constant calculation of leaky circular waveguide*. In Proceedings of 7th IEEE international symposium on microwave, antenna, propagation, and EMC technologies (MAPE) (pp. 303–308).
34. Zhang, J., et al. (2019). Efficient signal separation method based on antenna arrays for GNSS meaconing. *Tsinghua Science and Technology, 24*(2), 216–225.
35. Zhang, X., & Li, S. (2016). *Design of a novel circular waveguide TM_{02}-TE_{11} mode converter*. In 9th IEEE international conference on microwave and millimeter wave technology (ICMMT) (pp. 569–571).

36. Zhang, Y. (2011). Multiscale numerical algorithm for 3D Maxwell's equations with memory effects in composite materials. *International Journal of Numerical Analysis and Modeling, 1*, 41–57.
37. Zhao, A., & Ren, Z. (2019). Size reduction of self-isolated MIMO antenna system for 5G mobile phone applications. *IEEE Antennas and Wireless Propagation Letters, 18*(1), 152–156.
38. Zhao, D., et al. (2022). A miniaturized dual-band SIW filtering antenna with improved out-of-band suppression. *IEEE Transactions on Antennas and Propagation, 70*, 126–134.
39. Zhao, L., et al. (2016). An efficient algorithm for SAR evaluation from anatomically realistic human head model using DGTD with hybrid meshes. *Applied Computational Electromagnetics Society Journal, 31*(6), 629–635.
40. Zhao, L., et al. (2016). GPU accelerated discontinuous Galerkin time domain algorithm for electromagnetic problems of electrically large objects. *Progress in Electromagnetics Research B, 67*, 137–151.
41. Zhao, X., et al. (2018). Planar UWB MIMO antenna with pattern diversity and isolation improvement for mobile platform based on the theory of characteristic modes. *IEEE Transactions on Antennas and Propagation, 66*(1), 420–425.
42. Zheng, Z., et al. (2020). A compact waveguide slot filtering antenna based on mushroom-type surface. *IEEE Antennas and Wireless Propagation Letters, 19*(10), 1823–1827.
43. Zhou, E., et al. (2022). Low-profile high-gain wideband multi-resonance microstrip-fed slot antenna with anisotropic metasurface. *Progress in Electromagnetics Research, 175*, 91–104.
44. Zhou, P., et al. (2017). Design and fabrication of a broadband millimeter wave rectangular-metallic to dielectric rod-waveguide adaptor. *IEEE Transactions on Terahertz Science and Technology, 7*(1), 86–92.
45. Zhu, F., et al. (2021). Compact dual-mode bandpass filters based on half-mode substrate-integrated waveguide cavities. *IEEE Microwave and Wireless Components Letters, 31*, 441–444.

Index

I. Islamov, *Radio Engineering and Telecommunications Waveguide Systems in the Microwave Range*, https://doi.org/10.1007/978-3-031-37916-1

MIX
Papier aus verantwortungsvollen Quellen
Paper from responsible sources
FSC® C105338

If you have any concerns about our products,
you can contact us on
ProductSafety@springernature.com

In case Publisher is established outside the EU,
the EU authorized representative is:
Springer Nature Customer Service Center GmbH
Europaplatz 3, 69115 Heidelberg, Germany

Printed by Libri Plureos GmbH
in Hamburg, Germany